Radium and the Secret of Life

Radium and the Secret of Life

Luis A. Campos

The University of Chicago Press :: Chicago and London

The University of Chicago Press, Chicago 60637
The University of Chicago Press, Ltd., London

Paperback edition 2016
Printed in the United States of America

25 24 23 22 21 20 19 18 17 16   2 3 4 5 6

ISBN-13: 978-0-226-23827-2 (cloth)
ISBN-13: 978-0-226-41874-2 (paper)
ISBN-13: 978-0-226-23830-2 (e-book)
DOI: 10.7208/chicago/9780226238302.001.0001

Library of Congress Cataloging-in-Publication Data

Campos, Luis A., author.
Radium and the secret of life / Luis A. Campos.
pages cm
Includes bibliographical references and index.
ISBN 978-0-226-23827-2 (cloth: alk. paper) — ISBN 978-0-226-23830-2 (e-book) 1. Radium. 2. Genetics. I. Title.
QD181.R1C36 2015
572.8'38—dc23   2014032436

∞ This paper meets the requirements of ANSI/NISO Z39.48-1992 (Permanence of Paper).

To Life!

Contents

Introduction

The man of science must have been sleepy indeed who did not jump from his chair like a scared dog when, in 1898, Mme. Curie threw on his desk the metaphysical bomb she called radium. There remained no hole to hide in. Even metaphysics swept back over science with the green water of the deep-sea ocean and no one could longer hope to bar out the unknowable, for the unknowable was known.

—Henry Adams, *The Education of Henry Adams: An Autobiography*

Unfathomably rare and intensely powerful, glowing in the dark and utterly unaffected by any outside force of nature as it gave off rays of unprecedented energy, radium was perhaps the most wonderful and perplexing thing the modern world had ever seen—or had never seen, given that only the barest pinch of pure radium existed at the dawn of the twentieth century. The modern world had certainly heard about radium, however. Helping to overturn established ideas of atomic constitution and atomic behavior even as it gave birth to an immensely popular craze, radium challenged scientist and common man alike, and journalists scrambled to capture all its marvelous implications.

The eighty-eighth element in the periodic table was stunningly and starkly new. For melancholic man of letters Henry Adams, as for many others, the shock was indescribable: "Radium denied its God," he remarked, "or, what was . . . the same thing, denied the truths of . . . Science. The force was wholly new."[1] William James compared the upset caused by the discovery of radium to something like his beginning to "utter piercing shrieks and act like a maniac on this platform" and the doubts this behavior would sow in the minds of his students.[2] Adams, likewise, felt his "historical neck broken by the sudden irruption" of these forces that were both "anarchical" and "little short of parricidal in their wicked spirit towards science"—these rays were nothing like the wholesome, "harmless and beneficent" rays of the solar spectrum. Other phenomena could at least be measured—even "frozen air," if only "somebody could invent a thermometer adequate to the purpose," he said—but the new phenomena of X-rays and the radioactive properties of radium, the two of which seemed to be related in some as yet unknown way, brought about in Adams's mind a new "supersensual world" where nothing could be measured except by the imperceptibles themselves.[3] While the great mathematician Henri Poincaré had called radium a "great revolutionary," for Adams it was simply a sign of "physics stark mad in metaphysics."[4]

Depending on one's cast of mind, the discovery of radium could be said to illustrate the dawning of "a new epoch in chemistry," bringing investigators "nearer than ever before towards getting 'a glimpse of the nature of things,'" as the *Lancet* reported in 1903, or to so challenge preconceived understandings of the world that it was useful for explaining the pragmatic meaning of truth, as James believed.[5] Either way, one of the most striking features of those early years following the discovery of radium is the curious appearance of a metaphysics, and an attendant mode of metaphorical description, that suffused radioactivity with a peculiarly *biological* cast. Not only were radioactive phenomena characterized in quasi-biological ways from the earliest days, by their discoverers and by others, but radium itself—by far the most powerful and most popular of the radioactive elements—was often described as a "half-living" element in scientific and popular texts alike. Radium was sometimes even accorded vitalizing powers, an ascription that became part and parcel of the radium craze that swept the first decade of the twentieth century. While the earliest discoveries in radioactivity were immersed within rich sets of discourse that overdetermined radium's living aura, the popular craze that followed radium's discovery granted vitality and life a radioactive glow all its own.

Long before the hydrogen bomb indelibly associated radioactivity with death, many botanists and geneticists were eagerly remarking that radium held the key to the secret of life. No mere chunk of glowing earth, this most spectacular of elements was also, above all else, an ideal site for unexpected coincidences and fruitful confluences in the life sciences, for an overlapping of discourses and ontologies that persisted throughout the early twentieth century and proved as productive as it was provocative. Cross-fertilizing and recombining, these initially provocative connections between the radioactive and the living propagated over the decades, across disciplines, and between public and scientific discourses. These crossovers led to conceptual and experimental consequences and involved (at least in passing) many of the major biological questions of the day: the origin of life, the physiological effects of radiation, the nature of mutation, and the structure of the gene.

Emerging at a particular moment at the turn of the century and weaving together already-extant discursive strands and experimental traditions aimed at modifying and understanding life, this intersection between the physical and the biological—between the radioactive and the living—transmuted over the first half of the twentieth century, throwing off various new experimental systems and approaches. By critically engaging with the texts, narratives, and images generated both by scientists and by commentators of the day, I follow the varied and intertwining ways by which radium "came to life," how it played a significant role in the history of biology in the early twentieth century, and, in particular, how it contributed in surprisingly revealing and novel ways to the history of genetics.

Although it emerged in the context of the radium craze at the dawn of the century, this distinctive and provocative overlapping of metaphor and metaphysics, of terminology and technique, and of the living, nonliving, and even half-living proved remarkably productive in experimental terms and ultimately led to key insights into the origin of life, the nature of mutation, and the structure of the gene. Four revealing case studies form the core of my analysis as I examine how radium served for successive biological experimenters as vitalizer, stimulant, mutagen, and analytic tool.

This history does more than cast the established narratives of classical and radiation genetics in an entirely new light. In telling the story of how radium remained an epistemic tool, even as it eventually ceased to be an experimental one, I recount in later chapters how this powerful reworking of radium's role contributed to a crucial and widely recognized, but heretofore unanalyzed, shift in the meaning of mutation itself, from organism and chromosome to gene. Radium was thus not only a primary and vital part of the arsenal of early twentieth-century mutagens, but also played a constitutive role in the historical genetic "redefinition" of mutation, a redefinition that in turn, to date, has helped to obscure the central place of radium in the history of classical genetics.

Moreover, as the role of "atom of life" shifted from radium itself to microbes, mutant organisms, chromosomes, and finally genes, the trope of the "secret of life" moved ever inward. From the initial discovery of the new element in 1898 to the putative discovery of the "secret of life" with the elucidation of the structure of DNA in 1953—a midcentury moment by which the ties that had once bound metaphor and experimental practice together so tightly had decayed to mere discursive residues—this study traces the half-life of this transmuting connection between radium and life.

An introductory chapter sets the stage, finding the roots of this powerful association between radium and life in the earliest biological metaphors and metaphysics of early radioactivity research; in preexisting discursive traditions and popularization practices relating heat, light, electricity, thermodynamics, and notions of a "living atom" to life (all of which were easily subsumed under the new radioactive umbrella); in the popular radium craze of the first decade of the twentieth century; and in the aftermath of controversy regarding other types of rays supposedly produced by living things. Radium, in short order, became *the* living element: the element of choice not only for biological metaphors in a new realm of physics, but even for biological application.

Radium's presence in biology was to prove as striking as it would later be subtle. The first case study (chap. 2) examines the early apotheosis of these connections between radium and life in claims emanating from the Cavendish Laboratory at Cambridge that life had been *produced* from radium. John Butler Burke's controversial work, which comprised some of the first experimental work on the origin of life, wove radium into the history of life on the primordial earth and proved to be a key reworking of the history of spontaneous generation. In a series of sensational experiments that involved plunking a bit of radium into a petri dish of beef bouillon, Burke claimed to have produced cellular forms that were, if not quite living, at least *lifelike*. Appearing to grow and subdivide over a span of days and demonstrating other lifelike phenomena at the cytological level, they nevertheless decayed in sunlight and dissolved in water. Half radium and half microbe, these "radiobes" proved both immensely popular and immensely controversial. In that chapter I examine the ways in which Burke's work not only proved pivotal in the redefining of "spontaneous generation" in the Anglophone context, but also served as a founding moment in the history of experimental research into the origin of life that has to date been routinely overlooked. Burke's work explicitly linked the discourses of cosmic and organic evolution with concrete experiment for the first time with an element that appeared to bridge the two realms. Revealed at the height of the radium craze, Burke's findings also demonstrated the rapid sedimentation of the vitalistic metaphors surrounding radium. Not only reminiscent of life, radium itself, quite literally, vitalized matter.

Burke's spectacular claims offset the other, more respectable uses to which radium was also put in understanding basic biological phenomena. The second study (chap. 3) examines how botanical investigators in the early twentieth century used radium to induce or control biological evolution. Explicitly linking the transmutation of the physical species of radium with the transmutation of biological species, Daniel Trembly MacDougal and Charles Stuart Gager of the New York Botanical Garden and the Brooklyn Botanic Garden, respectively, independently irradiated plants with radium in an attempt to study its physiological effects as well as to provide experimental confirmation of Hugo de Vries's new "mutation theory." One of the dominant evolutionary accounts of the early twentieth century, de Vries's theory was widely seen as providing a mechanism for speciation where Darwinism had failed, and de Vries himself had suggested that radium and "the rays of Röntgen" might be useful in inducing mutations—a suggestion that was rapidly taken up. Metaphors of radium's powers were put to the experimental test at this moment and passed. Even those plants that happened not to mutate were seen to have been "stimulated" by radium, which "accelerated" their growth toward an "early senescence." What in a later nuclear age would be a clear sign of damage was—in the ongoing dynamic interplay between popular and scientific understandings of radium's biological effects—clear proof of radium's relevance in the novel early twentieth-century quest to induce and ultimately control evolution.

The third case (chap. 4) turns to further attempts along these lines made by two of the leaders of classical genetics: the Columbia University geneticist T. H. Morgan, best known for his work on the fruit fly *Drosophila melanogaster*, and Cold Spring Harbor investigator Albert F. Blakeslee, who would later become the second director of the Station for Experimental Evolution. Morgan focused on animals and on the use of radium in producing phenotypic mutants that could be attributed to mutant genes; Blakeslee, by contrast, focused on plants and on the use of radium in producing phenotypic mutants that were demonstrably shown to be chromosomal, and not simply genic, in nature. Morgan had turned to radium after having tried a number of other unsuccessful techniques to induce mutations in *Drosophila*, and he succeeded in his quest at nearly the same time as his friend and colleague, Jacques Loeb. Though Morgan later disowned his own claim that radium had been responsible for the mutants he discovered, his discounting of Loeb's mutants—and Loeb's dismissal of Morgan's findings in turn—presents a curious state of affairs that reveals how shifting ideas about radium's effects were inextricably related to ongoing shifts in the understandings of the artificial induction of "mutation." Blakeslee's experiments with radium, for example, established to widespread acclaim that new species could be produced by what he called *chromosomal mutation* (or "chromosomation"), and that this was as important a mechanism of evolutionary change as the *genic* mutation with which the drosophilists were more familiar. Blakeslee's work thus provides a key instance of the use of radium not only in attempts to confirm de Vries's mutation theory, but also to investigate in deeper cytological detail the ways in which induced mutation could occur in a suitable model organism. Although heretofore relatively unstudied by historians of genetics, Blakeslee's work shows how radium was instrumental not only in attempts to understand the physical nature of mutation, but also in what his contemporaries called "experimental evolution" or "evolutionary engineering" (Blakeslee himself would later refer to the emergence of a "genetics engineer").[6] The work of both Morgan and Blakeslee shows that radium remained a central experimental mutagen even as its precise role in inducing mutations—once so clear—came under increasing scrutiny as geneticists began their radium-inspired work on "artificial transmutation."

The fourth case (chap. 5) focuses on Hermann J. Muller's legendary "artificial transmutation of the gene," which has frequently been presented as the origin of the modern study of induced mutation. What is less well known is that Muller came to his researches through his fascination with the powerful metaphorical and metaphysical connections between radium and life that he encountered as a young man. Muller ultimately reworked these tropes to suggest that mutation and transmutation were fundamentally connected and that radium could be useful to produce not only phenotypic and chromosomal mutants, but mutations at the most fundamental level of all: the genes. In seeking to more precisely characterize the nature of mutation in *Drosophila*, Muller began with radium before shifting to X-rays as his mutagen of choice by the late 1920s. This shift was due to technological advances in the delivery of X-rays—the two radiations were increasingly being equated in their physical nature and in many of their biological effects by the mid-1920s—as well as to contingent circumstance, as Muller's vial of radium broke during a hot train ride through Texas in 1924. Muller's landmark 1927 announcement of his spectacularly precise and detailed new methods for the "artificial transmutation of the gene" ultimately earned him the Nobel Prize.

Radium was thus not only a primary and vital part of the arsenal of mutagens used by early twentieth-century researchers—from MacDougal, Gager, and Morgan to Blakeslee and Muller—but also played a constitutive role in the crucial and widely recognized, but heretofore unanalyzed, historical redefinition of "mutation." In his shift from radium to X-rays, and from transmission genetics to transmutation genetics, Muller ended up radically recharacterizing what had been a pluralistic set of understandings of "mutation" as a fundamentally genic phenomenon. This shift in the meaning and referent of "mutant" and "mutation"—from organism to chromosome to gene—not only marked the beginning of the end of a multilevel, nuanced understanding of mutation and its replacement by a fundamentally genic theory of mutation, but also ended up distancing radium from life in experimental terms. As the γ-rays of radium were increasingly understood by biologists to have the same effects as X-rays (physicists had long since equated the two), Muller's focus on the gene as the proper target for mutation and the X-ray as the proper tool for inducing it became a sentiment and a practice more widely shared. By the 1930s, X-ray-based "radiation genetics" had largely, but not entirely, replaced the use of radium in the study of the structure of the hereditary elements, and a larger "radiobiology" was still to come. This turn away from radium and toward other sources of ionizing radiation contributed in turn to the forgetting of the important role of radium in the successful earlier work of Gager, Blakeslee, and various others—work that Muller had encountered and studied on the path to his own remarkable experiments. Such was the aftermath of using radium as an epistemic, and not only an experimental, tool.

The fact that Muller's radioactive metaphysics of the gene, inherited from this earlier work with radium, could contribute to this process of historical rewriting shows further ways in which the powerful association between radium and life continued to transmute over the course of the first half of the twentieth century. The case of Muller is thus neither the zenith nor the end of the tale, but a fascinating inflection point: the collective forgetting of radium's early role is a consequence of the same processes of interpretation that permitted radium and life to become so closely associated in the first place. It is also reflective of the same historical processes that enable us to find and to trace this powerful association over the decades of this story. There is thus more to this account than the mere uncovering of the many and varied transmutations and disintegrations of the long-standing and powerful association between radium and life, or a series of disconnected musings on remarkable metaphors in a particular corner of biology.[7] In fact, studying the ongoing transmutations of this powerful association between radium and life across experimental systems, historical actors, and decades can reveal as much about the nature of history as about heredity.

This book is therefore structured to be read at two levels. At one level, it is a series of straightforward case studies on the applications of radium in biology—how and why these applications came to be, and how they were eventually lost to historical memory, as just described. These fascinating stories about radium—a kind of "prehistory of radiobiology"—not only uncover heretofore unknown but important dimensions of radium's life in biology, but also help to revise canonical moments in the early history of genetics. But at another level, the book is also a novel experiment in historiographical form in that it seeks to treat "radium" not only as the subject of the book and as an object that life scientists discussed and worked with, but also as the narrative conceit and immanent analytic for the book as a whole. In a manner broadly analogous to the ways in which the properties of radium inspired, structured, and sometimes disrupted the experiments of early physicists, botanists, and geneticists, I hope to reveal how reflexively taking "radium" seriously as an immanent analytic—tracing key moments of transmutation in the long half-life of radium's association with life—can inspire, structure, and ultimately challenge a historical argument through dynamics of transmutation and decay similar to those that were at work for the historical actors themselves. Their struggles with the intersection of radium-based radioactive discourse and experimentation are not fundamentally different from those encountered by the historian seeking to narrate the half-life of such a connection.

And so a word about half-lives. *Radium and the Secret of Life* is thus also an exploration of how to write about the intersection of the worlds of the radioactive and the living in the first half of the twentieth century without relying on all-too-familiar biographical tropes and metaphors such as the "life and death" of an element. Splitting the difference, the trope of the half-life serves as a narrative tool that suggests what it might mean for radium to serve as an immanent analytic in a historical account. Radium and life were powerfully and closely associated with each other as early as 1904, both in the public imagination and in scientific terminology and experimentation. How far did this association extend, and how long did it last? It began with a sharp initiatory moment (which was itself a perpetuation of and refraction of earlier entities and analogies) and has since intermittently decayed toward—but has never quite reached—a leaden state of complete dissociation. By tracing this asymptotic process of decay, and in ultimately coming to a point in the Conclusion where *it is no longer clear* whether the historical evidence speaks to a still-extant connection between radium and life, I hope to explore the possibility of a more consciously reflexive history—one in which the radiant narrative itself comes to test the limits of evidence and argument.[8]

In the final chapter (chap. 6), therefore, I explore the afterlife and persistence of radioactive residues in Muller's later work, in that of his contemporaries, and in the larger context of the study of heredity in the 1930s and 1940s. In these cases it becomes increasingly less clear whether there is any legitimate connection to be drawn to these further transmutations, decays, and disintegrations of what were once powerful associations between radium and life. In recounting this history, with its countless possible historical residues, *Radium and the Secret of Life* thereby challenges the very idea of any neat historical narrative of the "life and death" of radium's role in biology. In a theoretical coda, I suggest that this is what a hermeneutic of transmutation, seriously attempting to deploy "radium" as an epistemic tool for the historian as much as it was for the scientist, might look like in the form of historical narration. In short, as the experimental productivity of the once all-powerful metaphorical and metaphysical association between radium and life slowly decayed to trace residues (and tracers) in a generalized background of radiobiology, the once-pronounced clicking of the Geiger counter of historical narrative slowly merges into noise.

Throughout this study, I therefore consciously draw on Hans-Jörg Rheinberger's treatment of the *historial* and what he has called the "temporal structure of the production of a trace."[9] In his masterful empirical and theoretical analysis of "experimental systems," Rheinberger traced the emergence into scientific reality of "epistemic things" from the articulation and composition of "traces."[10] The traces in the story of radium and the secret of life, however, are inverted—they do not lead up to an epistemic thing, but rather away from a powerful originary moment when radium and life were clearly and commonly associated, when the secret of life clearly had something to do with radium. And so, after tracing a path from radium and its intersections with the spontaneous generation of life (Burke) to the cell (MacDougal) to the chromosome (Blakeslee) and to the gene (Muller), by way of a theoretical coda, I explore how, as discursive tropes and material agents alike continued to transition from radium to other sources of ionizing radiation, this initially powerful association of radium and life continued to disintegrate. Initially so rich with metaphor and metaphysical association, the connection between radium and life bifurcated into increasingly instrumentalized or metaphorical traces such that only distant but tantalizing echoes of its power remained. Just as investigators and commentators at the turn of the century had once held that the discovery of radioactivity entailed the unveiling of the "secret of matter," the ascription of the "secret of life" to the structure of DNA upon its discovery in 1953 can be viewed as one of many remaining radioactive residues in the mid-century disarticulation of radium and life some fifty years after their first powerful association—a well-known and convenient ending place for a story that, in approaching an asymptote, otherwise has no easy and neat narrative ending.

Interlinking metaphors and metaphysics, preexisting discourses and novel experimental ontologies, this story, then, in more ways than one, is the story of how radium came to life—and of how life came to radium. The element of choice for bringing together the realms of the radioactive and the living, radium was the atom of life and yet contained within itself the seeds of its own decay. This study thus seeks to reveal the changing particulars of this powerful association between radium and life over the decades and across experimental systems in order to illustrate how, as experimental productivity eventually outpaced metaphorical and discursive resonance, an initially unified coherence between "radium" and "life" was lost. Interacting with dominant conceptual frameworks, technological realities, and living organisms as this association generated a series of energetic and ever more productive experimental approaches, this initially powerful resonance between radium and life decayed to trace residues in a generalized background of radiobiology. Time after time, as the first half of the twentieth century unfolded, this nexus between radium and life—in a variety of directions and manners—transmuted.

1 The Birth of Living Radium

While uranium and thorium had already been known for decades, and while their newfound radioactivity catapulted them to greater prominence at the end of the nineteenth century, it was only with Marie Curie's discoveries of polonium and especially of radium, and with Ernest Rutherford and Frederick Soddy's subsequent theory of radioactive decay, that the new science of radioactivity took off—and with it an intense new culture of fascination with radium. The turn of the century saw the birth of a metaphorical (and sometimes more than metaphorical) understanding of "living radium."

For an age when chemistry and physics were thought to be closing in on the last few secrets of nature, the back-to-back discoveries of X-rays and of radioactivity came as a complete surprise. Curie's famed discovery had followed immediately after Wilhelm Roentgen's initial discovery of the penetrating power of X-rays in 1896 and Henri Becquerel's accidental discovery of the radioactive properties of uranium shortly thereafter. Dredging through tons of Joachimstal pitchblende to obtain the smallest fraction of radium in 1898, Curie had found with radium the radioactive element par excellence, some millions of times more radioactive than uranium. Incredibly rare and precious, even in minuscule amounts radium dazzled, glowing in the dark and shooting off rays in a seemingly endless blast of energy that came from nowhere in particular. As Rutherford later recounted, "The name radium was a very happy inspiration of the discoverers, for this substance in the pure state possesses the property of radio-activity to an astonishing degree."[1] In comparison, Roentgen's and Becquerel's discoveries had made nowhere near the impact on the public.[2] When Curie finally succeeded in isolating radium in a pure state in 1902, granting incontrovertible proof of its elemental status, radium was already well on the way to becoming the all-powerful and wondrous new element that could do everything—and that soon enough could do no wrong.

The peculiar connection between the phenomena of radioactivity and the properties of and discourses surrounding life first began to emerge in those earliest days of the science of radioactivity with Ernest Rutherford and Frederick Soddy's discovery that radioactivity, in fact, indicated the transmutation of the elements. Elements were supposed to be the fundamental building blocks of the physical world, the basic level of atomic composition of all things. "Atomic" literally meant that which could not be subdivided. A substance that had all the hallmarks of an element, that fit an empty spot on the periodic table, and yet came apart, spontaneously, was—prior to the discovery of radioactivity—almost inconceivable. "Elements" simply did not permit subdivision. The discovery of the transmutation of atomic species proved to be nearly as problematic a revelation for Rutherford and Soddy as the transmutation of biological species had once been for Darwin.

For all its mythical status, Rutherford and Soddy's legendary collaboration lasted only a year and a half. Beginning in 1901, when they both found themselves at McGill University—Rutherford an established professor of physics, Soddy an up-and-coming young chemist—their "local and intense" collaboration resulted in the production of nine papers, the last of which, "Radioactive Change," appeared in May 1903 and presented their theory in its final form. Their famed collaboration not only brought forth the first solidified account of elemental transmutation—the "disintegration theory of radioactive transformation"—but also served to explain many of the other radioactive properties of the radioelements and to advance the idea of an evolutionary history of the universe told by its elements.

The *experimentum crucis* that led to the birth of the disintegration theory of radioactive substances took place in April 1902, when Rutherford and Soddy observed thorium X spontaneously changing within the confines of their laboratory setup into the noble gas argon. Soddy, though no alchemical adept, had nevertheless always been interested in the connections, historical and otherwise, between alchemy and chemistry and had even lectured on alchemy in his course on the history of chemistry: "I made that goal [of transmutation] quite clear," he said. The appearance of alchemical transmutation before his very eyes, however, was almost "too devastatingly simple." He recalled himself "standing there transfixed as though stunned by the colossal import of the thing."[3]

> I remember when I interpreted my first experiment I could not wait to tell Rutherford, but words would not come. I could feel my heart throbbing, and as though propelled by some outside force I heard myself utter unbelievable words: "Rutherford, this is transmutation!"

Rutherford, "in his breezy manner," is said to have shouted back: "For Mike's sake, Soddy, don't call it *transmutation*. They'll have our heads off as alchemists."[4]

Soddy, however, remained transfixed by the idea of elemental transmutation. Once disparaging of earlier attempts at alchemical transmutation, he now found himself converted. In short order, he publicly declaimed in a lecture at McGill that "alchemy must be regarded as the true beginning of the science of chemistry." Accordingly, he said, transmutation "is, as it has always been, the real goal of the chemist." From doubtful practicing chemist-cum-historian of alchemy to firm adherent, Soddy came completely around[5] and found himself "entirely engrossed" in interpreting his newfound transmutation:

> The atoms were disintegrating, so disposing of the chemists' cherished theories of its immutability. I began to consider the state of the disintegrated atom. Was it now a smaller atom of the same element? By its integration would it have assumed another character, become another element? By further possible emissions would it further disintegrate and if so, at what rate? How long would such a disintegrating atom live? Since it seemed obvious that most of the atoms in the element would at some time be suffering disintegration it followed that the element would be composed of atoms in various stages of disintegration.[6]

Soddy recollected Rutherford afterward "taking me to task because people were saying that what we were saying was tantamount to 'transmutation,' and I had to convince him that it *was* transmutation and put him *au fait* with the chemical evidence to confute anyone who disputed it." Transmutation still smacked too much of the alchemical for a respectable scientific report, however, and in the first published account of their discoveries in April 1902, the word "transmutation" was replaced with the more benign "transformation."[7] Yet the excitement of their alchemical discovery still bubbled beneath the surface: in an effort to get their first paper published in the *Transactions of the Chemical Society*, Rutherford had written privately to Sir William Crookes, saying that "although of course it is not advisable to put the case too bluntly to a chemical society, I believe that in the radio-active elements we have a process of disintegration or transmutation steadily going on which is the source of the energy dissipated in radioactivity."[8] Nevertheless, the shift to "transformation" as the term of choice was rapid—by September of that same year, Rutherford and Soddy reported in the *Philosophical Magazine* that radioactivity was "a manifestation of sub-atomic chemical change" and, as such, "the radioactive elements must be undergoing spontaneous *transformation*."[9] From here on out, a distinction emerged between rather more scientific references to "disintegration" and "transformation" and what were clearly more popular references to "transmutation"—although Soddy continued to blur the lines from time to time.

With their different disciplinary interests, it was only natural that Rutherford and Soddy would pursue different paths after their discovery. Rutherford, the physicist with "a most radiating smile," focused on further experimentation aimed at discovering the nature of the α-particles produced in moments of radioactive transformation.[10] Soddy, the chemist, focused more on the chemical implications of the new discovery and looked for further proof of transmutation. While Rutherford remained at McGill until leaving for Manchester in 1907, Soddy had already transferred to William Ramsay's laboratory at Cambridge by 1903. Soddy found his first samples of radium by chance one day in April of that year as he walked "past a store [Isenthal's] on Mortimer Street off Upper Regent Street" in London. A sign in the window read: "Pure radium compounds on sale here." At a time when radium was available only "by favour of the Curies," as Soddy recalled, this was an exceptional find: "Here it was to be bought in a London shop at some eight shillings a milligram of pure radium bromide," the product of a German production firm (Geisel of the Chinin Fabrik of Brunswick) that had begun to manufacture radium compounds on a commercial basis.[11]

The final proof of transmutation thus came in Ramsay's laboratory with the production of helium from the radium sample on April 27,

1903.[12] The result of Ramsay and Soddy's collaboration, one commentator noted, was nothing less than "the chemical sensation of the summer of 1903."[13] The presentation of the proof of transmutation at the annual meeting of the British Association for the Advancement of Science in Southport in 1903 came at a time when Lord Kelvin was still espousing the idea that it was the ether that carried energy to radioactive substances, rather than seeing such energy as something inexplicably inherent in the atom.[14] By 1903, however, general agreement was beginning to fall in favor of the Rutherford-Soddy account of radioactive transformation. The only significant holdouts against the theory of radioactive transformation in the British context, it turns out, were Kelvin and Henry E. Armstrong. Armstrong attacked the disintegration theory, "which assumes that nature has endowed radium alone of all the elements with incurable suicidal monomania," but both men were largely silenced after Rutherford's presentation.[15] The physics of the new phenomenon of radioactivity was beginning to come together.[16] The broader cultural and biological import of radium, however, was just beginning.

"Physics Stark Mad in Metaphysics"

"In pre-radium days," W. Hanna Thomson remarked in his popular 1909 *What Is Physical Life? Its Origin and Nature*, "we took the diverse chemical elements for granted, with vague speculations as to their possible evolution from some primitive kind of stuff out of which the fabric of the world has been spun." But the discovery of radioactivity, he went on, "has made it certain that one element can be evolved from another, or, in other cases, legitimately thought of as evolved from another, by the addition or separation of certain components."[17] Whether or not Thomson's description of the phenomena of radioactivity is technically accurate, what is certain is that the work of Rutherford and Soddy took what had previously been a merely suggestive connection between the processes of cosmic and biological evolution and linked the two much more closely. From Robert Chambers's all-encompassing *Vestiges of the Natural History of Creation* of 1844 (which continued to outsell Darwin's *Origin* even years after the latter's publication in 1859) to the work of Herbert Spencer and others, many in the nineteenth century readily viewed evolution as a simultaneously cosmic and biological process.[18] While elements in everyday experience may have been stable, the idea that elements—much like living things—at some point in the history of the cosmos underwent an evolutionary process was considered only mildly far-fetched and not entirely beyond reasoned imagination.[19] Astrophysicist Norman Lockyer's 1900 book *Inorganic Evolution* is perhaps the most notable indicator of the idea being already "in the air" just prior to Rutherford and Soddy's work.[20] It was only following their proof of transmutation, however, that these earliest evolutionary links between the radioactive and the living could, and did, become much more closely and provocatively established.

Given these preexisting traditions that linked cosmic with biological evolution, Rutherford and Soddy's new quasi-alchemical talk of the "transmutation" of the elements could not help but resonate with talk of *biological* transmutation. Elemental transmutation seemed to imply cosmic *evolution* of some sort, just as biological transmutation implied biological evolution. Radioactivity and life were thus linked not only from the dawn of research into radioactivity, but from the very dawn of time. As such, the terms of the one could be applied with ease to the other—which is precisely what Soddy proceeded to do.

In his Wilde lecture of February 23, 1904, "The Evolution of Matter as Revealed by the Radio-Active Elements," Soddy remarked at greater length on the nature of the relationship between the process of radioactive change, the "evolution of the elements," and cosmic evolution more generally, conjecturing whether one could ever "regard the universe . . . [as] proceeding through continuous cycles of evolution" and discussing the period of average life for the first time.[21] To an observer getting only a glimpse into a vast cosmic process of evolution "going on for indefinite ages," he wrote, the currently recognized "atoms of the periodic law" were probably only a subset of the original constitution of the universe and its "parent-element[s]." These elements as we now know them were, Soddy wagered, but "the forms with longest life, which exist to-day because they have survived a long process of evolution in which those physically unfit have disappeared." Or, as he elsewhere characterized cosmic history, "Matter has passed to its present position of apparent immutability by a long process of natural selection. The elements known to the chemist are stable because they exist and have survived. On the other hand, it is now possible to examine some excessively unstable forms of matter," or, in other words, the radioactive elements.[22] The radioactive elements—uranium, thorium, and radium, and by no mere coincidence the three heaviest elements in the periodic table thus far—were for Soddy "transition forms" or "elementary forms of matter physically unfitted to survive, but which are brought within our powers of knowledge because they constitute the temporary halting places through which matter is passing in a scheme of slow continuous evolution from the heavier to the lighter forms."[23] As Soddy said in his annual summary of the year's findings for the British Chemical Society, "We have here the introduction into chemistry of a conception analogous to that of evolution in the biological sciences."[24]

The choice of language is striking: not only was there a kind of cosmic evolution taking place—by natural selection, no less—but atoms were said to have *parents*, to have *lives*, to *survive*. ("Its simple existence is eloquent of its fitness to survive," Soddy wrote.)[25] As Rutherford wrote to Jacques Loeb in 1907, "I'm feeling very fit and hard at work examining the recent evidence of the parentage of radium—latest report still uncertain though no doubt this is a productive parent."[26] Parents, grandparents, and even great-grandparents all made an appearance in Soddy's thinking: "Radioactive children frequently resemble their great-grandparents with such complete fidelity that no known means of separating them by chemical analysis exists."[27] (Even Marie Curie would refer in her 1911 Nobel lecture to "the atom of radium [that] gives birth to a train of atoms of smaller and smaller weights.")

Rutherford and Soddy needed a term to describe these species of unstable atoms transmuting their way from one element to another, a term that highlighted these kinds of particular and as yet unsung connections between the animate and inanimate that would grant the inorganic a particular kind of half-living status. The name they settled on was deliberately evocative of one of the most basic of living processes: these radioelements were to be called *metabolons*.

Metabolons: Half-Living Elements

The term "metabolon" first appears in Rutherford and Soddy's final paper, "Radioactive Change," published in the *Philosophical Magazine* in 1903. Here, aware that other corpuscles were being expelled at the same time as a ray was produced, they remarked:

> It seems advisable to possess a special name for these now numerous atom-fragments, or new atoms, which result from the original atom after the ray has been expelled, and which remain in existence only a limited time, continually undergoing further change. . . . We would therefore suggest the term *metabolon* for this purpose.

Some metabolons—and here Soddy explicitly mentions radium—are "metabolon[s] in the full sense of having been formed by disintegration of one of the other elements present in the mineral." As such, he said, they form "the common ground between metabolons and atoms, possessing the properties of both." Like living things, metabolons were curiously self-reproducing entities that reproduced differentially—never reproducing themselves, but always some other element further down the chain, either one, and only one, other metabolon or one of the other stable elements. Though not living, these metabolons clearly had a "life" all their own—and Soddy first used the term "life" cautiously, in quotation marks. By 1904, however, the quotation marks were gone: a metabolon was one of "a certain number of short-lived transition-forms of matter intermediate between the initial and the final atoms," or simply, as Soddy summarized it, "an atom with a limited life."[28]

Early on, others had similarly used quotation marks when referring to radium's "decay."[29] But this analogy, too, went from being merely suggestive to being provocatively real within a short time. As Soddy noted, "At first sight it seems the atomic theory, which bears out and is borne out so strikingly by atomic disintegration, opposes a barrier to any conception of atomic up-building. . . . But the atomic theory appears to demand equally with a *per saltum* degradation, a *per saltum* accretion."[30] And already by 1904, one contemporary noted, "The atoms are crumbling and decaying. Must they not also be forming and coming to birth? Decay only, without birth and cumulation, cannot be the last word!"[31] And in the same year, the American physicist Robert Millikan could refer to decay in such a way and remark without irony that

> the only change of this kind thus far discovered to be going on in the structure of the atom is in some respects similar to the changes that are incessantly occurring in the organic world in the structure of molecules. By the ordinary process of decay, the more complex molecules are continually disintegrating into simpler ones, and in so doing are setting free the energy that was originally put into them when the processes of life first built them up into their complex forms. . . . The analogy suggests a profoundly interesting question. Is there any process which does among the atoms what the life process does among the molecules, which takes the simple forms and builds them up again into more complex ones?[32]

A profusion of biological terms rapidly emerged in short order and came to be widely adopted to describe the process of radioactive change.[33]

From the earliest days in the Curies' laboratory, radioactivity was—like a disease—described as "catching."[34] In fact, noted one popularizer, "After working with radium for some time an experimenter finds that everything in his laboratory, the walls of the room, and worst of all, he himself, has become radio-active."[35] Still others referred to material "being all the time *exhaled* or emanated from the radium."[36]

Biological metaphors ran unchecked. In Soddy's grand scheme of cosmic evolution by natural selection, radioactive phenomena were said to be the result of a process of radioactive *decay* that produced *daughter elements* from *parent elements* through the passage of radioactive *generations*, a process that could be measured with the idea of a "half-life"—a term first coined in 1907 and first used with respect to radium.[37] By 1906 Soddy was referring to radium as "the missing 'big brother' of the alkaline-earth family of elements" and to helium as "the lightest member, or 'baby,' of a whole family of gaseous elements exactly similar in chemical nature."[38]

The table of contents to Soddy's popular *The Interpretation of Radium*, published in 1909, reads like a summary of the vitalized radioactive discourse, referring to the "decay of the emanation and its reproduction by radium," "its expectation of life," the "average life of radium," "average life of a disintegrating atom," "the parent of radium," the "period of average life of uranium," the "growth of radium by uranium," "the direct parent of radium," "the stately procession of elementary evolution," "survival of the fittest or most stable atoms," "radium and the struggle for existence," and the "universality of the conception of evolution to the material universe, animate and inanimate." Is it any wonder that by 1912 Soddy felt compelled to publish an article entitled "Transmutation: The Vital Problem of the Future"?

Soddy was not alone. Rutherford seemed equally transfixed by the biological valences of radium, and he devoted a whole chapter of his 1906 *Radioactive Transformations* to the "origin and life of radium," describing its growth and decay, albeit in terms less florid than Soddy's.[39] Others picked up on the terminology rapidly, promoting their own further vitalization of the discourse surrounding radium—a report in the *Lancet* even referred to radium's "native habitat."[40] For still others, radium was not only alive, but even had "a short life and a merry one."[41] One commentator even referred to "these fascinating problems of the ultimate state of *extinct* radio-active matter."[42]

Soddy had described the evolution of the elements as a "struggle for existence between non-living substances" and as a "contest of stability, the unstable constantly breaking down into more stable forms better fitted to survive in the conditions under which they find themselves." The Darwinian resonances of Soddy's account were not lost on his contemporaries: one popular writer held that just as "Darwin and Wallace revealed to us the evolution of living organisms; it seems possible that Thomson, Larmor, and Rutherford may enable us to trace the corresponding process in *inorganic* matter."[43] T. C. Chamberlin concluded in 1909 that "if the atom shall show an authenticated pedigree, it will easily take its place in the procession of the derived, with the plant, the animal, the earth, and the stars."[44] Even a decade later radium rays were still being described as "of the same category and obey[ing] the same laws as the forms which before have nourished and embellished life."[45] Still others picked up on the vitalized terminology and referred to radium as having to be "aged" (as if it were a fine wine) before reaching maximum activity,[46] or, even more confusedly, to its mortality and the "death of a molecule" of radium.[47] One even went biblical:

> Having discovered that the atoms are not immortal, chemists are now hard at work constructing family trees. Like other genealogical tables there are some discrepancies between them, and in places one or more generations may be skipped, but they read somewhat as follows: Now Thorium begat Uranium, and Uranium begat Radium, and Radium begat Helium and Polonium, and Polonium begat Lead.[48]

Radium Regnant

This vitalized radioactive discourse necessarily applies to all radioactive elements in general. Why, then, was the claim made for a specific connection between *radium* and life? The answer, in brief, is that as went radioactivity, so went radium—but even more strongly so. Radium, the quintessential metabolon, held pride of place among the radioelements, even as slippage between "radium" and "radioactivity" was everywhere apparent in the literature.[49]

As early as November 1903, for example, the index of *Nature* already had separate headings for "Radiography" and "Radium." Similarly, while "Radioactivity" had become a subject category in the 1903 *International Catalog of Scientific Literature*, growing out of the earlier heat-and-light subject category ("The Emission of Radiation, Phosphorescence, etc."), by 1906 the subheaders had multiplied across the page as the category was renamed "Radioactivity (Radium, Etc.)." Even the popularizer William J. Hammer noted that of the three radioactive substances he most frequently mentioned, "radium is by far the most important and is of extraordinary interest."[50] Years later, Eve Curie would write:

> Radium, radium, radium! The magic word came up ten or twenty times, passed from tongue to tongue, and sometimes provoked a regret in Marie: chance had arranged things badly in making radium such a prodigious substance and polonium—the first element the Curies had discovered—an unstable body of secondary interest. The patriotic Marie could have wished that polonium, with its symbolic name, had drawn fame upon itself.[51]

Soddy had explicit reasons for holding radium above the other radioactive elements, reporting to the British Chemical Society in 1904 that of the eighteen radioactive "substances" then known, only in the case of radium was there direct evidence that an *element* had been found.[52] In 1911 he continued to hold that radium was "the most important of the new radioactive elements"—not only was it "by far the most completely investigated of the shorter-lived radio-elements," it was also conveniently short-lived enough to show its radioactivity and long-lived enough to allow it to be studied.[53]

Radium was not only the first radioactive element to be isolated in pure form, but also the most intensely radioactive. In bromide form it literally glowed in the dark with a bluish hue that reminded many scientists of the glowworm. Moreover, Soddy claimed, while all metabolons behaved similarly, the sheer intensity of *radium's* radioactivity meant that by all rights it was *not* some leftover product from earlier universal-historical processes of decay, but had to be actively "*being reproduced*"—not just produced—"as fast as it disintegrates." This revelation led Soddy to further experiments of a vitalized cast to see "whether a quantity of uranium, originally free from radium, would not *grow a crop* of the latter element in the course of time."[54] He began to speculate on the necessity of some kind of "regeneration process" of "atomic up-building"—the same thing that Millikan had wondered about.[55] The special aura of the living that imbued radium even led Soddy to remark that "this one element has clothed with its own dignity the whole empire of common matter. Its ultra-potentialities are the common possession of that world which, in our ignorance, we used to refer to as mere inanimate matter."[56] Radium was special, in other words, not least because it had made the inanimate world *alive*.

Radium was quite the wonder, becoming so rapidly associated with "puzzles" and "riddles" of all sorts that "the mystery of radium" became a stock phrase in wide circulation during the first decade of the twentieth century. (Life itself was also routinely and widely described as an "enigma," "riddle," "mystery," and "secret" at this time.) Charles Vernon Boys, as president of Section A of the BAAS, in reviewing the scientific accomplishments of the year 1903, "characterized the discovery of the properties of radium as transcending all others in their intrinsic importance and revolutionary possibilities." Boys was said to have thought the claims for radium so remarkable that "if the half of it were true, the term 'mystery of radium' was inadequate; the 'miracle of radium' was the only expression that could be employed."[57] As a consequence of all these factors, much of the vital discourse surrounding the metabolons could be, and in fact *was*, concentrated on the wonder element itself—radium.[58]

The Metaphysics of Metaphor

This was physics stark mad in a peculiar kind of metaphysics: a metaphysics of metaphor. Actively linking the organic and the inorganic in an array of mutually constituting and mutually reinforcing ways, key living metaphors—from the merely descriptive to the pregnant turn of phrase—aided the development of the underlying metaphysics of radioactivity. Conversely, metaphysical considerations, from ontological assumptions about the nature of the universe to the life histories of its constituent elements, helped drive the selection of the appropriate (lifelike) descriptive metaphors.

A full spectrum of possibilities for relating metaphysics to metaphor was at play in Soddy's and other early radioactivists' reports. On some occasions, for example, Soddy and others seem to have held that there were clear ontological parallels—or even identical processes—going on in the world of the living and the world of the radioactive, and they called for the same terms to be used to describe both (the metaphors chosen reflected a deeper metaphysical identity). At other times, the sheer conceptual productivity and playfulness provided by such analogies and metaphorical overlaps between the two realms seemed to be the attraction, as was undoubtedly the case for radium's popularizers. At still other moments, the biological phrasing for radioactive phenomena was simply thought to be the most convenient to retain, as when

Soddy later in the decade recommended continued use of the phrase "the parent of radium," rather than any other complicated term, until the radioactive nomenclature could be systematized.[59]

And then there were those intriguing moments when despite obvious difficulties in applying the terminology of life to radioactive phenomena, its use nevertheless seemed compelling. At moments like these, Soddy and other writers seemed not so much confused by the putatively living or lifelike status of radioactive phenomena as much as carried along and entranced by the all-encompassing discursive comfort, convenient overlap, and novel prospects such terminology provided. In this way, metaphors and metaphysics served to coproduce each other. Soddy himself was well aware of the obvious difficulties that clearly attended the description of a radioactive element as "living"—for example, one atom of a radioactive element could last for centuries while a neighboring atom disintegrated in moments. Indeed, he noted that the "average life period of the atom is totally different from that of any living creature" and thought that this could be easily demonstrated by experiment: "If you contrast very old with new-born radioactive atoms of the same kind each sort will have the same period of average life. . . . Both lots are quite indistinguishable and change at the same rate." Nevertheless, Soddy felt the appeal of a vitalized terminology and regularly chose to describe radioactive phenomena in living terms.[60]

An obvious difficulty in applying vitalized terminology to radioactive phenomena concerned the alleged process of "evolution" itself: while many of Soddy's contemporaries conceived of organic evolution as generally progressive, inorganic "evolution" was unabashedly a story of the decomposition of weightier elements, of dissolution and of *decay*. How could this story be reconciled with the dominant vitalized discourse of radioactivity? A variety of approaches ensued. One immediately apparent resolution of an otherwise indeterminate oscillation from the radioelements having a "life" (in quotes), to having a *life*, to being recognizably distinct from life was the development of the term "half-life." This term, now so familiar to us in its delimited scientific denotation, first emerged out of attempts to "get at" the curious persistence of these half-living, decaying elements within a vitalized discourse of radioactive phenomena. Curiously appropriate as a physical measure (the amount of time until only half of a sample was still radioactively "alive") while simultaneously pregnant with biological resonances, the term "half-life" also served as a convenient way to reconcile the putatively biological affinities of the radioelements (half-*life*) with other physical realities that provided grist for disanalogy (*half*-life).

This vitalized radioactive discourse existed side by side with—and at times even co-opted the terms of—other modes of reasoning that helped propel it onward. Take, for example, the case of the *measure* of radioactivity, λ. While Curie had previously characterized *each atom* of a radioactive element as steadily supplying the observed radiated energy, Rutherford and Soddy—aware of "transition-forms" of elements not known to Curie, and aware that at any given moment only a few atoms were unpredictably and individually undergoing spontaneous transformation—appealed to the idea that "a property which is contributed by a *constant* fraction of the total is indistinguishable from a property possessed by each atom in common," thereby redefining radioactivity as the property of a population.[61] This reinterpretation of radioactivity as the property of a *population* of radium atoms rather than an atomic property per se—a reinterpretation that involved the establishment of the technical measure of a "half-life"—is one means by which statistical and biological modes of discourse came to be inseparably united. In the process of reinterpreting radioactivity, Rutherford and Soddy not only produced a new measure of radioactivity, λ (which stood for "the proportionate fraction changing per second"), but returned to the idea of an atom of radium having a life by holding that the inverse, 1/λ, represented "the *average life* of the metabolon in seconds."[62] The statistical treatment of atomic populations—at the heart of the development of the technical concept of a half-life—was clearly a physicalist and statistical mode of reasoning, and yet so long as one was already committed to speaking of "parent-elements" and "metabolons" having a "life," it was a mode of reasoning that was steeped in biological metaphors.

Not every radioactive term with an apparent biological tenor was the result of a conscious decision, however. Many putatively "living" terms for radioactive phenomena came from other traditions with their own particular historical trajectories but found a suitable home in the realm of the radioactive. The concept of "decay" in radiation, for example, was named by Rutherford in 1897, and the concept of a "lifetime" came at least in part from the pre-radioactive study of the "lifetimes" of phosphorescent phenomena. Both of these terms predated the discovery of radioactivity (discussed below).[63] And yet, just as the clear alchemical roots of Soddy's talk of "transmutation" had a perhaps unsuspected but nonetheless strong and immensely productive biological valence (cosmic evolution already having been linked to biological evolution), the often immediate resonance of such initially unaffiliated phrases as "decay" and "lifetime" led to their ready incorporation within a larger vitalized radioactive discourse. That the physical realities of the atom when combined with choices made for ease of measurement could lead to the idea of a "half-life"—a term that resonates simultaneously in both physical and biological idioms—shows how the incorporation of unintended felicities can and often did serve to contribute to the furthering of a newly vitalized radioactive discourse.

Indeed, at the far end of the spectrum of possibilities in the metaphysics of metaphor, past even the serendipitous coexistence of discourses and their ready appropriation, lay the merely suggestive, that which had a "biological flavor," even as it came from other roots and was never fully adopted into the vitalized discourse of radioactivity. The concept of the positively charged central "nucleus" of an atom, for example, postulated by Thomson in 1904[64] and named by Rutherford in 1912, emerged at a time of intense interest in nuclear cytology, where the term had already long been in use. Similarly, Henry Fairfield Osborn coined the phrase "adaptive radiation" to describe a cluster of speciation events in evolutionary history at nearly the same time that experimenters began to systematically investigate the effects of radiation on evolutionary processes. By the time of the centennial of Darwin's birth in 1909, even his achievements came to be cast in a radioactive mode: they were described as "a radiant influence so penetrating and so stimulating that it has been felt in every field of thought."[65] These uses are, by all rights, totally divorced from any genetic link or intentional carriage from the radioactive to the living. Such happenstance connections may, nevertheless, have had important roles to play in the development of long-lasting associations between radium and life: they may have helped to expand the realm of vitalized radioactive discourse by inadvertently enlarging perceptions of the breadth of such discourse (discourse is as discourse says); they may have helped to encourage those who found them discursively resonant to make connections that turned out to be experimentally productive (with suggestive terminology leading to the exploration of new frontiers—this was arguably the case with Soddy, as it was also with Burke, as we will see in the next chapter); and finally, and most importantly for the historian, such "merely suggestive" connections may provide glimpses into the previously existing discursive traditions that were drawn on in the establishment of a connection between the radioactive and the living in the first place. These varied intersections of metaphysics and metaphor—intentionally applied biological metaphors, unintended felicitous reappropriations, and a broader realm of unrelated but potentially resonant discourse—were all means by which radioactivity "came to life."

"A Happy Association": Radium and Preexisting Discursive Traditions

The living roots of technical radioactive terminology reflect only part of a larger set of already powerful extant discursive traditions that radium came to be grafted onto and which it ultimately transformed. Glowing in the dark, producing heat seemingly from nothing, and even producing—with the right equipment—sparks of light, radium's unusual properties enabled it to weave together varied tropes relating life to heat, light, and even electricity. These were already familiar themes of philosophical musings about the nature of life, such as in German *Naturphilosophie* and English romanticism, which envisioned the physical and living worlds as part of one harmonious and interconnected system. At a time when larger cultural narratives of decay were dominant, however, radium's seemingly endless supply of heat and light combined with its purported similarities to the ways of living things to bring about a new questioning of theories of thermodynamics. Finally, the discourses surrounding radium drew on long-standing tropes of "living atoms" and late nineteenth-century theories of particulate heredity to contribute to novel forms of thinking about the very "atoms of life." These powerful kaleidoscopic reworkings of tropes and discoveries, and the multiple and proliferating registers of scientific and cultural concerns they intersected with, further served to bring radium and life closer together.

Heat and Light. Heat had long been associated with life—indeed, it was often considered to be "indistinguishable from life itself." The idea that animals owed their heat to a phenomenon similar to the phosphorescence of an element dates back to at least the eighteenth century.[66] One account in the *Philosophical Transactions* of 1745 regarded phosphorus as the "animal sulphur." It held that "all animals contain some phosphoreal principle" and "that phosphorus exists, at least in the dormant state, in the animal fluids. All that is necessary is that the phosphoreal particles be brought into contact with aerial particles and as a consequence heat must be produced."[67] But phosphorus, in the course of its combustion with atmospheric oxygen, not only produced heat—it also produced light. Many breathless accounts of the shoemaker Casciarlo's famous "Bologna stone" of 1602 attest to this power and the amazement it wrought.[68]

Enter radium. Its remarkable ability to produce what Marie Curie called an "astonishing discharge of heat"—one of the first major revelations made about the new element, and a fact easily enough established without the need for any complicated theoretical framework about atomic change—readily assisted its grafting onto an earlier tradition linking heat and life.[69] Radium not only produced vast quantities of heat, of course, but also produced light—a phenomenon noted very early on by the Curies, who sometimes chose to visit their laboratory at night to witness the glow. Strikingly, many early popular accounts held that radium's light "was like that of phosphorus."[70] At the same time that Soddy held that radium was the new philosophers' stone, it was also assuming the mantle of a modern-day Bologna stone. This is perhaps unsurprising, given that it was Henri Becquerel's own study of phosphorescence—and the idea that naturally phosphorescent bodies might, under the influence of light, emit radiation like the newly discovered X-rays—that had inadvertently led him to the discovery of radioactivity in uranium in the first place.[71] The narrative of radium's properties from its earliest years thus began to weave together longstanding discourses associating life with heat and with light in a potent mix drawing on both the alchemical and the preternatural, and in its own way, with its own scintillations, contributed something to the maintenance of the "spark of life" trope.

New possibilities for the enlivening of radioactive discourse also emerged from this encounter. Most obvious and most notable was the transfer of the idea of a "lifetime" from descriptions of the macroscale phenomena of traditional phosphorescence (which radium did not fit perfectly—its light had no lifetime and never faded) to descriptions of the "life" of the constituent atoms of radium that radioactive theory held to be responsible for such phosphorescence. The lifetime of the phosphorescence thus moved inward into the lives of atoms even as Soddy sought to describe radium as a quasi-living element. Together, these descriptions provided yet further means for the association of the radioactive and the living and for the radioactive concept of a "half-life."

By 1900 there were already many deep-rooted discursive links between light and life. The study of bioluminescence had been of perennial interest to naturalists. The following passages from Alexander von Humboldt's *Cosmos* help capture some of these deep-rooted discursive links between light and life:

> The still radiance of the vault of heaven is for a moment animated with life and movement. In the mild radiance left on the track of the shooting star. . . .
>
> Here the magical effect of light is owing to the forces of organic nature. Foaming with light, the eddying waves flash in

> phosphorescent sparks over the wide expanse of waters, where every scintillation is the vital manifestation of an invisible animal world. . . .
>
> The waters swarm with countless hosts of small luminiferous animalcules . . . which, when attracted to the surface by peculiar meteorological conditions, convert every wave into a foaming band of flashing light. . . .
>
> It is not only at particular points in inland seas, or in the vicinity of the land, that the ocean is densely inhabited by living atoms, invisible to the naked eye.[72]

Remarks like these not only resonated with long-standing traditions of German *Naturphilosophie*—which often sought to use "the latest findings in biological research to argue for a continuum between the world perceived and the human consciousness that perceived it," as Laura Otis has noted—but were also readily mined by later writers, romantic and otherwise, searching for descriptions that would capture the new phenomena of radium in terms familiar to them.[73] Sir William Crookes did so in his description of the discovery of the scintillation of a phosphorescent screen when a sample of radium was brought near it—"one of the most impressive spectacles which we had for a long time," as one commentator remarked in *Science*.[74] Crookes evoked lifelike descriptions and called on the same kinds of tropes of light and life that Humboldt had used in describing the glories of the living ocean at night, describing to the Royal Society of London in 1903 how the phenomena he saw reminded him of phosphorescent plankton:

> On bringing the radium nearer the screen the scintillations become more numerous and brighter, until when close together the flashes follow each other so quickly that the surface looks like a turbulent luminous sea. When the scintillating points are few there is no residual phosphorescence to be seen, and the sparks succeeding each other appear like stars on a black sky. When, however, the bombardment exceeds a certain intensity, the residual phosphorescent glow spreads over the screen without, however, interfering with the scintillations.[75]

Crookes invented a device to share the phenomenon more widely, the spinthariscope (named after the Greek word for "scintillation"). Peer into the spinthariscope, remarked one writer a few years later, and you will find "'dead radium' displaying what suggests eternal life."[76] A French scientist, writing in the appropriately named *La Matière: Sa Vie et Ses Transformations*, even described each scintillation as due to "une particule d'hélium qui, dégagée du radium et arrêtée par une substance phosphorescente, transforme sa force vive en lumière."[77] And a scientist at Johns Hopkins would similarly report in *Science*:

> If one sits for several minutes in an absolutely dark room, and then examines the plate with a powerful pocket magnifying glass, the appearance reminds one of an enormous star cluster as seen in a telescope, the individual stars lighting up and disappearing in rapid succession, producing an impression which has been likened to that produced by moonlight on rippling water. . . . On carefully scrutinizing the screen it is almost impossible to avoid forming the opinion that the points of light are in motion, the whole field squirming with light, like a colony of infusoria under the microscope.[78]

The literature of the early radium years is thus dotted with references linking radium and life through the phenomena of living luminescence in ways ranging from the Romantic to the technical. As one journalist noted early on, "The problem of the glow-worm and firefly, the problem of light without heat, is vexing the soul of the scientist."[79] Sir Oliver Lodge wondered aloud whether the glowworm emitted light "because the insect has learnt to control the breaking down of the atoms, so as to enable their internal energy, in the act of transmutation, to take the form of useful light."[80] Or, as a 1904 book proposed, "In this cosmic process, the gradual break-up into simpler forms of a complex arrangement built up untold ages ago, we may possibly find the explanation of the light of the glow-worm and the firefly; that may be using the energy of the world's break-up to produce their light."[81] Even Millikan thought that looking at the spinthariscope was like "viewing a swamp full of fireflies."[82] H. G. Wells, not one to be left out of the trend, immortalized the connection in *The World Set Free* in his description of the childhood of his fictional scientist Holsten:

> He was to tell afterwards in his reminiscences how he watched the fireflies drifting and glowing among the dark trees in the garden of the villa under the warm blue night sky of Italy; how he caught and kept them in cages, dissected them, first studying the

> general anatomy of insects very elaborately, and how he began to experiment with the effect of various gases and varying temperature upon their light. Then the chance present of a little scientific toy invented by Sir William Crookes, a toy called the spinthariscope, on which radium particles impinge upon sulphide of zinc and make it luminous, induced him to associate the two sets of phenomena. It was a happy association for his enquiries.[83]

Radium's luminescence was never affected by temperature or other gases, of course, but Holsten's serendipitous linking of fireflies with radium epitomizes the contingent but historically implicated way in which the phenomena of phosphorescence, and then luminescence, helped to link the realms of the living and the radioactive.

Radium rapidly assumed the place of a powerful new element reordering the fin de siècle world. While an earlier generation had speculated that a sort of "electrical decomposing and secreting operation" was "inherent and necessary to the development, growth, constitution, and vital career of the identity we call 'our' globe,"[84] by the early twentieth century radium—with its remarkable production of heat—had replaced electricity. Radium even lay at the heart of the major dispute between Rutherford and Kelvin on the ages of the sun and the earth. Kelvin's original calculations, from which he concluded that "the inhabitants of the earth cannot continue to enjoy the light and heat essential to their life for many million years longer, unless sources now unknown to us are prepared in the great storehouses of creation," found their disproof in radium. And Rutherford calculated in 1904 that a radioactive earth would imply a life span much longer than Kelvin allowed—thereby helping to preserve Darwin's slow brand of evolution by natural selection as a viable account. This, too, helped strengthen in yet another way the link between the radioactive and the living.[85] As Rutherford described the occasion:

> I came into the room, which was half dark, and presently spotted Lord Kelvin in the audience and realized that I was in for trouble at the last part of my speech dealing with the age of the earth, where my views conflicted with him. To my relief, Kelvin fell asleep, but as I came to the important point, I saw the old bird sit up, open an eye and cock a baleful glance at me! Then a sudden inspiration came, and I said Lord Kelvin had limited the age of the earth, provided no new source was discovered.

> That prophetic utterance refers to what we are now considering tonight, radium! Behold! The old boy beamed upon me.[86]

Thermodynamics, Decay, and Perpetual Motion. As radium's endless production of heat led to wondrous claims for perpetual motion, scientists were forced to confront substantial issues in prevailing narratives of thermodynamics. As the *Los Angeles Times* noted, radium "apparently violates one of the fundamental laws of physics, namely, that of the conservation of energy."[87] By invoking new sources of energy not previously known, radium provided ways out of the pessimistic fin de siècle thermodynamic narratives of the end of the world. Indeed, according to historian of physics Abraham Pais, the discovery of radioactivity was the first of three times in which "prominent physicists waver[ed] in their faith in the universal validity of the law of conservation of energy."[88] The effort to determine where radium's energy came from eventually brought not only the first but also the second law of thermodynamics into question.[89] Radium's boundless energy rapidly caused more than one onlooker to remark that its discovery "could barely be distinguished from that of perpetual motion, which it was an axiom of science to call impossible."[90] Even *Punch* published a rhyme beginning, "Radium, very expensive, the source of perpetual motion. . . ."[91]

For many in the late nineteenth century, even well before the discovery of radium, speculation already abounded whether living things might somehow be able to resist the second law of thermodynamics—as well as whether the origin of life itself might be related to thermodynamic considerations. One 1885 account illustrated these sorts of resonances between the thermodynamic and the living quite clearly:

> This is the dawn of Life. . . . The tiny globules are each and every one an arena of warfare perpetual, incessant, between opposite irreconcilable forces of Nature,—the centrifugal, the radiant, the spiritual, and the centripetal, concentrative, physical. Life is not that which energizes, but the index of the energizing tension of inter-atomic repulsion and attraction. The one is within: it is the life-principle, the vital energy, the *vis viva*, the efficient cause, which would drive the atoms apart, dissolve every organism, destroy a world, were not such radiant energy counteracted and held in quivering æquipoises by that which is without,—exterior physical and chemical resistances, which press the particles into closer union, and, like the balance-wheel

> of a mechanism, slow the vibration of the atoms to a rate compatible with vital processes.[92]

This account, from a book entitled *The Daemon of Darwin*, places life in deliberate engagement with nineteenth-century issues of thermodynamics. Its title is an implicit reference to Maxwell's demon, that fictional entity and microscopic intelligence able to arrest the second law of thermodynamics and prevent the inexorable heat death of the universe by ferrying molecules and atoms back against the grain of entropy loss. *The Daemon of Darwin* transports Maxwell's demon to the realm of the living: while Maxwell's demon initially posed a problem for thermodynamic theory, Darwin's "daemon" is a reveling in the living escape from thermodynamic constraints, a realization that life might similarly exist eudaimonically against the grain of the energy flow of the universe. Firmly a nineteenth-century piece, *The Daemon of Darwin* illustrates how radioactive discourse could draw on and transform preexisting thermodynamic strands of discourse as it sought to relate one form of perpetual motion (organic) to another (inorganic). Within the discourse of thermodynamic concerns, radioactivity could thus not help but be allied with life. Demons, organisms, and radium all did the same thing thermodynamically: they all appeared to violate the second law. As such, the modes of description of one seemed as readily applicable to the modes of description of another. Radium proved to be both Darwin's saving grace, as in Rutherford's battle against Kelvin, and Darwin's daemon at once.[93]

Thermodynamic considerations thus proved to be yet another way in which the realms of the living and the radioactive came to be discursively linked. Nineteenth-century thermodynamically inflected talk of life-principles and vital energies within the earliest form of life on earth striving to get out was easily interwoven with later descriptions of the energy contained within an atom of radium, waiting to release its energizing power before decaying. The thermodynamic reference to atoms being driven apart, the dissolution of every organism, and the destruction of a world would again come to resonate, albeit in a different register, within a world newly made radioactive. Indeed, in a fin de siècle period full of cultural narratives of decay, the observation that living things somehow seemed able to resist the relentless and universal tendency toward entropy complemented claims that energy itself was not only enlivening, but in some sense alive. Oswald Spengler, in a chapter of his *Decline of the West*, spoke of the possibility of turning dead energy "once more into living . . . through the simultaneous binding of a further quantum of living energy in some second process," as in the combustion of coal, but added a footnote later in the chapter that made explicit reference both to radium and to radioactive "lifetimes."[94]

On the other hand, even as radium was able to connect concerns over the thermodynamic dissolution of the world with the thermodynamically puzzling non-dissolution of organisms, it was itself, once properly enlivened, subjected to the same sorts of late nineteenth-century cultural narratives of degeneration and decay. In fact, the concept of radioactive "decay" probably has as much to do with these cultural fears as with any technical undertones from the concept of "lifetime" carried over from the study of phosphorescence. H. G. Wells—referring to the "secret of vigour, Tono-Bungay" in his 1908 book of the same name, and referring to "quap . . . a festering mass of earths and heavy metals, polonium, radium, ythorium, thorium, cerium, and new things too"—captures some of the cultural valences of these radioactive atoms:

> Those are just little molecular centres of disintegration, of that mysterious decay and rotting of those elements, elements once regarded as the most stable things in nature. But there is something—the only word that comes near it is *cancerous*—and that is not very near, about the whole of quap, something that creeps and lives as a disease lives by destroying; an elemental stirring and disarrangement, incalculably maleficent and strange.
>
> This is no imaginative comparison of mine. To my mind radioactivity is a real disease of matter. Moreover it is a contagious disease. It spreads. You bring those debased and crumbling atoms near others and those too presently catch the trick of swinging themselves out of coherent existence. It is in matter exactly what the decay of our old culture is in society, a loss of traditions and distinctions and assured reactions. . . . I mention this here as a queer persistent fancy. Suppose indeed that is to be the end of our planet; no splendid climax and finale, no towering accumulation of achievements but just—atomic decay![95]

Living Atoms. The idea that atoms could decay is of course only part of a tradition that could think of atoms as somehow *living* in the first place. Radium was far from the only inanimate entity viewed as potentially living: crystals, viruses, the *arbor Dianae*, and other forms also readily fit this role.[96] But while the full history of the "living atom" tradition has yet to be written, some preliminary observations can be made. One of its obvious features is that discourse on the "living atom" goes back to ancient times, dating back at least to Lucretius, and has been reworked multiple times over its history (often under the label of "monads" or "monism"). In recent centuries, the claimants for the role of living atom were not only actively contested, but recognized as constantly shifting in light of new discoveries. Were the living atoms to be understood in the first instance as microscopic animalcules or particles? Anton van Leeuwenhoek, for example, had referred to the living, moving structures he had seen through the microscope as "living atoms." Enlightenment thinkers such as d'Holbach and Diderot, on the other hand, held that "the viable units of living matter were the same as those of matter, namely 'atoms.'"[97] Buffon readily referred to the "life" of organic molecules, and Maupertuis drew on the concept of living atoms to explain hereditary processes in his popular and controversial *Vénus physique*. The distinction between "living atoms" (atoms that lived) and "atoms of life" (the smallest particles of life)—a distinction unknown to the ancients—began to make itself more apparent at this time. By the mid-nineteenth century, the work of Schwann and Schleiden had placed the *cell* at the forefront of the living atom tradition, while by 1875 Lionel Beale was arguing that it was not the cell, but the subcellular *bioplasm* and its constituent parts, that must be the atoms of life—and yet these alone, he argued, were not enough to account for or to be all that life was. Writing just before the great storm of late nineteenth-century particulate theories of heredity, Beale claimed that the problem with materialist accounts of the spontaneous generation of life was that they also required as the last step in their recipe that "the groups of atoms must be made to live." This, however, was impossible:

> It is remarkable that those who have spoken of *living atoms* should have thought so little about the matter as to have permitted themselves to suggest an absolute impossibility,—for it is obvious that *no single atom can be thought of as alive*. The idea of a living atom of oxygen or hydrogen or carbon or nitrogen is clearly untenable. We might as well talk of a living atom of sulphur or iron or lead. And it is absurd to talk of *dead* carbon, oxygen, or hydrogen atoms, since a living state of the atoms of these and other bodies is thereby implied; but such living state is inconceivable.[98]

The discovery of radioactivity, of course, was to throw precisely this assumption—that such a living state of an element was "inconceivable"—into question, precisely because the discursive commonalities maintained from earlier traditions of the living atom were now applied and used to characterize the new radioactive phenomena. Perhaps atoms could live after all. As C. W. Saleeby, the great British popularizer, wrote in *Harper's* in 1906: "That is the position of thinking men to-day. More and more do they hesitate to believe that there is a difference in kind between living and so-called lifeless matter. If anything in the world is alive, is not radium alive?"[99]

Beale's railing against materialism and his ardent critiques of attempts to produce recipes for the spontaneous generation of life echoed Enlightenment debates, and, indeed, the discussion of "living atoms" arose especially frequently in the context of debates over vitalism and materialism. As part of a nineteenth-century vitalist tradition that referred to "vital centers" and often associated these centers with the nuclei of cells, Beale's rhetoric spoke of a vitality kept deep within, which the outside environment could never impinge on and from which effects went outward:[100]

> Within every centre of every one of the thousands of minute molecules of which I conceive every particle of living matter is constituted, is a more central centre in which the matter is a degree nearer the point where it began to live, and where new powers were first communicated to it . . . *it is to the power acting from the centre within, and welling up, as it were, from a yet more central source of power which seems inexhaustible, that vital phenomena must be attributed* . . . and that, therefore, the central life-communicating power plays a far more important part in the phenomena of life than anything in the environment acting from without.[101]

Beale's vitalist tropes placing the agent of life at the radiating center of the cell resonated both with contemporaneous thermodynamic considerations—Darwin's daemon is in evidence here—and with later discourses that described radium as having its own secret storehouse of power radiating outward from its center. (H. J. Muller's arguments for the nature of the gene a half century later would sound eerily similar; see chap. 5.) Even just before the discovery of radium, a lecturer at the Marine Biological Laboratory at Woods Hole would tell his students that "life is an affair of atoms and molecules rather than of large and visible masses of them," and that the particular movements of living things were "due likewise to harmonic changes of energy inseparable from the atoms themselves."[102]

If such accounts seemed to many by the last quarter of the nineteenth century to be without proof, many other investigators, in their quest to explain "organism" and "life," were just as keen to find fundamental "life-units" and "life-processes." And in their quest to keep the living atom tradition alive, they were not above inventing hypothetical units to serve their purposes. Charles Darwin's "gemmules," Hugo de Vries's "pangens," and the numerous particulate inventions of August Weismann's fertile mind stand witness not only to the late nineteenth-century concern with "particle biology," but also to the increasing *hereditarian* cast the living atom tradition attained by the turn of the century. This shift was one of the major steps in the transmutation of the "living atom" tradition into an "atoms of life" tradition.[103] Even Ernst Haeckel's fanciful theory of "the perigenesis of the plastidule" called for the phenomenon of heredity to be explained by the vibration of *Lebensteilchen.*

Dovetailing with the almost exclusively hereditarian cast that the living atom tradition took on by the first years of the twentieth century, the vitalized discourse surrounding radioactivity—the genealogies and kinship relations of the radioelements and the idea of radioactive atoms having a life—and the whole slew of new atomic terms radioactivity brought with it helped to further this reworking of the tropes of the living atom tradition. This hereditarian resolution of the essence of life into particles of one kind or another readily facilitated the equation of the respective hereditary "quanta" of life with energy—an equation that was simply obvious to many contemporaries.[104] Indeed, concepts of energy had been used to delimit the fundamental unit of life just a year before the discovery of radioactivity: according to the *Oxford English Dictionary*, the term "energid" was coined in 1897 to describe "the nucleus of a cell together with its active cytoplasm regarded as a vital unit. . . . The distinguishing characteristic of an energid is the living element (protoplasm and nucleus)." The discovery of radium a year later meant that a new and different sort of "living element," whose vital power also resided tucked away in its nucleus, had now been found to which this kind of discourse could be applied. As the *Medical News* put it, "The mysterious 'vital spark' may be identical with some of the physical agents which have recently loomed into the ken of the physicist."[105]

Earlier talk by Ernst Wilhelm von Brücke and others of possible *Elementarorganismen*, both in the nucleus and elsewhere, similarly helped to set up discursive commonalities with that other "elemental organism," radium.[106] While the evidentiary strength of any one of these connections may be debatable, what should be clear through the complexity of the associations presented here is that the emergence of a powerful association between radium and life depended on the appropriation of previously existing discursive traditions, as well as on acts of novel improvisation with and within these traditions to better fit and characterize the novel phenomena associated with radioactivity. The ongoing transformation of the "living atoms" tradition into an "atoms of life" tradition resonated so well with the new discoveries in radioactivity and with the biological modes of describing its behavior that Henry Fairfield Osborn was led to speculate in the late 1910s as to the existence of a life "element" so far unknown, but which might be "like radium . . . wrapped up in living matter but remain as yet undetected, owing to its suffusion or presence in excessively small quantities or to its possession of properties that have escaped notice."[107] No longer was radium questionably alive; living things, at least for one prominent scientist, were now in need of some element like radium to account for their ability to live. The move from living atoms to centers of vitality to the discovery of a living atom culminated in a near-complete reversal for Osborn: radium was arguably more alive than the living matter that surrounded it!

Small wonder, then, that radium—scintillating on zinc sulfide screens on both sides of the Atlantic, glowing in the dark, and producing heat of its own accord; reinventing the spark of life trope by bringing it into conjunction with electrical discourse and discourse of the living atom; and dovetailing with nineteenth-century thermodynamics, perpetual motion, and vitalist-materialist debates on the origin and nature of life—should rapidly become the *Ur*-substance of life. Radium was not simply the eighty-eighth element and a remarkably energetic one at that. Radium was above all the most serendipitous of sites for the reworking and confluence of a host of earlier traditions. Bringing together fermentation, heat, light, sparks, change, a discourse of living atoms, and the very atoms of life, radium did it all: it was all of these things associated with life, and all at once. No wonder the wondrous element stole the popular limelight.

Radium Reaches the Public

Word of the miracle of radium spread rapidly, and public lectures on radium became immensely popular. Some lectures were even repeated on the following day, "owing to several persons being turned away from the doors."[108] Soddy became well known for giving a number of well-attended public lectures in London and Cambridge, whose recurring theme was the immense storehouse of energy within the atom that radioactivity had revealed and which could one day be put to good use.

The transmutation of uranium, thorium, and radium had been going on since the creation of the universe, Soddy said, which meant that life could never have survived in the intensely radioactive environment of the early earth. But, he thought, the time was now ripe: "Radioactivity would decrease with time and leave to us in this generation just a suitable amount—if we could but find it and control it—for the use and benefit of all peoples of the world."[109]

Soddy transferred many of the themes of his lectures, especially those he gave at Glasgow University in 1908, where he had been a lecturer in chemistry since 1904, into his book *The Interpretation of Radium*.[110] One reviewer, having read the "forty pages of speculation," criticized Soddy for the way he "injects metaphorically into his veins pint after pint of radium emanation; and enjoys, and makes his readers enjoy, a wild, extravagant dream of infinitely controllable infinite energy."[111] While Soddy had once remarked on the mystery of being able to identify helium from 100 million miles away—"what a theme for, say, Mr. H. G. Wells," he once exclaimed—Wells more than returned the compliment with his fascination for Soddy's findings.[112] Wells's 1914 novel, *The World Set Free*, described smoke-free industry, the artificial production of radioactive elements, and even the explosive force that could be released from energy hidden within radioactive atoms. Wells dedicated his book to Soddy.[113]

In his novel, Wells captured the excitement of public lectures on radioactivity, surely drawing on his own experiences as a young man:

> A certain professor of physics named Rufus was giving a course of afternoon lectures upon Radium and Radio-Activity in Edinburgh. They were lectures that had attracted a very considerable amount of attention. He gave them in a small lecture-theatre that had become more and more congested as his course proceeded. At his concluding discussion it was crowded right up to the ceiling at the back, and there people were standing, standing without any sense of fatigue, so fascinating did they find his suggestions. One youngster in particular, a chuckle-headed, scrub-haired lad from the Highlands, sat hugging his knee with great sand-red hands and drinking in every word, eyes aglow, cheeks flushed, and ears burning.[114]

As if paraphrasing Soddy's own lectures, Professor Rufus continued: radium was the source of "the intensest force" that could "keep Edinburgh brightly lit for a week," the discovery of which—like early man's discovery of fire—brought "the dawn of a new day in human living" and "the possibility of an entirely new civilisation." And, in an echo of Soddy's earlier narrative of cosmic evolution, Wells inserted radium into a narrative of natural selection, but with a new twist: this freeing up of uncountable stores of energy, Professor Rufus said, would mean that "that perpetual struggle for existence, that perpetual struggle to live on the bare surplus of Nature's energies will cease to be the lot of Man."[115] Soddy's own description of radium as the by-product of an elemental struggle for existence here appropriately meets the claim of his literary reflection that radium could help humans escape from their own struggle for existence. Perhaps most striking is the professor's comment, "We know now that the atom, that once we thought hard and impenetrable, and indivisible and final and—lifeless—lifeless, is really a reservoir of immense energy." Once again, the discovery of radioactivity and of the energy lying latent within the atom was a mode of bringing the atom to life.

The "chuckle-headed, scrub-haired lad" left the crowd after the lecture, Wells concluded, anxious that no one "should invade his glowing sphere of enthusiasm" as he wandered "through the streets with a rapt face, like a saint who sees visions." Soddy's lectures undoubtedly had something of a similar effect on his listeners. It was with such glowing visions of endless power waiting to be tapped within the living atomic storehouse—images largely derived from Soddy's public lectures—that radium first thoroughly captured the public imagination on both sides of the Atlantic.

Although radium was a discovery of European science, it was the American knowledge of radium—"based more on enthusiasm than experiments," according to Carolyn de la Peña—that carried the element's "cultural narrative . . . beyond subjects covered in journals and laboratories":

> In the United States, radium knowledge evolved from a series of trickle-down information networks, the majority of which were begun by scientists and laymen who had little direct experience with the element. This led to a dramatically different vision of radium than European science could support. Instead of focusing on radium's fragility and volatility, properties that were apparent to those who understood the element scientifically, Americans concentrated on its power and malleability. It was this "discovery" that inspired an American following unparalleled by any previous scientific discovery.[116]

American public lectures on the properties of radium were thus more often situated within the context of a history of *showmanship*, with the public paying to see samples of the wondrous element and learn of its many marvelous effects. "It was these individuals," de la Peña says, "among them common charlatans and renowned scientists, who served as primary disseminators of radium knowledge."

Prime among these disseminators was William J. Hammer, an electrical engineer by training and a onetime assistant to Thomas Edison, who left Edison in 1902 to join Marie Curie in Paris for a time before bringing back nine tubes of radium with him on his return—the first radium to enter the United States. Once back in the United States, Hammer spent most of his time "experimenting with radium-derived luminous substances, promoting radium therapy for physicians, and giving paid lectures across the country . . . more showman than medical professional."[117] On February 19, 1903, he first exhibited the tubes at a meeting of the New York Academy of Medicine. Reports at the time remarked that the specimens, estimated to be some eight thousand times more powerful than uranium, "glowed visibly in the dark."[118]

From 1902 to 1907, Hammer was undoubtedly the most famous of the scores of radium experts on the circuit. One image even granted him a glowing radioactive aura.[119] Hammer talked about all aspects of radium, from its use in fighting tuberculosis, to its immense expense to the multitude of practical applications that could be dreamt up, comparing its penetrating power with that of Roentgen rays (X-rays). As he wrote in his 1903 book *Radium, and Other Radio-Active Substances*, "It is doubtful whether any substance has been discovered in the history of the world of such stupendous interest and importance and possessing such puzzling characteristics as radium, which seems to at variance with well-established scientific theories as to the constitution of matter."[120] Hammer's personal experience with the element, one review of his lectures seemed to imply, added to his status and distinguished him from his lecturing contemporaries: "Mr. Hammer is well informed on the subjects of which he treats, and the dignified auspices under which his lecture was delivered guarantee freedom from dependence upon the scissors and paste pot in the making of it."[121]

Hammer's career shifts—from Edison to Curie, from electricity to radium, and from assistant researcher to popular lecturer—exemplify how the popular discourse surrounding radium and its marvelous powers, and the institutional contexts in which public knowledge of the element circulated, were linked to already existing modes of discourse and institutions associated with electricity. Radium may have seemed to have appeared out of nowhere, but the ways in which it was first popularized rehearsed tropes parallel to and disseminated along discursive paths first trod by novel electrical phenomena a century before. Radioactivity was both *physically* and *institutionally* linked to the properties of electrons and electricity: the theory of radioactive transformation was often referred to as the "electronic theory of matter,"[122] for example, and Soddy gave a series of lectures, entitled "The Internal Energy of the Elements," to the Institute of Electrical Engineers (which were transcribed and published in the Institute's journal, *The Electrician*, in 1904).[123] On the other side of the Atlantic, well-attended lectures on radioactivity were held at the Electro-Chemical Society in New York City. Although radioactivity was clearly a distinct phenomenon with its own particularities, it was thus both discursively and performatively linked with earlier traditions of electrical showmanship. The very modes of description, means of public display, and institutions available for the popularization of radium meant that radium was grafted onto an electrical stem.[124]

Moreover, in the ongoing American story of technological energy enhancement, as de la Peña has shown, "radium entered into and flourished within a popular culture of energy fantasy well established by previous mechanical and electrical energy devices." The radium craze fell within the broader discourses of human energy prevalent at the turn of the century: within the human body, radium "seemed capable of creating energy by causing a cellular change that rendered the body infinitely able to renew its own energy supply."[125] It was, in other words, the "new electricity." Similarly, as Spencer Weart has noted in his masterful history of atomic discourse, "If electricity was becoming a humdrum household matter, then the mysteries once associated with it could be transferred to radioactivity, where they seemed to fit even better."[126] What for an earlier generation in the early nineteenth century would have been described in terms of electrical or magnetic phenomena was by the dawn of the twentieth century increasingly likely to be described in terms that glowed radioactive.[127] As an article in *Everybody's Magazine* reported in 1903, "it is believed that the discovery of Radium will answer the question, 'What is Electricity?'"[128]

While a lesser and shorter-lived kind of "entrepreneurial showman's and artisan's culture" surrounded the American popularization of radium than surrounded the earlier popularization of electrical phenomena, still, just as with electricity, where "shocks and sparks were well calculated to impress a paying audience," radium attracted its oglers.[129] With street marketers and other quacks in the late nineteenth century already heralding that "electricity is life," the shift to speaking of radium and life was all but guaranteed.[130] Radium had at last reached its public—and the public was all too eagerly reaching back in return.

"An Indecent Curie-osity": The Pop Radium Craze

Hammer's book, published in 1903, may have been the first book on radium published in the United States, but it was certainly not to be the last. "Never, perhaps, was so much ado made over so small a quantity of any substance," one reporter noted in the same year.[131] Hammer's book and lectures helped unleash a deluge of interest that rose throughout that epoch-making year, bringing the marvels of radium to ever wider audiences. The first decade of the twentieth century witnessed a wave of tremendous worldwide interest in radium—the so-called radium craze. Fascination with radium pervaded every level of popular culture, from the learned societies and their journals and the upper-crust literary digests of the time, to the pages of the *New York Times*, the more popular press, and other hoi-polloi literary and trade journals of the day, to popular novels, nonfiction pamphlets, advertisements, and even church sermons.

As historian of radioactivity Lawrence Badash has noted, "The subject of radioactivity firmly engaged the public's attention by 1903, and far more articles were printed in that year than in the three preceding years combined. Newspapers shared in this trend, the *New York Daily Tribune*, for example, running thirteen editorials and twenty-three stories on radium during 1903, having printed virtually nothing on it earlier."[132] In fact, so many articles were being published on the amazing properties of radioactivity and its presence in even the most mundane things, such as snowfalls and waterfalls, that the secretary of the Royal Society, Joseph Larmor, exclaimed to Rutherford that "the newspapers have become radioactive."[133]

Following the radium story in the newspapers, the *New York Times* reported in 1903, "has become a matter of absorbing and almost exciting public interest. The average man in every civilized country knows as much about radium as do the most advanced physicists, and the daily progress of investigation is watched with interest by the well-informed newspaper reader everywhere."[134] Newspapers were "the real educators in applied science"—articles on radioactivity were not "'caviare to the general,' but read with appreciative interest by multitudes."[135] Or, as historian Alex Keller described it:

> That was the springtime of radium. Those who before had scarcely noticed radioactivity, now went overboard. The *Illustrated London News* published pages of sketches of its wonderful works, as explained by William Crookes. The *Engineer* has one brief mention of the discovery of radium, in a summary of chemical progress in 1899, and then not a word until 1903, when there are seven in the first half of the year, ten in the second half. Radium received the honor of the main editorial on 10 April: what is this stuff which gives out heat without ostensible cause?[136]

After tracing the astonishing increase in references to radium in *Nature*, Keller noted, "If that was the response of sober science, the public was even more entranced." Along with other radium-themed cartoons, *Punch* published a frontispiece of the year that involved "the good fairy Radium [who] make[s] everybody happy with her magical powers."[137]

Most of the popular literary magazines at the turn of the century "made efforts to educate their readers in matters radioactive," as Badash has noted. One article in *McClure's Magazine* described radium as "not one discovery but a dozen that we were contemplating," while the *Times Literary Supplement* chimed in by publishing an article on the disintegration theory. As a consequence, Badash noted, "by the middle of this century's first decade scarcely a person in the civilized world was unfamiliar with the word 'radium' or the name of its discoverer." The Austrian embargo on the export of pitchblende ore and the imposition of a 25 percent import duty by American customs only "added to the public's fascination with radium," this "natural Roman candle."[138]

Some of the earliest American accounts of radium's powers were tempered with doubts that it would ever be anything but "a laboratory metal" and that it could ever live up to the extravagant claims made for it.[139] One critic opined that while radium compounds could and would be of great help to science, "from its extreme rarity it can never be of corporal use to man."[140] The question of the uses of radium was frequently asked in American journals: while one report noted that "at present radium grinds no axes," there was still early hope that further study might lead someday, for example, to "a method of obtaining light in a cheaper and more convenient manner than any now known."[141]

Fantastic proposals for the use of radium were expounded, from perpetual light to inexhaustible sources of heat, or that half a pound would be enough to keep a room warm for hundreds of generations. As Hampson recounted in his popular 1905 book, *Radium Explained*, "With such words from such men in his ears, no one need fear to be called stupid if he listens with opened-mouthed astonishment to the tale of the marvels of radium, or to be called a wonder-monger if he repeats them."[142]

Speculation as to radium's properties ran rampant. A farmer in 1903 wondered about the effects of "radium mixed with chicken-feed. . . . The radio-eggs would either hard-boil themselves upon being laid, or would hatch the chicks without need for an incubator."[143] Either result was obviously a positive effect. A Russian doctor even experimented with creating "radium energised wool" that might readily "become part of pharmaceutical stock and at no great expense"—though he cautioned that "before making 'emanated wool' an article of pharmaceutical commerce we must know how and in what particular cases the commodity would be useful—and that is still a question for the future."[144]

Even established professors followed their fancy wherever it led. Sir William Crookes reportedly held early on that the emanations that arose from the radioactive elements gave rise to the sense of smell. As the *Literary Digest* reported in 1902, "Whatever may be the nature of the emanation from radium, uranium, and the other so-called radioactive bodies . . . for the first time in the history of physics, the physical cause of odor seems to have been connected with the other physical phenomena known to science."[145] One early critic complained in the *New York Times* that with all the misleading statements, deceptions, and "vivid descriptions of all the things which radium is supposed to do . . . there has not been a single article written by an authority on the subject who has come to the front with definite statements and undeniable facts of what radium *cannot* do." In fact, he felt obligated himself to counter claims from the winter before that the blind could be made to see, saying that "it is quite as impossible as it is to restore life to a dead body."[146]

Another *New York Times* article, reflecting on the fantastic claims made for radium and asking whether men might "shine brilliantly in the dark," noted that "nowadays, what with Roentgen and other rays, wireless telegraphy, and radium, it takes a good deal to give us a start."[147] Nevertheless, by the end of 1903, radium was all the rage.[148] As another radium popularizer of the time noted, "For more than twelve months radium has received an amount of public attention which is not often bestowed on a strictly scientific subject. Everyone is now familiar with the word at least, which has obtained such wide-spread recognition, that, besides seeing radium dances in ballets, we can buy radium collars, radium stoves, and radium polish."[149]

"Radium was fun." There were radium chocolates, crucifixes, watches, toothpastes, clothing, toys, jewelry, doorbells, tonics, sweets, and, according to some sources, even "radium-spiked diet bread" and "sipping séances."[150] Suppositories, ointments, and vitalizers and tablets all followed in short order, accompanying

> endless medical products claiming to cure everything from the mundane to the exotic. . . . At one point . . . shares in dubious uranium mines were given away free with hamburgers and packets of toothpaste. . . . Quite apart from radium-water and medical lotions and potions, every conceivable product, from ventilation systems to chocolate to clothing seemed to be available in a form containing "health-giving radium."[151]

Entrepreneurs and charlatans alike took advantage of the situation, and it often remained unclear at times whether these products contained any radium at all. Marie Curie herself was involved in exposing the low levels of radioactivity in a particular fertilizer. No doubt some products did contain radium—with telling results in later years, as we will see in chapter 6. In other cases, businessmen were eager to vouch for the authenticity of their claims: one supplier of phosphorescent costumes, in an indignant letter to the *New York Times*, insisted, contrary to an offending article, that his props were in fact painted with radium paint, which "does not contain any phosphorus [the cheaper alternative for glowing paint] whatsoever."[152] One can only imagine today the untold radium dances and plays with "pretty, but invisible, girls, tripping noiselessly about in an absolutely darkened theater, and yet glowingly illuminated in spots by reason of the chemical mixture upon their costumes."[153]

Radium lived a complex life in the early decades of the twentieth century. It was hawked as the food preservative of the future: "The time is coming when radio activity will entirely supplant the chemicals now used for preservatives," reported one article.[154] It was used in advertising rhymes to sell shoes, and even in manure to be spread on fields and flowers (figs. 1 and 2). Songs, stories, medicine, games, parties, and meals all involved radium.[155] There were even games of radium roulette. As George Bernard Shaw commented, "The world has run raving mad

FIGURE 1. "Radium makes things grow." (Advertisement, *Literary Digest*, April 3, 1915, 775.)

FIGURE 2. Giant (*right*) versus untreated (*left*) flowers allegedly resulting from radium treatment. (Courtesy of Paul Frame, Oak Ridge Associated Universities, and Patrick McDermott, Rutgers University.)

on the subject of radium, which has excited our credulity precisely as the apparition at Lourdes excited the credulity of Roman Catholics."[156] It is hardly any wonder that Marie Curie was in short order referred to as "Our Lady of Radium."[157]

The religious valences of radium also multiplied. Not only could radium give sight to the blind, or so it was said; some even claimed that radium could resurrect the dead. In the creation story of the universe, radium was both the new Adam and the new atom. Religious disciples of Swedenborg even claimed that his theory of vortices granted him priority in enunciating the principles of radioactivity and the attendant corpuscular theory.[158] Others, like Dr. Louis Albert Banks, the preacher at Grace Methodist Episcopal Church on the Upper West Side of New York City, gave sermons on "spiritual radium" (other clergy still do).[159] Even Henry Adams sought to relate radium to the divine, claiming that the new rays "were a revelation of mysterious energy like that of the Cross; they were what, in terms of medieval science, were called immediate modes of the divine substance."[160] By the late 1910s, others were still happy to claim that "there is a deposit, an infinitesimal deposit it may be, of the radium of romance in the slag of all souls," which is "harder to separate from the spiritual dross of us than radium from its carnotite; a kind of atomic property of the spirit which breaks up its substance; which ionizes, energizes, and illumines it."[161]

In more academic settings some spoke of "that intangible something which is transmitted from person to person by association and contact, but can not be written or spoken—we may term it inspiration, or personal magnetism, or perhaps the radium of the soul."[162] Even in the theatrical world, radium found its place. One reviewer described how the "radium of Shakespeare" would be "release[d] . . . from the vessel of tradition,"[163] while another claimed that

> there has always been some kind of magnetism or radiation recognized in the charm of a great singer or a great actor has for an audience [*sic*], and it may be that it is a form of radioactivity. One may not be able to analyze it, so to speak, and reduce the psychological radium to practical demonstration . . . [but] why should the greatest art even not be a still more impalpable and unmeasurable force or a variation of the same inscrutable mystery? . . . So, through the whole scheme of nature radioactivity seems to be the characteristic of man and matter alike.[164]

In their own peculiar way, even the words "radium" and "radio-active" attained popular status in the idiom as early as 1904, when, according to the *Oxford English Dictionary*, "radium" referred to "a smooth, plain fabric with a sheen of silk," on through 1905 ("Eliza has found that London is radio-active, hence enjoyable"), and certainly at least through 1909 ("She did not begin to live, socially, till her body was at rest. . . . Then her individuality would be radioactive.")

One particularly radioactive personality was Marie Corelli, the best-selling novelist of her generation and the author of thirty-one novels from 1886 to 1925. Even the titles of some of Corelli's books—*The Mighty Atom* (1896), *The Secret Power* (1921), and *The Life Everlasting* (1911)—hint at a kind of literary connection linking radium and life, a connection that Corelli herself was quite explicit and adamant about defending. In the prologue to *The Life Everlasting*, Corelli laid out precisely how she intended her "Electric Theory of the Universe" to stand in for radium. Of her earlier *A Romance of the Two Worlds* (1886), she said:

> I was forbidden, for example, to write of *radium*, that wonderful "discovery" of the immediate hour, though it was then, and had

> been for a long period, perfectly well known to my instructors, who possessed all the means of extracting it from substances as yet undreamed of by latter-day scientists. I was only permitted to hint at it under the guise of the word "Electricity"—which, after all, was not so much of a misnomer, seeing that electric force displays itself in countless millions of form.[165]

Corelli had no doubt that "this vital radio-active force" existed—she even claimed to have hinted at radium prior to its discovery (a claim that placed her squarely among her fellow alchemically oriented writers—all that was lacking, she implied, was a "fitting name"):

> *This was precisely my teaching in the first book I ever wrote.* I was ridiculed for it, of course,—and I was told that there was no "spiritual" force in electricity. I differ from this view; but "radio-activity" is perhaps the better, because the truer term to employ in seeking to describe the Germ or Embryo of the Soul.[166]

With radioactivity more spiritual than electricity could ever be, Corelli transformed others' earlier fascination with "the embryology of the soul"—the title of chapter 8 of Haeckel's best-selling *The Riddle of the Universe* (1898)—into an idiosyncratic mix that cited contemporary scientists and Paracelsus alike. Linking light and heat with what she called the "Soul or 'Radia' of a human being," Corelli reflected the craze of her times, identifying radioactivity variously with the Fountain of Youth, the elixir of life, and the divine. The "secret power" of that "mighty atom," radium was part and parcel of the "life everlasting."

Radium was also part of everyday mortal life back on earth, where it was actively promoted as healthful, restorative, and rejuvenating—a stimulus to all life (not just chicken eggs) and conduct. The *New York Times* held, for example, that in matters of international diplomacy, "there are times when silence is more than golden; it has the value of radium." The *Wall Street Journal* commented that "as radium is used to cure cancer so the radium of publicity may be relied upon to cure the cancer of corruption."[167] By 1923 the popular discourse itself was being termed "radioactive," according to the *Oxford English Dictionary*, even as it continued without any sign of stopping ("this radio-active quality of popular idiom, this power to give out life and never lose it"). Well before the better-known craze of the 1950s, an atomic euphoria had set in.[168]

The fascination with radium was total: "There seems to be hardly any limit to the marvels of radium," one *New York Times* article reported, even going so far as to find H. G. Wells remarkably prescient for his *War of the Worlds*, which was thought to give "an appalling picture of radium heat rays in full operation, although they were still to be discovered by our scientists."[169] Others saw closer parallels to Wells's story of the stolen cholera bacillus when a tube of radium was lost in the Paris Métro in late December 1911.[170]

As men like William J. Hammer were hawking the wonders of radium to paying audiences, public museums staged immensely popular exhibits of the rare element. The American Museum of Natural History in New York City received a gift of radium, which it immediately proceeded to exhibit in September 1903.[171] Donated to aid in scientific experiments and initially intended to help the gem curator in assessing the effects of radium on various minerals, the element was in demand by physicians for medicinal uses almost immediately. The public merely wanted to see the two grains (about 125 milligrams) of "yellowish powder" as they rested on cotton in a protected glass case. Crowds flocked to the museum "to see a little capsule of the new wonder of science," newspapers reported, and a policeman even had to be employed to keep the crowds moving along on Labor Day.[172] As one *New York Times* society gossip columnist remarked, "Apropos of this, as it is not generally known, clubmen and society generally are taking much interest in the lectures at the Museum of Natural History, especially those on radium and the exposition of that new property. . . . An afternoon at the museum sometimes resembles a society function."[173]

At least partly due to its sheer expense and scarcity, radium rapidly took on a rarefied air and became the must-have item of high society.[174] By the time radium was on exhibit at the 1904 World's Fair in St. Louis—where U.S. Geological Survey employees delivered lectures on it twice a day—it was already at the heart of popular discourse and on everyone's lips, both metaphorically and literally.[175] A "liquid sunshine cocktail" had been invented for the alumni of MIT to toast with, having been made to glow in the dark by dipping a tube of radium in a glass filled with water. One report that described the glimmering event, with phosphorescent cigars and human skeletons covered in glowing paint brought out to dance in the dark, noted that after the toast, "none of the self-sacrificing scientists who drank the liquid became transparent afterward, although all were assured that, as a matter of fact, their interiors were thoroughly illuminated."[176] One of the physicians who invented the cocktail, William J. Morton, held that his liquid sunshine, in lighting a patient up from the inside out, could serve more medicinal purposes: "We know of the value of sunshine on the outside," the doctor remarked, "particularly where bald heads are concerned, and we believe it will have a similar effect on the inside."[177] Or, as the *Literary Digest* declared in 1912, "Our brains are especially radio-active, the heart less so, and the kidneys still less."[178]

Morton was but one of many American doctors who either capitalized on the situation or who, "trapped by the hysteria, sold their hard-won supplies to the popular dilettantes," and who, just a few years later, facing an "insatiable demand for luminous materials . . . began selling their meager stocks of radium to the military" in order to provide luminous watches for each soldier while European sources were unavailable (some 6,500 watches in 1913, up to 2.5 million in 1919). As David I. Harvie has noted, "Ironically, radium—in cautious use in medicine—was widely available indirectly to society at large." Claims emanating from the medical journal *Radium* that "radium has absolutely no toxic effects, it being accepted as harmoniously by the human system as sunlight is by the plant," were "happily received in the USA," Harvie notes, "where at one time the efficacy of radium was measured in homely 'Sunshine Units,'" and where newspapers reported that radium permitted the "bottling of sunshines."[179]

By 1905 radon inhalation therapy was established in what was termed an "emanatorium." Radium became inextricably and essentially linked to conceptions of good health and vitality, which made it seem no more sinister than a frolic in the park on a sunny day.[180] (Debates about whether the sun was composed at least in part of radium were also carried on at this time.[181]) While X-rays, with their bony revelations, connoted death almost from their very inception, radium rapidly came to connote life: as one radium therapist noted in 1910, "This metal is, so far, the only radioactive body used for therapeutic purposes."[182] As Bettyann Kevles has noted, "It did not make the visible translucent, but it glowed by itself and seemed a different kind of miracle."[183] Even as late as 1921, the director of the Radium Institute of New York still proclaimed that "radium is the most remarkable therapeutic agent emanating from the laboratory of the Almighty."[184]

More than simply exhibiting life*like* qualities, as Soddy had proposed—which would be a fascinating enough story in its own right—the half-living element was viewed by a wider public as life-*giving*. While Soddy had employed a discourse of living radium in no small measure to help conceptualize what was going on in the physical world, popular writers and commentators brought about their own transmutation of radioactive discourse to ultimately grant radium a vital*izing*, and not just a vital, character. Soddy had succeeded in bringing radium to life;

others, with the help of these kinds of popular transmutations, were soon to use their understandings of radium to irradiate life itself, both figuratively and experimentally.

Once radium reached the popular realm, the metaphors that Soddy had used freely, but with a modicum of circumspection, were taken as reflecting realities and used to *prove* previously existing hypotheses. Just as Millikan had previously sedimented the metaphor of radioactive decay, the popularizer C. W. Saleeby wrote in *Harper's* in 1904 that radium proved that Herbert Spencer—"the greatest thinker whom the Anglo-Saxon race has ever produced"—was right: "Radium . . . proves that his great formula of evolution is as applicable to atoms as it is to societies of solar systems." Radium "proves the truth of *atomic evolution*" in that it was "not manufactured, but evolved."[185]

Radium was both vitalizer and vitalized. As one popularizer put it, "The great peculiarity of radium is that it is a metal that is practically an inexhaustible reservoir of energy, which not only imparts vitality to another body, but does not appreciably lose any of its own in the process."[186] Soddy wasn't the only one to describe the radioactive realm in living terms: Gustave Le Bon literally equated the transmutations of radioactivity with changes in living things in his popular book *The Evolution of Matter* of 1907.[187]

Radium was stimulating, effervescent, life-giving, and omnipresent—if not omnipotent. A 1904 London review had announced: "All Nature Now Alive!" as all sorts of ordinary things were found to be radioactive—not just newspapers, but air, water, and trees. The *Los Angeles Herald* notably related such discoveries under the headline: "Secret of Life." As Weart has noted, from the very beginning, "newspapers hinted that radium might 'solve the problem of life.' . . . For there was already a connection between radiation and life, a connection found everywhere from ancient transmutation myths to modern science news. . . . Radioactivity somehow reminded people irresistibly of life."[188]

End Rays

Life also somehow reminded people irresistibly of radioactivity. The dissemination of new ideas about radium's vitalizing power spread far beyond the newspapers, the pulpit, and the popular novel—it couldn't help but reach the ears of practicing scientists themselves, eventually conditioning the kinds of experiments that were undertaken and the modes of interpretation employed in examining their results. The *Lancet* posed and answered the following question in 1902: "Does the human body possess properties akin to radio-activity? The researches of M. Charpentier would appear to indicate clearly that it does."[189] Augustin Charpentier, professor of medical physics at the University of Nancy, had claimed to find a new form of radiation emanating from living things. The discovery of a biological origin for these rays—first christened "N-rays" by their discoverer, Charpentier's physicist colleague René Blondlot—was tremendously exciting.[190] While Blondlot had earlier concluded that N-rays were "not without influence on certain phenomena of animal life or vegetable life," it was Charpentier who reported to the Académie des Sciences between December 1903 and July 1904 that N-rays were emitted particularly strongly by living tissue—in particular, by nerves and muscles.[191] N-rays were even held to improve visual acuity and heighten the senses, and a putative link was found between levels of "psychic activity" and the N-rays detected. Moreover, *Nature* reported, "the emission of the n-rays by living bodies is not peculiar to man; it has been found in rabbits, frogs, and other animals."[192] One author even linked phosphorescent insects to N-rays, while another noted that N-rays "appear to flow from the nerves of the body, and are increased by the contraction of a muscle, indicat[ing] the radio-activity of the physical organism."[193]

Coming in a decade when "the discovery of a new radiation was not a surprising event," as historian of the N-ray controversy Mary Joe Nye has noted, "many of the properties that Blondlot reported for N-rays were similar to ones established for radioactive rays."[194] As news of the N-rays' stimulating effects and their putative biological origin brought them alongside the indisputable wonders of radium and the other radioactive emanations in the *Lancet*'s review of recent discoveries, N-rays—as the new "rays of life"—were readily brought into the radioactive fold, sharing the limelight with radium throughout 1904.[195]

According to Nye, at least forty people all told had observed N-rays, and between 1903 and 1906, the rays were discussed in some three hundred separate articles by a hundred different scientists and physicians.[196] N-rays rapidly became a favored topic for research. As another N-ray historian has noted:

> In the year and a half following Blondlot's announcement of his discovery the number of publications on the subject grew almost explosively. In the first half of 1903 four papers on the subject appeared in *Comptes rendus*; in the first half of 1904 the number had risen to 54. (It is interesting to note that in the latter period *Comptes rendus* carried only three papers on X rays.)[197]

Outpacing even X-rays in popularity, N-rays were on the make—until, that is, a visit to Blondlot's laboratory by the American physicist Robert W. Wood and the subsequent reports of his visit published in *Nature*.[198] Within months, N-rays would vanish from legitimate scientific concern nearly as mysteriously as they had first emerged.

Difficulties in replicating N-ray phenomena led to mounting criticism. While Blondlot had largely attributed differences in the ability to detect the effects of N-rays on a screen to individual visual acuity, there was also the matter of peripheral vision to consider, and some critics—including a young researcher at the Cavendish Laboratory by the name of John Butler Burke—took Blondlot to task for failing to consider other obvious but important objections. Published in *Nature* in February 1904, Burke's criticisms became part of an increasing flood as the year 1904 progressed. "I have endeavoured to repeat M. Blondlot's experiments," Burke reported in deadpan, "but quite without effect."[199] In fact, 1904 was to be the last year in which *Comptes rendus* published any papers on N-rays. While controversy still raged in some quarters, scientists increasingly turned their attentions elsewhere as confirmation of the existence of N-rays became increasingly localized to Nancy.[200]

Once a competitor for the mantle of the radiation-life connection that was everywhere abundant in those early years, N-rays rapidly gave way to the dominance of radium. Often, however, it was the same researchers who worked on both phenomena: Rutherford, for example, had been exploring the putative properties of N-rays in Montreal even as he continued his investigations into the disintegration products of radium.[201] And Burke, at the Cavendish, soon found his attention moving further and further from the study of fluorescence—his first research love, and the one that had led him to the N-ray controversy in the first place—and toward radium.

"By their rays ye shall know them," Soddy had once remarked in a prophetic tone. The connections between the radiant and the living fleetingly claimed by N-rays were instead rapidly parlayed into and supported the growing connection between radium and life.[202] Soddy's passion for enlivening the discourse of the radioactive world therefore contributed to the spread of a vital radium-life connection not only through the popular realm, but among scientists as well—and that connection was soon to prove experimentally productive. While lecturing on radioactivity at University College, London, Soddy had been invited by J. J. Thomson, discoverer of the electron and director of the Cavendish Laboratory in Cambridge, to give a "short course of experimentally illustrated lectures" on the stuff of his researches with Rutherford.[203]

Having just recently opened its doors to young researchers from other countries eager to pursue advanced degrees, the Cavendish was a hotbed of young blood and cutting-edge research. As Soddy discussed his research on the chemistry of radioactivity, his discourse of living radium and his references to a radioactive early earth may well have sparked a connection in the ears of his listeners. Very likely sitting in Soddy's audience at the Cavendish was the same young Anglo-Irishman who had criticized Blondlot's N-rays: John Benjamin Butler Burke, one of the first to enter the newly internationalized laboratory and resident there since at least 1900.

The stage was set for an intensification of the associations between radium and life. Radium had been discovered and described in lifelike terms, drawing on and transforming earlier discursive traditions that made it into *the* living element, the element of choice for biological connection. Burke, on the other hand, had come to radium from his earlier studies in fluorescence and phosphorescence through the intervening medium of the N-ray controversy. Well placed at the Cavendish Laboratory, hearing of Soddy's work with radium and the living discourse surrounding it, and having yet to make a name for himself, Burke found that the time was ripe for him to rework Soddy's metaphysics of metaphor in ways that were experimentally productive. While Soddy remained focused on the chemistry of radioactivity and the potential uses of radioactive energy, Burke was entranced by Soddy's story of radioactive "transition-forms" in the great saga of cosmic evolution and—in the wake of Rutherford's victory over Kelvin—took intense interest in the idea of a more intensely radioactive early earth. Soddy had provocatively concluded his Wilde Lecture in 1904 by saying:

> As science has advanced, the limitations of the universe with reference to the extent of its past and future existence have been forced steadily back. By the discovery of radioactivity and the revelation that has followed, of the vast and hitherto unsuspected stores of energy associated with the atomic structure of the more complex forms of matter, the possibilities of life in both directions have been enormously extended.[204]

Could radium produce life—or at least account for its origin? Soddy didn't think so, but Burke intended to find out.

2 Radium and the Origin of Life

On June 20th [1905] the scientific world was startled by the sensational announcement that a momentous discovery concerning the origin of life had been made by an English scientist. Working experimentally at the famous Cavendish laboratory in Cambridge, Mr. John Butler Burke, a young man in the prime of life . . . succeeded in producing cultures bearing all the semblance of vitality.
—William Ramsay, "Can Life Be Produced by Radium?"

In 1905 the world at large first learned of an epoch-making discovery from blaring double-columned headlines on the front page of the London *Daily Chronicle*:

> A wonderful discovery is stated to have been made by Mr. J. B. Burke, in the Cavendish Laboratory at Cambridge. By means of radium and sterilised bouillon, placed together in a test tube, he has developed cultures that present an appearance of life, and sub-cultures of which possess the power of sub-division. "They suggest vitality," says Mr. Burke, who is still conducting experiments.[1]

Word of Burke's discovery crossed the pond almost instantaneously and hit the front page of the *New York Times* with an equally sensational effect: "Generation by Radium: Cambridge Professor Reported to Have Produced Artificial Life." Burke seemed to have discovered the means of creating life at will in the laboratory. As a *Times* headline asked three weeks later, "Has Radium Revealed the Secret of Life? Recent Investigations Held by Some to Justify Such a Belief."[2] For its part, the *Daily Chronicle* repeated Burke's suggestion that the strange forms he had discovered were "in the critical state of matter between the mineral and the vegetable, in fact, on the borderline between the living and the dead."[3] The *New York Times* breathlessly declared that the "new creatures" stood "on the frontiers of life, where they tremble between the inertia of inanimate existence and the strange throb of incipient vitality."[4] Neither radium nor microbe, but partaking of some of the properties of both, Burke thus christened his newfound growths *radiobes*.[5]

Burke, like Soddy, was working out his own distinctive synthesis of the realms of the radioactive and the living. While Soddy's focus had been on the characterization of radioactive phenomena by means of biological metaphors, Burke took Soddy's metaphysics of metaphor one step further. Where Soddy had posited a process of natural selection among radioactive "metabolons," Burke posited an even more provocative theory of "physical metabolism" spanning the cosmic and organic worlds. Burke's radiobes thus soldered together a narrative of evolution that extended from the vast reaches of nebulae in outer space to the very origin of life itself, linking the cosmic with the organic through the half-living element and the half-living organisms it produced.

In some respects, Burke's work was thus entirely of a piece with various long-standing traditions that by the nineteenth century frequently viewed cosmic and organic evolution as operating by the same means and functioning as essentially the same process.[6] Indeed, it had been commonplace throughout the nineteenth century to speak of cosmic evolution as essentially a "developmental" or unfolding process. Just how organic evolution might have illustrated cosmic processes was never entirely clear, but Burke's experiments with radium, sitting at the precise juncture of these two discourses of inorganic and organic evolution, were ideally situated to rework these older traditions. As Burke put it:

> Germ-plasm . . . would thus consist of a nebula of uncondensed matter, of corpuscles or electrons in the state of formation, as planets and solar systems are evolved from atoms. This is not an absurdity, as may at first sight be imagined, but merely an extension, and a logical extension, as I venture to think, of the theory that atoms are such miniature planetary systems.[7]

Burke's work also fit well with the intractable series of debates and experiments on "spontaneous generation" that spanned much of the end of the nineteenth century and remained contested terrain well afterward. Despite his lack of prior interest in engaging in such debates, Burke initially framed his work in just these terms.[8] Matching what James Strick has called a "late nineteenth-century gestalt" that held that there was "some single *sine qua non*" of life, Burke's use of radium paralleled many other efforts to identify the ultimate "stuff of life": "protoplasm" (the physical basis of life that had been popular in theories about the nature of life ever since Thomas Henry Huxley's *The Physical Basis of Life* of 1869), "colloids," "gemmules," "plastidules," "micelles," "biogen," or some other entity, depending on the theorist.[9] As Strick has noted, "From the 1860s through the first decade of the twentieth century, theories abounded as to the simplest 'living unit.'"[10] In the context of the long nineteenth-century discourse of spontaneous generation, Burke's work thus contributed not only to the physical reworking of the "living atoms" tradition, but also to the re-envisioning and biological reworking of a much older and long-standing tradition of "atoms of life" speculations, from Leibniz's monads and Buffon's and Maupertuis's *molécules organiques* to Liebig's monads of the mid-nineteenth century, and arguably even Haeckel's more recent reflections on the connections between crystals and life (and his own brand of "monism").[11]

Moreover, Burke's work also fits into a diverse series of attempts around the turn of the century "to create life-like artifacts from inorganic matter"—a set of efforts that would later come to be called "synthetic biology."[12] Researchers in this tradition often claimed to produce useful mineral, crystalline, or other models for living systems rather than actual living beings themselves. Such experimentation was intended as an aid to speculation about the nature of life and perhaps even its provenance.[13] Burke's work fit with this tradition as well. Uninterested in merely rehearsing familiar claims to have created life in the test tube, he ultimately came to claim not that his experiments had produced life, but that they had produced something far more ancestral. By ultimately disavowing the label of "spontaneous generation" for his work, and by weaving together radium with the history of life on the primordial earth, Burke would soon come to characterize his work as the one of the earliest experimental attempts to get at the question of the historical origin of life on the early earth.

This chapter thus builds on the repeated suggestion in the literature that there are deeper continuities between the spontaneous generation of the 1870s and the later emergence of origin-of-life studies in the early twentieth century. While the larger tale of this decades-long shift from spontaneous generation to the emerging field of the origin of life extends far beyond the immediate case at hand, radium was a heretofore unacknowledged keystone bridging these worlds of the animate and the inanimate. Moreover, the Cavendish physicist's radiobes present a story of how radium came to life in a second way, through the weaving together of three different legacies. Baptized in the fiery controversies of spontaneous generation, inheritor to the rich legacies of cosmic evolution, and matching thoughtful speculation with real-world experiment, radium would turn out to have a central role to play in reworking all three of these legacies. The element with a half-life and the half-living element would produce a half-living thing.

A Bit of Beef Tea

As Burke himself later recounted, it was at the start of the Michaelmas term in October 1904 that he "exhibited to a host of people at the Cavendish and Pathological Laboratories at Cambridge these first experiments made on the action of radium salts on sterilised bouillon. The bodies thus observed were very curious indeed, and some of their properties very remarkable."[14] Plunk a bit of radium chloride or bromide from an ordinary test tube in sterilized bouillon, Burke said (fig. 3)—nothing more than "beef tea," as one critic opined[15]—and this was the result:

> After 24 hours or so in the case of the bromide, and about three or four days in that of the chloride, a peculiar culture-like growth appeared on the surface, and gradually made its way downward, until after a fortnight, in some cases, it had grown fully a centimeter beneath the surface.[16]

Burke conducted more experiments on May 10, and he published his findings in a letter to *Nature* on May 25, 1905, in which he described his technique in further detail and repeated his statement of surprise: "Bouillon is acted upon by radium salts and some other slightly radioactive bodies in a most remarkable manner."[17] (One commentator wondered if Burke might "have been a little premature in writing to *Nature*" in order to secure priority, but excused the fault as "nowadays a common one, particularly where radium is involved."[18])

Burke was called on a few months later to describe the results of his experiments before the Ordinary General Meeting of the Röntgen

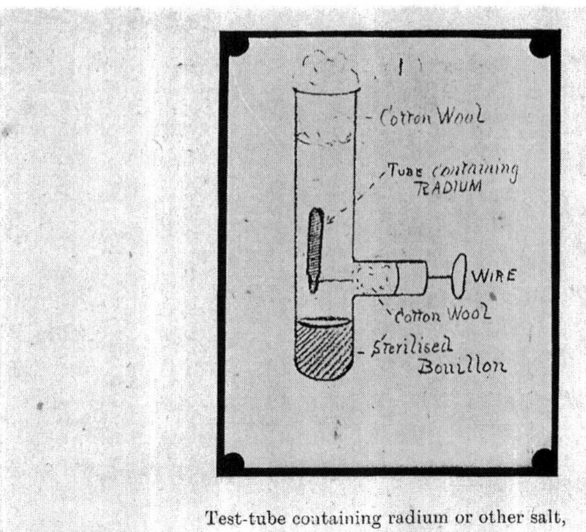

Test-tube containing radium or other salt, for purpose of sterilisation.

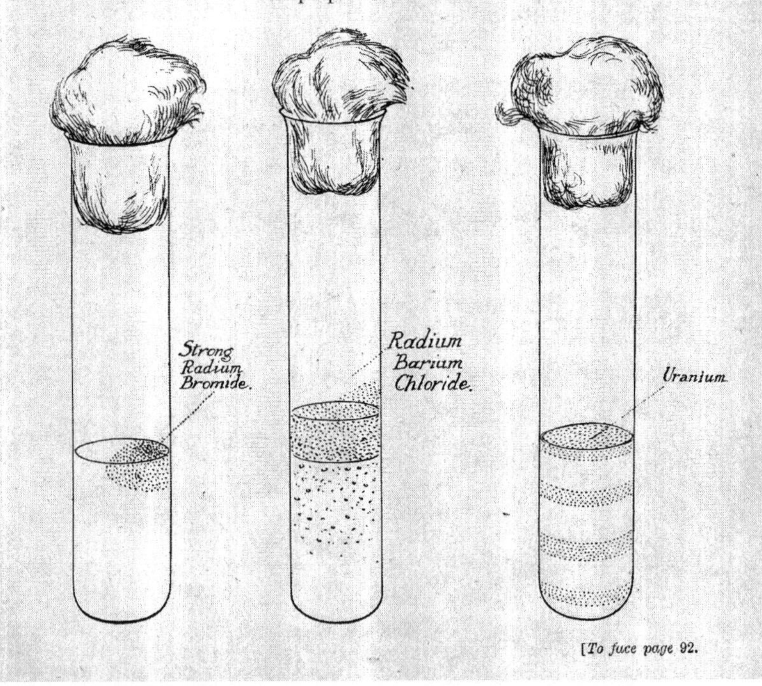

FIGURE 3. Burke's experimental setup with radium and bouillon. (From Burke, *The Origin of Life: Its Physical Basis and Definition*, 1906, 92.)

Society in London. Burke was introduced on December 7, 1905, to the "unusually large" assembled crowd by the newly installed president of the society—none other than Frederick Soddy—and "was received with applause."[19] According to the minutes of the meeting, Burke made preliminary references to the work done by others on the creation of artificial cells and on the actions of various salts on beef gelatine—a typical medium for growing microorganisms in vitro at this time—before turning to the more remarkable substance of his own experiments. Burke's strange growths "possessed a peculiar structure," the minutes noted, and were observed to grow in size over the course of days, showing an "evolution of shapes":

> They started as mere dots, and in the course of time they showed signs of development and aggregation. They expanded and assumed the dumb-bell shape. All this could be explained on the assumption that they were mere air bubbles. Later on, however, the appearance of nuclei became manifest, and it was these nuclei which seemed to him so peculiar. The nuclei began to divide, and the divisions, developed from the original, certainly did not present the appearance of air bubbles. They became more of the nature of organic crystals and were divided into distinct segments. [Burke] did not think that bubbles would produce that effect. . . . The form started as a dot, then took on the appearance of two dots, then the dumb-bell shape, and finally the biscuit shape, bearing a very close resemblance to genuine bacilli.

Or, as Burke himself noted:

> The growth is from the minutest visible speck to two dots, then a dumb-bell shaped appearance, later more like a frog's spawn, and so on through various stages until it reaches a shape largely different from its previous forms when it divides and loses its individuality, and ultimately becomes resolved into minute crystals, possibly of uric acid. This is a development which no crystal has yet been known to make, and forces upon the mind the idea that they must be organisms; the fact, however, that they are soluble in water seems, on the other hand, to disprove the suggestion that they can be bacteria.[20]

Existing at the limits of microscopic vision, Burke's growths were at times extraordinarily difficult to see, ranging in size from about 0.3 micrometers (or less than 1/60,000 of an inch by another account) to the merest of specks. Burke reported that they were "only visible with a very high power of the instruments—scarcely visible, indeed, with one-sixth-inch focus power, plainly visible with one-twelfth."[21] Resolving the nature of these growths thus required frequent reference to Burke's "photographs," although some of them looked more like hand drawings than photographs. A few of these images were published along with Burke's initial report in *Nature* (additional images accompanied his later 1906 book *The Origin of Life*; see fig. 4).[22] Burke even showed a photograph of his growths "in which the nucleate structure could be seen" to the Röntgen Society. Burke's images were of enormous interest; they also, however, required constant interpretation. As the *Daily Chronicle* remarked, photographs of the "magnified growths" showed "globular dots signifying nothing to the unexpert eye, but much to the trained intelligence of the scientist."[23]

Though they might look superficially like bacteria, it was clear to Burke that his growths were *not* bacteria or recognizably "life as we know it." Subcultures did *not* form when inoculated into a fresh gelatine; the growth patterns were peculiar (with cultures developing only some significant time after exposure, and growing only "slightly"); and the new forms had a perplexing tendency to grow only in the dark and to vanish in sunlight, to be soluble in water, and to simply vanish at 35°C rather than to coagulate as bacteria would. These were all strong reasons to doubt any attempt to grant full vitality to the new forms. Sims Woodhead, a Cambridge pathologist, had pronounced that Burke's radiobes were not bacteria. As one of the most popular science writers of the day, C. W. Saleeby, noted:

> It looked like a growth of bacteria, but on microscopical examination it proved to consist of minute rounded, nucleated (?!) bodies, such as Professor Woodhead had never seen. (I once had a bad quarter of an hour with Professor Woodhead, which suggests to me that the bacteria he has not seen no one has seen.)[24]

The growths lasted only "a fortnight or so" before proceeding to "break up," "dissolve," and "disappear" of their own accord at the end of their unnatural life. "Yet," as Saleeby remarked, "they seemed to be alive."[25] Or, if not quite living, Burke's new forms at least exhibited some of the main characteristics essential to life. Burke also described the forms, with a distinctively atomic terminology and a vitalistic cast, as growing atoms.[26] Or, as he wrote in his letter to *Nature*: "They are clearly

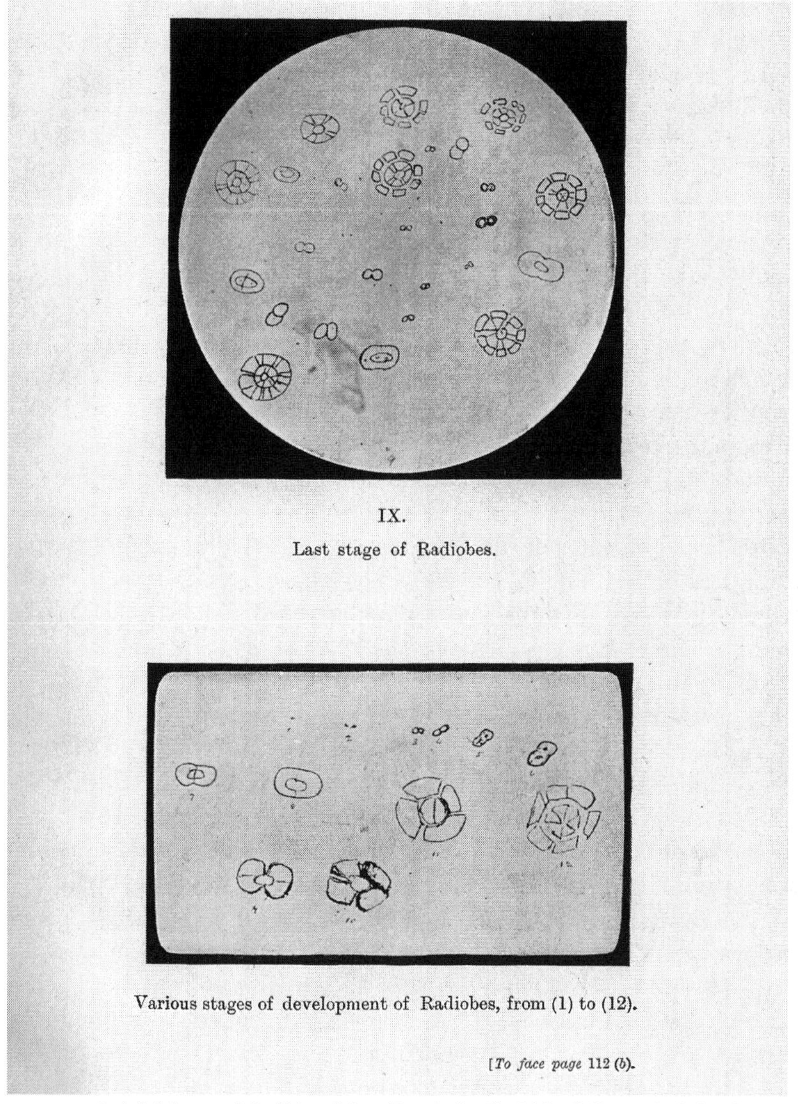

FIGURE 4. Burke's "photographs" of his radiobes. (From Burke, *The Origin of Life: Its Physical Basis and Definition*, 1906, 112.)

something more than mere aggregates in so far as they are not merely capable of growth, but also of subdivision, possibly of reproduction, and certainly of decay."[27] Burke's lifelike growths—said to be the result of his having "vitalized" gelatine with radium[28]—were also portrayed in the popular press almost immediately after their discovery as "short-lived but independent little atoms."[29] The strange new forms even—or so Burke reported—underwent cellular division akin to mitosis. Trained solely as a physicist, Burke may have been more than a little ignorant of the fundamentals of cytology, but he certainly knew that finding such fissioning in these inorganic forms was a point well worth advertising.

Far enough from truly living things, and yet just as far from being mere crystals, Burke's growths showed a kind of "physical metabolism" somewhere between *accretion* (an inorganic process) and *assimilation* (an organic process). And so Burke readily extended the living metaphors that Soddy and others were already using to describe and conceptualize novel radioactive phenomena to his curious new growths, forms that seemed even more obviously "alive" than the radium compounds themselves. Paralleling Soddy's vitalized discourse of radioactive "transition-forms" or "metabolons" in the grand story of cosmic evolution, Burke took his radium-induced growths to be new particular and peculiar instantiations of a larger *physical* metabolism that "is everywhere present" and which could be "controlled by certain types of inorganic bodies; but most of all by the vital units which form the basis of all life."[30] The growths, Burke thus suggested, although they "obviously lie altogether outside the beaten track of living things," were still properly to be considered "within the realm of biology." He held that they were "suggestive" of both the nature and the origin of life, and perhaps qualified even as transitional forms of life, as "they appear to possess many of the qualities and properties which enable them to be placed in the borderland, so to speak, between crystals and bacteria—organisms in the sense in which we have employed the word, and possibly the missing link between the animate and the inanimate."[31]

The widespread attention the radiobes drew and the debates they helped engender, the provocative connections of the radioelements with the hot early earth on which life may have originated, and the slipping of a vitalized discourse of radioactivity across the realms of the physical and the organic with several overlapping senses of "half-life"[32]—all these factors came together to permit Burke's radiobes to resonate powerfully and simultaneously in several different registers and realms.

How Experiments Begin

An active researcher at the Cavendish Laboratory in Cambridge, Burke was well placed but far from well known at the time of his sensational discoveries. As a popular lecturer on the radium circuit at the height of the radium craze, Burke was a young, respectable, and perhaps respected (though no one was quite sure) scientific investigator. (He was such an unknown quantity in 1905 that the *New York Daily Tribune*, evaluating his meteoric rise to fame, was forced to ask, "Now, who is this young man and what has he done?"[33])

This much was clear to his contemporaries: born in Manila on November 4, 1873, and raised in Dublin, John Benjamin Butler Burke matriculated at Trinity College before becoming a lecturer at Birmingham and later a fellow at Manchester. By 1898 Burke had moved on to Trinity College, Cambridge, where he found himself granted access to the prestigious Cavendish Laboratory, and where he received his M.A. in research under the eminent physicist J. J. Thomson in 1902. Burke's biological training was minimal at best, but at the Cavendish, where he was surrounded by fellow physicists, this proved to be no obstacle to undertaking physicalist researches with a potentially fundamental biological import. His institutional situation, the nature of his experiments, and the nature of his claims all served to reinforce a link between the transmuting processes of radioactivity and the potential uses of that transmutation for crossing the border between the lifeless and the living.

The Cavendish was an exciting place to be in the first years of the new century. Founded in 1871, it was almost immediately recognized as the best laboratory in Britain for new advances in physics and chemistry, which included the discovery of the electron, the characterization of cathode rays, and the study of "the discharge of electricity through gases."[34] Under Thomson's leadership, a number of "advanced students" from other universities (and selected countries) who had not completed their undergraduate work at Cambridge were permitted into the laboratory.[35] The first of these advanced students, including Ernest Rutherford, arrived at the Cavendish in 1895.[36] Arriving the following year, as one of 39 other students who worked in the lab during the period 1896–1900, Burke aimed to study the production and propagation of the phosphorescence of gases in the presence of an electrical discharge.[37]

He began his research career ordinarily enough, writing a paper entitled "Luminosity and Kinetic Energy" for the British Association in 1902, and another, "On a Modification of FitzGerald's Model of the Ether," in 1904, and contributing a note for the *Proceedings of the Royal Society* on fluorescence and absorption, among other publications.[38] Just as fluorescent and phosphorescent phenomena had induced Henri Becquerel to investigate and eventually discover radioactivity, however, Burke's interests led him to the study of the properties and effects of radium that was soon to make him famous.[39] In the first popular account of Burke's experiments, the *Daily Chronicle* reported that "it had not occurred to anyone, however, that even the workers of the Cavendish Laboratory would detect any connection between the phenomena of radium and the problem of the origin of life."[40] It had, however, occurred to Burke.

Open to "any Cambridge man who wanted to research physics," the Cavendish was a place where Burke would have been largely left to his own devices.[41] The extent of this laissez-faire tradition was such that researchers "freely made" shifts of research focus that were "sometimes casual, sometimes dramatic." This tradition does much to account for the freewheeling nature of Burke's research. Even though studies of radioactivity were not among the laboratory's primary interests, the Cavendish was sufficiently well endowed to have ready access to radioactive materials, and Burke was already familiar with the radioactive work of Rutherford and Soddy. This institutional environment enabled Burke to pursue his own rather unorthodox interest in radium.[42]

"All Matter Is Alive—That Is My Thesis"

Burke's wholesale adoption of the discourse of living radioactivity is readily apparent. No stranger to Ernest Rutherford and Frederick Soddy's work on atomic disintegration, he had in fact already summarized their work in the popular digest the *Monthly Review* in 1903, shortly before beginning his own experiments with radium. Describing with biological metaphors how "radio-activity is . . . infectious, but the infected body recovers in the course of time," Burke noted that the "continual formation" and "gradual destruction" of "radioactive stuff" caused him to "perceive a striking analogy which appears to exist between such a process and that of *metabolism*, although the two phenomena, so far as our knowledge at present goes, are distinct." A ready comparison with processes known to take place with "complex molecules of albumen" was possible, Burke knew, "but here once more we must be careful lest our imagination should carry us away, and lead us into regions of pure fancy, to a height beyond the support of experimental facts." But the process of "perpetual change" in matter nevertheless intrigued him:

> It changes its substance—in a limited sense lives—and yet it is ever the same.[43] Is it not so with the cell? Is the atom an elementary cell, a living thing? Our own view is that the atom preserves its identity in the same manner that a cell does. . . . The distinction apparently insuperable, that the biologist holds to exist between living and so-called dead matter, should thus pass away

> as a false distinction, and all Nature appear as a manifestation of Life; this being the play of *units* of we know not what, save that it is what we call *electricity*. Atoms and molecules would be elementary living cells, possessing some of the properties, but not all, of the more highly organised cell, the unit with which the biologist has to deal. These are not idle thoughts. Heterodox or orthodox they are to us the logical outcome of all that we have had to say. All matter is alive—that is my thesis.

Burke concluded by noting that at long last, perhaps a clue was available "as to the ultimate constitution, perhaps also as to the ultimate destiny, not only of Life as we know it, but of a simpler Life, that of matter too."[44] Burke's interest in the "life" of matter and the processes of "physical metabolism" is strikingly apparent here a full year before he began his experiments on the effects of radium on gelatine.

Burke's claim that "all matter is alive" itself has a rich history dating back at least to the philosophical musings of nineteenth-century *Naturphilosophen* such as Novalis or Schelling. But in using the new phenomena of radioactivity to characterize life, Burke's approach dovetailed nicely with Soddy's relation of life to radioactivity—his ongoing discourse of living radium, his references to a radioactive early earth, his discussion of "transition-forms" and metabolons in the great saga of cosmic evolution. "The 'radium' may be, and I am at present not loath to think is, that state of matter that separates, or perhaps unites the organic and the inorganic worlds," Burke noted. Moreover, this vitalized terminology of radioactive phenomena also served as a provocative sign pointing to the nature and perhaps even the origin of life: "In the radioactivity of matter and the products produced by it, in their growth, disintegration and decay, may be sought that vital principle and source of vital energy which in the beginning withstood that high temperature at which most assuredly our planet must have once existed."[45]

Soddy's framing, the broader contemporaneous radium craze, Burke's thesis that "all matter is alive," and his belief that radium had vitalizing powers were all factors leading toward his radiobe experiments. But so, too, were his years of study: by the time of his 1904 experiments, Burke had been regularly publishing papers on fluorescence and phosphorescence in the *Philosophical Transactions* and elsewhere for several years.[46] He was particularly intrigued by the luminous glow produced on a screen when a radioactive object was brought near. Drawing perhaps on Rutherford and Soddy's disintegration theory, Burke supposed that the scintillations on the fluorescing screen were the result of the formation and decay of molecular aggregates. Partly as a result of these and other experiments on phosphorescent bodies, Burke proposed a connection between light and life and sought to explore "whether such dynamically unstable groupings could be produced by the action of radium upon certain organic substances."[47] Dynamically unstable groupings might prove to be responsible not only for phosphorescence, but also for some of the properties of life.

As already noted, phosphorescence had long seemed a natural link between the realms of the inorganic and the organic, and the supposition that both phosphorescence and life involved the molecular aggregation and disaggregation of molecules was far from new. That phosphorescence might parallel, or in some way even generate, a basic kind of living metabolism seemed obvious not only to Burke, but also to several of his predecessors. Indeed, Burke himself readily acknowledged that the inspiration for his first attempts to study phosphorescence and internal energy transformation came from Eduard Pflüger's earlier praise of cyanogens and Max Verworn's focus on metabolism.

Both Pflüger and Verworn had found numerous provocative analogies between cyanogens and what they termed "living proteid." In searching for the molecular basis of life, Verworn had wondered whether it might be cyanogen itself that granted to "living proteid molecules its characteristic properties." Earlier, Pflüger had gone even further in a remarkable statement: "This similarity is so great that I might term cyanic acid a *half-living* molecule."[48]

Burke first encountered Pflüger's claim that the phosphorescence of the cyanogens reflected certain elements of life while he was a fellow at Owens College in Manchester. Although at first he thought the theory was "a rather wild one," it ultimately seemed to him "a very reasonable thing that if cyanogen was a living thing it ought to grow in culture media." After failing in his first attempts to grow it, and "observing that radium had several qualities in common with cyanogens—it is highly excitable and contains a vast store of energy," he turned to radium.[49]

Whatever the merit and internal coherence of Pflüger's and Verworn's positions, Burke found an undeniable resonance between their accounts and the properties of and discourses surrounding radium. Into the thorny debate over spontaneous generation; wrapped up with Pflüger's talk of half-living cyanogens and the necessity of a labile chemical compound whose internal metabolism could itself be held responsible for the characteristics of life; an intimate heir to Verworn's focus on metabolism as the root of all vital phenomena and his account of the history of the vital process as a complex motion partaking in some sense of the history of the cosmos—into all this entered radium. A number of other different threads also came together for Burke by the time of his 1904 experiments: the laissez-faire, vaguely radioactive atmosphere of the Cavendish; his role as a popular lecturer on the properties and wonders of radium; his thorough immersion in the living discourse of radioactivity that Soddy and others were helping to popularize; his interest in the properties of fluorescent and phosphorescent gases in the presence of electrical discharge; his acquaintance with the work of Pflüger and Verworn and their reference to the "half-living" molecule, cyanogen; and his attempts to culture cyanogen and then radium—all these were part of the pathway leading to the experimental discovery of half-living forms, his radiobes. Was Burke an innovative genius creating half-living forms or a well-placed crackpot? His reputation was made and unmade in the most public of ways.

"He Has Taken the World by Surprise"

Emerging contemporaneously with all the swirl and hubbub of the radium craze, the results of the experiments in Burke's radioactive crucible "startled every scientist and immensely interested the public at large," drawing both "overwhelming correspondence from many quarters of the globe" and "embarrassing publicity."[50] One journalist noted that Burke had "taken the world by surprise."[51] On the day of its big scoop, the *Daily Chronicle* had already captured an element of Burke's diffidence, describing him as "unaffectedly modest and somewhat reticent about his remarkable discovery," reporting that "he discussed the matter as calmly as he might have described the merits of a new lamp or a water-tap."[52]

Burke's experiments not only revitalized the "spark of life" trope, but were rapidly re-inscribed within the frame of the life-as-electricity connection: "The manifest intimate connection between vital and electrical phenomena is all in favour of the validity of Mr. Burke's conclusions,"[53] noted one commentator, while the *New York Times* remarked, "We leave to the next generation the task of ferreting out the genealogy of the radiobe and the electron."[54] Indeed, Burke's results were taken to drive what was known at the time as "the electrical theory of matter" in new directions: "The new theory of the composition of an atom and the possibility that one element may be transmuted into another relate only to the nurture of inanimate matter. A hint that life may spring out of the latter without the exercise of will or intelligence is vastly more startling," reported the *New York Daily Tribune*.[55] "It is more than probable," said another commentator, "that radium is one of the A B C stepping-stones to such a law of cosmic life."[56]

Saleeby, after initially confessing that it took him "many readings of Mr. Burke's letter to persuade me that it would not be possible to detect some fallacy, some experimental error somewhere," shortly afterward waxed eloquent about Burke's findings, telling "tales of cells of gelatin all but alive."[57] Like many others, Saleeby had already speculated in 1904 that radium was the new philosophers' stone, as it led from one element to another.[58] But by 1906, Saleeby went so far as to enthusiastically popularize Burke's claim that the radiobes might constitute a "missing link" between the inorganic and organic realms. Such work, he said, was "profoundly altering our view of matter and utterly disorganizing the accepted definitions of life." If even so-called "lifeless" matter, "the seat of incessant, manifold, potent, and seemingly self-caused activities . . . must undergo a profound alteration," he asked, was there any meaningful "difference in kind between living and so-called lifeless matter"? "If anything in the world is alive," he asked, "is not radium alive?" Saleeby concluded that the question of whether Burke's radiobes were alive depended on considering "the reputed behavior of an atom of radium."[59] Burke agreed: "The question of whether the microscopic forms therein described are, or are not, living things depends altogether upon what our conception of life is, and upon how broad or how narrow is the definition we are willing to accept of the phenomena of vital processes."[60] For Burke, the answer was clear: the "disintegration and decay of inorganic substances is one of the most remarkable analogies between them and living matter."[61]

By the end of the summer, Burke's discovery was still making headlines as his results were being considered everything from provocative to a prank. W. P. Pycraft called out overly eager journalists as "ill-informed enthusiasts [who] ecstatically assured us that the greatest of all mysteries had now indeed been laid bare, and this by the aid of radium and a little beef-tea!" Pycraft was merciless in his criticism: it was "a pity" that the radiobes were incapable of long-term survival, but even their disintegration, he noted ironically, was said to show signs "so closely suggestive of death" as to signify "the crowning glory of the discovery."[62]

But by September another commentator was calling the radiobes "the most surprising discovery since the first isolation of radium by M. Curie," even though—in light of the "contradictory characteristics" of the radiobes—it was "evidently premature to speak of Mr. Burke's discovery as in any way throwing light upon the chemical origin of life."[63] By November, others were touting Burke's findings as "a discovery that has provoked more discussion, perhaps, than any event in the history of science since the publication of the 'Origin of Species,' for it has a direct bearing on all speculative theories of life."[64]

If the discovery of radioactivity had uncovered the secret of matter, then Burke's radiobes—not fully living but certainly having some of the properties of life and emerging in highly radioactive circumstances similar to those thought to exist on the early earth—bolstered the notion that the discovery of the secret of life was at hand. Indeed, newspaper accounts referred to Burke's discovery in just these terms.[65] The minutes of the Röntgen Society meeting pushed the seat of this secret of life still further: "There was probably something in the nucleus itself which had not yet been discovered. It might be that there was some element far more unstable than radium, and which possessed its properties in a more marked degree."[66] If radium itself was not the secret of life, then this radium of radium deeper within must surely be it.

Indeed, Burke's experiments drew enormous attention. "The interest attached to them," Burke said, "has been such that the brief note communicated to *Nature*, May 25th, 1905, and the few words uttered to a representative of the *Daily Chronicle* . . . have resounded from the remotest corners of the earth to an extent quite beyond the expectation even of my most apprehensive friends."[67] Radium was Burke's ticket to fame, his experiments taking the press by storm and promoting him from his status as a dime-a-dozen lecturer on the radium circuit to "the most talked of man of science in the United Kingdom."[68] As Burke himself noted, "The interest in the question of spontaneous generation has scarcely surpassed, nor in some respects does it appear to have equaled, the enthusiasm which these experiments, whatever their ultimate bearing on the question, have aroused."[69]

Mind the Gap: Redefining Spontaneous Generation

Burke's experiments were clearly relevant to the unsettled debate over spontaneous generation in Britain, and his findings were regularly cast by the press in just these terms. At first, Burke was quite comfortable with this characterization. (In his 1906 book detailing the results of his experiments, *The Origin of Life*, Burke continued to use the discourse of spontaneous generation and even used the phrase as the title of his eleventh chapter.)[70] He was well aware that he was reopening the spontaneous generation controversy, and he seemed to revel in bringing the materialist implications of his science to the forefront, along with the theological consequences they might provoke. As he remarked to the

Chronicle, "Should my experiments prove the possibility of 'spontaneous generation,' it is a principle not in the least destructive of the deistic conception of the universe. In fact, if it can be shown that dust and earth can produce life on account of radio activity, it would only confirm the truth of Biblical teaching."[71] (On hearing this, one wit remarked that "he knew his Bible pretty well, but did not remember that it even hinted that life was produced by spontaneous generation, or that radio-activity acting on dust and earth produced it."[72]) Others, in a deist vein, felt the discoveries were a "gain to religious thought."[73]

When scientists were asked to give preliminary commentary on Burke's findings, however, they were less sanguine about Burke's intended revival of the debate. Gathering initial comment, the *Daily Chronicle* reported Sir William Ramsay as saying, "There may or there may not be something in the discovery. . . . I have seen what has appeared in the papers on the subject, and I think it is a pity that so much has been made of the demonstration." Ilya Metchnikoff agreed that it was too early for comment and that more corroboration would be needed, while a professor of public health at University College, London, stated, "He is a bold man who would revive the theory of spontaneous generation, unless it is based upon work of the highest order." Marie Curie, by contrast, was said to be "profoundly interested," but as she was not "a biological expert," the *Chronicle* reported, she "therefore hesitated to express any opinion at the present juncture as to the possibility or otherwise of spontaneous generation by the direct agency of radium."[74]

Some newspapers had reported that Burke's experiments implied that the creation of life was just around the corner—and some even reported that he had already accomplished it. For all his dabbling with the label "spontaneous generation," however, Burke soon stated explicitly that his experiments did *not* prove spontaneous generation: "We do not claim to have produced spontaneous generation," he said, "if by this term is to be understood the appearance of life from the absolutely lifeless."[75] The most he said he hoped to indicate was that "we have arrived at a method of structural organic synthesis of artificial cells, which, if it does not give us natural organic life such as we see existing around us, gives us at least something which admits of being placed in the gap, or, as it might preferably be called, the borderland, between living and dead matter, as familiarly understood."[76]

Indeed, for all the journalistic hype—with writers leaping all over the discovery and energetically spreading the news among the populace on both sides of the Atlantic—many of the first reported scientific responses to Burke's discovery were distinctly guarded. If corroborated, the radiobes would be a "sensational" find—the word was used time and again to describe Burke's findings—but first reports were not to be taken instantly at face value.[77] The *Daily Chronicle* captured the mood: "There is a general tendency in the scientific world to await further results before pronouncing definitely on the question."[78] But that didn't stop the paper from continuing to run its lead story—potentially one of its biggest stories since its breaking the news of Roentgen's discovery of X-rays to the British public in 1896—or from continuing to search out additional scientists with something to say about the radiobes. Many saw in Burke's discoveries echoes of the discovery of similar bodies or artificial cells decades earlier by different means. But some, like Sir Oliver Lodge, saw in Burke's work a confirmation of some of their own speculations. Lodge had published speculative comments only a month earlier on the "complex molecular aggregates" that would "probably be found on the road towards organic evolution."[79] By these, Lodge meant none other than the radioelements.

Having attended and been inspired by what he took to be a "brilliant" lecture by Rutherford at University College, London, Lodge had published an essay in the *North American Review* only a month before Burke's experiments, asking, "What is life?" and "What is an element?" Burke's work seemed to Lodge to be an experimental verification of a promising idea he himself had already been considering: that the evolution of matter in the elements might lead to the evolution of life. In fact, Lodge wondered whether "all effort at spontaneous generation has been a failure" either "because some essential ingredient or condition was omitted," perhaps such as radium, or "because great lapse of time was necessary."[80]

But if radium could theoretically bridge the gap, as Soddy, Lodge, and innumerable independent others dared to wonder, and if radium were the element of choice not only to bridge the living and nonliving worlds, but also to end the debate over spontaneous generation, then the discovery of Burke's radiobes went one better. No longer was it necessary to claim that radium was "alive" in any substantive or provocative sense, or even that one had produced something living by means of radium. One could argue instead, as Burke did, for the critical role played by radium in the production of half-living forms, of a new kind of *artificial* life unlike the natural variety: "The vital processes in the radiobe and other such bodies constitute merely artificial life, as distinct from that natural life we see around us, and which it is beyond our wildest hopes to imitate, much less to create."[81] And elsewhere, "These cells are not alive in the familiar sense of the word. In fact they do not show more than the rudiments of vitality, when the word is used in its more extended sense; but they help to illustrate the manner in which cellular bodies may be formed from protoplasmic substance."[82] Indeed, Burke never claimed that his radiobes were living in the ordinarily understood sense: "To expect to make a full-blown bacillus at the present day," he soon wrote, "would not be more absurd than to try to manufacture a man!"[83] As the president of the New York Academy of Medicine summarized it, Burke's aim was to do nothing less than "enlarge enormously our conception of what life is. He denies that protoplasm is its sole basis, and he thinks that practically all physical phenomena are vital phenomena."[84]

Though Burke had at first happily endorsed the label "spontaneous generation" for his findings, by 1906 he realized the misunderstandings to which he was prone by continuing to use such terminology—not to mention his unwanted association with colleagues like Henry Charles Bastian, with whom he often disagreed.[85] Through further careful reflections on the nature of life, the source of vitality, the place of the radioelements, and his own earlier work with cyanogens, Burke increasingly sought to distinguish his findings from those of spontaneous generation generally. While the production of "living organisms from inorganic matter would be without question a case" of spontaneous generation, Burke proposed instead that deep within inorganic substances there may be "some germ, or germs, still hitherto unknown, and of a nature quite distinct from any we have yet had reason to regard as living," and yet which might have "the principle of vital process, in an elementary form, as a part and parcel of their being. It is so with the dynamically unstable substances which of their own account manifest radio-activity."

Expanding on Soddy's living metaphysics of radioactivity, Burke noted that the "dynamically unstable bodies" of the radioelements "have to some extent some of the properties of life . . . the products of radio-active bodies manifest not merely instability and decay but growth, sub-division, reproduction, and adjustment of their internal functions to their surroundings, a circumstance which I think will be found to be equivalent to nutrition."[86] Fully aware that this was "merely analogy," and quoting Darwin's line that "analogy is a deceitful guide," Burke nevertheless felt that if analogy led to verifiable results, as his had done, then "its utility should have a greater claim to our attention than to be passed over with indifference and ignored." In other words, Burke proposed, the analogy was more than merely suggestive.

Burke thus ultimately set out to redefine the focus of his investigation as something *other* than spontaneous generation as traditionally conceived. While appealing to, say, Louis Pasteur or John Tyndall and their arguments against spontaneous generation "sounds very simple, very clear, and very forcible," Burke asked, "Has it really any bearing on the question as to whether radio-activity can afford the internal energy of vital processes?" Burke here recast the question into one that could be answered purely in a physical mode: physics could provide answers to questions in biology that biologists using their own methods had been stumped by. After all, "there are questions in biology of the deepest interest to the physicist."[87] Burke was apparently keen to cross treacherous disciplinary divides:

> I think . . . that there are a great many problems really common to physics and physiology that should be taken up and studied thoroughly. It would be a great deal better if they were not supposed to be divided up into watertight compartments, and each looking with suspicion on the other. But if the origin of life is to be solved it will be by a physicist, for the physiologist seems to have given up the task as hopeless.[88]

Burke's confidence in this quest came in no small measure from his willingness to confront the eternal bugbear of biology: the question "What is life?" In a remarkable passage, Burke declared that biologists should feel free to pronounce on the question rather than constantly shy away from it—chemistry and physics had found their respective grounds in atoms, why not biology? "What is this unit or atom of life; whether it is a man, an acorn, a microscopic cell, or the atom of radium, or something else that we know nothing of?"[89] While physiologists assumed that life came from life, one of the older and most established dicta of biology, Burke contended that "we physicists think that life is such an elementary form that its origin is to be found elsewhere. Radio active bodies disintegrate and decay, and therefore there is an analogy between the organic and the inorganic."[90] Moreover, physics would succeed where biology had failed, and would do so in precisely in the gap that Burke was so fascinated by. After all, he said, "Life-activity is a phenomenon of matter as much as radio-activity."[91] It was no longer enough to give rise to something merely *like* life. Rather, physics could help get at the first emergence of something that was like life because it was *analogous* to life, and perhaps even because it was a potential *precursor* to life. Burke was confident that physics would lead the way: "The problem of the origin of life, which the biologist for the time being has almost, if not altogether, abandoned, if not in despair, at least in quiet resignation, would thus appear to present more hopeful signs of yielding a solution in the hands of the physicist than in his own. I say this with all due deference."[92]

Evolving Life

If radium had the ability to produce vaguely living forms, and if the early earth had higher levels of radium than it does at present—in what seemed an obvious corollary of Rutherford and Soddy's theories of radioactive decay—then these radiobes might be much more than mere examples of spontaneous generation, and contested examples at that. They might instead be missing links in a larger chain of organic evolution, indicative not just of disputable life created in the here and now, but of the very first kinds of living things to have historically emerged on earth and that no longer existed.

With his petri dishes pregnant with possibility, Burke's experiments with radium thus served as midwife to the nascent field of the study of the origin of life. Redefining terms such as "spontaneous generation" and "artificial life" (even as he continued to use them), Burke worked a fundamental transformation in the terms of the debate in the first decade of the twentieth century. No longer would scientists have to continue to endlessly debate the age-old question of the origin of life; Burke would instead provide them with the first form of experimental access to what he labeled an "obliterated page in the history of our planet."[93] Abiogenesis was not only a problem in evolution; it was a problem ripe for experimentation. Although Burke had claimed early on to be somewhat wary of extending the implications of his findings as far as the historical origin of life—publicly expressing misgivings at the Röntgen Society about such an extension—he proceeded to do precisely that, even at that very meeting. Burke's radiobes thus not only bridged the gap between the living and the nonliving in the here and now, but also enabled Burke to at last bring together for the first time two separate discourses of evolution, one cosmic and the other organic, in experimental fashion.

Even if his radiobes could not properly be called living, he said, they were nevertheless "suggestive" of the origin of life in that they might correspond "to some simple form of life that existed in a far distant age" and which was unknown at present due to the action of natural selection over the eons.[94] Just as Soddy had theorized that the radioactive elements were unstable "transition forms" in a grand process of cosmic evolution, with the most unstable already selected out, so Burke proposed that his radiobes might best be considered unstable transition forms in the origin of life and indicators of an evolutionary process from inorganic to organic that no longer existed on the earth: "Possibly they are a primitive form of life," he concluded. "Nearly everything is radio-active. The earth itself is, and in some suitable medium life may have originated on the earth in that way."[95]

Burke's radiobes thus ultimately depended for their privileged status not on their demonstrated growth, division, or reproduction, but on their place within a narrative of the terrestrial evolution of life from nonlife and the half-living stage in between. The *New York Times* ran with this story line under the headline "The Microbe's Ancestor," saying that Burke had discovered "in a radium product . . . the 'prehistoric ancestor' of the microbe." Breathlessly exclaiming that "the chemically dead compound is observed to give birth to living organisms," the reporter described how "minute creatures revealing a highly organized structure push their way to the surface, display phenomena of 'budding' and reproduction, and after a slight development in the segregated state decay and are resolved into minute crystals. Mr. Burke is not prepared to affirm with positiveness that these organisms are quite alive. They may be *half alive*."[96] And just as Burke's radiobes sat poised halfway between life and nonlife, Burke's own position hovered somewhere between the discourse he had come from and the discourse he was helping to inaugurate. Burke's discourse of the half-living was, fittingly, only half-born.

When dealing with experiments on the edge of life, words matter as much as things.[97] While the story of the progression from the half-living *element glowing* in the dark to a half-living *radiobe growing* in the dark might seem at first a mere rhetorical pleasantry, radium and radiobes found a real connection in Burke's mind, and in the minds of his contemporaries, through their common half-living status. And this connection was made possible by Burke's peculiar redefinition of spontaneous generation and his marriage of the two seemingly independent discourses of organic and inorganic evolution, not only through a common radioactive metaphysics but through experiment itself.

"Something Queer Has Happened in His 'Bouillon'"

While some reports had held that Burke understood that "radium was the life-producer" and that a radiobe was "actually a living organism,"[98] others claimed that Burke had even been able to obtain subcultures, and that the growth of the radiobes continued even in the absence of radium. A rash of newspaper headlines trumpeted, "Generation by Radium," "The Secret of Life," and "The Microbe's Ancestor."[99] Burke had largely shied away from overt sensationalization and frequently underplayed his findings in the popular press: the *New York Times* had even quoted him as saying, "What has been done has suggested vitality. Do not put it higher than that."[100] As another reporter noted early on, "What Prof. Burke says is not so much more than that something queer has happened in his 'bouillon,' and that it seems to have developed life out of unlife."[101] Burke's statements were actually quite modest when compared with the enormous stir his experiments aroused. But just what he had claimed he had done changed in the reportage of the time as his work increasingly fell from favor.

Burke's vaunted position at the Cavendish and his academic pedigree initially protected him from some of the worst of the attacks. His work deserved "the most respectful consideration," opined the *New York Daily Tribune*: "The simple fact that such leaders in science as Lord Kelvin and Professor J. J. Thomson have faith in his character and capacity, while it is not a verdict on the young Irishman's experiments themselves, indicates his title to a hearing."[102] Even an unfavorable review of his later book detailing the experiments held that "Mr. Burke has been the victim" of "the sensational announcements of his discoveries or observations,"[103] though another source saw Burke as "willing prey, to an enterprising journalist . . . [who] made the most of his time."[104] Another review held that "Mr. Burke has had the misfortune to be heralded by a particularly loud fanfare, and if the result causes disappointment he must thank his trumpeters. When the world is informed that a mountain is in labour the mouse that issues, though quite a good mouse, is rendered ridiculous."[105] Other commentators reprimanded the purveyors of such yellow journalism: "A study of the original authorities is always advisable before publishing their contents. For the sake of scientific journalists who may wish to publish Mr. Burke's future discoveries to a wider circle, we may perhaps state that the price of *Nature* is sixpence, and that it can usually be obtained at the larger railway bookstalls."[106]

Once the initial tempest had subsided, cooler heads were called on to investigate the phenomenon more fully and pronounce on it. One witty Fellow of the Royal Society remarked that while there may have been plenty of radium on the early earth there certainly wasn't any beef-tea.[107] Another critic similarly took issue with Burke's use of bouillon, calling it a "highly developed proteid" and animal product: "If he has obtained organisms, his discovery will do much to show how matter that has once lived may be made to live again, but it will not in the least explain the origin of life on this planet for the proteid essential to the experiment is itself a highly specialised product of animal activity."[108] In the same vein, the remarkably named systematic theologian Agar Best remarked that "unquestionably our planet was once destitute even of bouillon, *i.e.*, of the marvelous and complicated carbon compounds which are the constant garb of life, and are found only where there is or has been life."[109]

Even Saleeby, one of Burke's foremost popular advocates, claimed that Burke's experiments "offer no correspondence at all to the conditions which must have obtained on this planet, hundreds of millions of years ago" and that "there is no evidence . . . that salts of radium were present upon this cooling earth of aeons ago, in any proportion comparable to that of the radium in Mr. Burke's test tubes." Moreover, he noted, Burke's experiment "would be irrelevant, since not only the experimenter but also his beef gelatin are themselves products of life":

> Beef-gelatin is itself a product of living matter and . . . even though it could be artificially produced by the chemist, yet there were no laboratories on the cooling earth a hundred million years ago, and if there was no life to produce gelatin, it tells us little of the *original* origin of life to know that it may now be produced by the submitting of organic compounds to the action of radio-active substances.[110]

Saleeby concluded, nevertheless, that Burke's accomplishment was "signal enough," and he praised Burke for, like his rival Bastian, having "gone far to show that spontaneous generation occurs in the world to-day."[111]

Burke quickly countered his critics by saying that any experiment would have to make use of conditions not available on the early earth, but that this need not invalidate his findings:

> How could ever we imagine the existence of beef-tea? I should add, or of a laboratory, retort stands, and test tubes? Obviously, if we are to imitate Nature in the laboratory, the processes of artificial synthesis will differ to some extent if not widely from the original. In the original the constituents of protoplasm were present. But are they not each and all also present in the gelatine culture medium?[112]

Nevertheless, the tide of opinion began to turn against Burke. The *New York Times* went on record scandalously associating his radiobes with the very N-rays he had years before been instrumental in criticizing: "His 'radiobes' will be best seen by the 'N-rays,' about which there was not long ago a hot dispute, now quite died away, but as the professor is not too sure one way, other people will be most safe if they refrain from being too sure the other."[113] Others immediately asked the obvious question: Was the bouillon used thoroughly sterilized? Was contamination a possibility? The *Daily Tribune* called for "independent verification from competent experts." And indeed, as alternative interpretations of Burke's findings went hand in hand with stronger criticism, attention focused primarily on those critics who attempted to replicate Burke's results.

Among Burke's top critics was Sir William Ramsay, a discoverer of the noble gases and an authority on the products of radioactive disintegration.[114] Ramsay not only disparaged Burke's attempt to reopen the spontaneous generation controversy, but also—in a letter to *Nature* of his own—called Burke's account "mad," the unworthy reporting of a "mad experiment."[115] Ramsay swiftly explained the radiobes away in purely physicalist terms as the mere aftereffects of the emanations of radium acting on the bouillon. In "a well-considered and frankly skeptical and sensible article"[116] first published in the *Independent*, Ramsay argued that the heat and gases given off by the radium as a by-product of its decay would undoubtedly disturb the bouillon just as a solution of gas in water could coagulate the white of an egg. As Ramsay wrote:

> Mr. Burke made use of solid radium bromide in fine powder. He sprinkled a few minute grains on a gelatine broth medium, possibly somewhat soft, so that the granules would sink slowly below the surface. Once there they would dissolve in and decompose the water, liberating oxygen and hydrogen, together with emanations, which would remain mixed with these gases. The gases would form minute bubbles, probably of microscopic dimensions, and the coagulating action of the emanation on the albumen of the liquor would surround each with a skin, so that the product would appear like a cell.

For Ramsay, the life span of a given bubble—which "would resemble a yeast cell" and, by implication, constitute one of Burke's "radiobes"—along with its quasi-reproduction by "budding" or death by "bursting," would last only so long as the radium remained active ("the best part of a thousand years"). "The 'life,'" Ramsay noted, "therefore, would be a long one, and the 'budding' would impress itself on an observer as equally continuous with that of a living organism."[117] Despite having written eloquently just a few months before about the potential connections between the discovery of radioactivity ("the philosophers' stone") and the *elixir vitae*, Ramsay was unconvinced by Burke's work.[118] In fact, Ramsay was instrumental in increasing doubt surrounding the existence of the radiobes. By the following spring of 1906, the *New York Times* noted that it was largely thanks to Ramsay's explanation that "the quietus is put upon the theory of Burke that he has created a cell, or the beginning of organic life, through radium."[119]

Ramsay was far from the only critic, however. The experimental nail in the radiobes' coffin came from W. A. Douglas Rudge, one of Burke's former colleagues at the Cavendish from 1900 to 1902, who performed a series of experiments explicitly designed to figure out just what had happened at Burke's lab bench. Rudge soon became Burke's nemesis in the radiobe controversy. Rudge designed his experiments carefully, attempting to replicate Burke's experiments in similar media, and was also committed to using photographs to convince his readers that Burke's "radiobes" were in fact nothing but radium precipitates, remarking that his own work "deals chiefly with the results obtained by the aid of photography, which obviously is a much more satisfactory method of recording than mere drawing."[120]

In a communication to the Royal Society made on his behalf by J. J. Thomson—who may have been trying to distance the Cavendish and its reputation from Burke—Rudge concluded from his systematic examination of all kinds of metallic salts that only those of strontium, lead, radium, and barium had any effect akin to what Burke had found: "As these metals are those which form insoluble sulphates, it seemed likely that the growth originated about the precipitates which form with the sulphur compounds present in the gelatin."[121] Rudge also found that gelatin made with distilled water produced no precipitates, but that gelatin made with tap water produced a "*very dense* growth." "It was thus quite evident," he concluded, "that the presence of a sulphate was necessary for the formation of the growth."[122] The radiobes, in other words, were nothing but sulfate precipitates.

Rudge retraced his experiment step-by-step: "The first effect of the action of radium salt was to cause an evolution of gas in the form of minute bubbles, owing to the decomposition of the water; the evolution soon ceased, but simultaneously a nebulous growth was seen to proceed from the point of contact of the salt with the gelatin," which continued rapidly for a time before slowing down and then ceasing. "This precipitate," Rudge concluded, "has, undoubtedly, a sort of cellular structure." Nevertheless, any further resemblance to Burke's radiobes failed to materialize. Rudge noted, for example, that "many 'pairs' of cells" could be seen, but that their "grouping is purely fortuitous," and moreover, that his constant photographing revealed nothing of "the nature of 'cell division' or growth, in the usual sense, taking place." Perhaps most tellingly, Rudge wrote, "there is no *trace of a nucleus*, even on pushing the magnifying power by projection up to 12,000!!, this figure being, of course, a long way past the limit of 'useful' magnifications." Rudge tested again for the formation of radiobes without sulfate, and obtained none: "It thus seems to be quite clear that the cellular growth cannot be produced by radium or barium unless a sulphate is present."[123]

As impure radium was often found associated with barium, Rudge interpreted Burke's and his own failure to carry out inoculation of subcultures as consistent with the interpretation of the radiobes as precipitates of barium sulfate. Curiously, what for Burke had indicated that the radiobes were *not quite living*—their inability to develop a culture on fresh medium as real bacteria would—served for Rudge as evidence for the *purely physical* nature of the radiobes. (Burke denied the precipitate argument altogether, saying that he had found the radiobes to be soluble in warm water, whereas barium sulfate, quite plainly, was not.)

Rudge concluded from his experiments that radium had "*no specific action in forming cells*" and that any observed effect was caused by the barium often associated with radium. Pure radium salt would probably produce only the evolution of gas, he concluded, since "radium salts are less satisfactory as cell-formers than the impurer ones." Most damningly, Rudge concluded from his photographs that "the cells do not divide or bud or show anything resembling 'karyokinesis,' the growth very quickly reaches a maximum, and they do not decay or split up, save as a consequence of the drying of the gelatin." All in all, he concluded, "radio-active substances, unless they contain barium, do not give rise to the formation of cells."[124] Rudge's experimental "disproof" of the radiobes' existence, reducing them to mere physical precipitate, was by far the strongest criticism of Burke's findings.[125]

Still others, like Jacques Loeb, criticized Burke's radiobes for having only "an external resemblance to living cells": experiments that produced colloidal precipitates that "imitate the structures in the cell" were common, but such precipitates routinely lacked "the characteristic synthetic chemical processes" central to life.[126] Artificially producing life, Loeb thought, required the production of a "substance capable of development, growth, and reproduction" and that synthesized the chemicals it needed for growth: "Whoever claims to have succeeded in making living matter from inanimate will have to prove that he has succeeded in producing nuclear material which acts as a ferment for its own synthesis and thus reproduces itself. Nobody has thus far succeeded in this, although nothing warrants us in taking it for granted that this task is beyond the power of science."[127] To make autosynthesis a requirement for life, however, was precisely to deny Burke's claim that life may have originated in many different ways and, moreover, that primitive life might look distinctly different from contemporary life. (It seems ironic that Loeb, famous for his engineering approach to life, was more concerned with the historical characteristics and trajectory of living systems than Burke.) Loeb was also bothered by the loose use of words and metaphors: "The purely morphological imitations of bacteria or cells which physicists have now and then proclaimed as artificially produced living beings, or the plays on words by which, e.g., the regeneration of broken crystals and the regeneration of lost limbs by a crustacean were declared identical will not appeal to the biologist."[128] Nevertheless, Burke and Loeb were routinely lumped together in the popular press as proponents of "artificial life," though their techniques, and even their ideas of what "artificial life" could possibly be, were to some degree distinguishable: Burke was after the artificial production of life; Loeb, its artificial control.

"Biology Is Decidedly Not His Forte*"*

Burke summarized his many experimental findings in *The Origin of Life: Its Physical Basis and Definition* (1906), which united and expanded on his earlier publications. Strangely enough, reference to this fascinating founding text in the origin of life literature has by and large disappeared (as has any awareness of Burke's role more generally). But this is perhaps not without reason: contemporary reviews of Burke's book were distinctly less than flattering.

The *Dublin Review*, while calling the book "highly interesting" and acknowledging the "wide circle of readers" it would undoubtedly reach, called it an "unconvincing work, marred by some curious errors and rendered exceedingly difficult of comprehension in divers places by the singularly involved style in which it is written." Burke's work was also cytologically rather naïve and was dependent on a somewhat idiosyncratic understanding of "organism" and "life": "We more than doubt whether Mr Burke would find any biologist willing to adopt his definition as anything like an adequate or satisfactory summation of the facts." Equally troubling for many readers was Burke's crossing of disciplinary divides and his "unwarrant[ed]" mixing up of "physical questions" with "biological considerations." But the most egregious way in which Burke failed his cause was his demonstrated lack of proficiency in biological terminology. For all his knowledge of the elements of physics, one reviewer noted, Burke displayed a "fundamental ignorance of the elements of biology. . . . This is a strong statement, but we think we can justify it." The reviewer pointed out Burke's errors in thrice misidentifying chlorophyll as chromatin, his failure to acknowledge that the primitive non-nucleated living cell (or "*Monera*") "probably does not exist and never did exist," his failure to understand that protoplasm was no longer generally considered to be crystalline in nature, his misunderstandings concerning the nature of fertilization, his equation of the nucleolus with the centrosome, his misunderstanding (and misspelling) of "mytosis," and more.[129] Burke's abysmal understanding of cytology and the phenomena of karyokinesis had even led him to state that his radiobes—contrary to the reigning biological state of affairs—divided cells before they divided nuclei.[130] "These are errors which one ought not to be confronted with in a book which professes to deal with the fundamental laws of life and living things."[131] Burke was a physicist through and through, and his claims, while potentially of great interest and fascination to biologists, at times revealed a basic ignorance of biological fundamentals—a fact that his critics pointed out with glee. (So much for physics paving the way for biology.)

Burke's competence was clearly under attack. According to one reviewer, Burke readily "demonstrate[d] that biology is decidedly not his *forte*," while another commented on his several "errors indicative of haste, and [the] disconcerting lack of correspondence between some of the figures and the references to them in the text."[132] Another criticized Burke for the poor structure of the book—which discussed his experiments in only one of its nineteen chapters, and waited until the sixth chapter at that, with too many "preliminary considerations"—as well as its style: "It is to be hoped that he is more skilful with the test-tube than with the pen. His style is extraordinarily loose and awkward. . . . [Some of his sentences] have subjects without predicates, predicates without subjects, and sometimes neither subject nor predicate. Sometimes the construction is not English at all."[133]

Editorial problems aside, the relevance of Burke's experiments to the question of the origin of life remained equally contested. Sir Bertram Windle complained that Burke's radiobes "at all times . . . appear to be soluble in warm water, and they end up as crystals. It is hard to see how objects of this kind can be held to throw any light upon the origin of life." His radiobes seemed "more like some aberrant process of crystallization than the behaviour of a living organization." And yet Burke's novel reconceptualization of life was, time and again, noted front and center, with striking passages quoted in full, as when Windle quoted Burke's statement that the radiobes were "analogous to living types and may, as we say, be called artificial forms of life, but they are not the same as life as we know it to-day. . . . If these artificial things are alive, it is not life as we know it in nature. It is not life which can claim descent from the remote past, and it is not life which will hand on its own type to the distant future."[134] This was a subtle point, hard for many to grasp, even when so clearly stated and prominently placed.

A reviewer of Burke's book in *Nature* delivered another scathing assessment. Although noting that Burke spoke of his radiobes as "possessing $n - 1$ of the n properties of living bacilli," the reviewer went on to complain that Burke went "soaring in a region where verification and contradiction are alike impossible." Vigor without rigor was almost enough, but not quite: "The author is so enthusiastic over his radiobes and with nuclei that we almost wish we could believe more in the importance of either of them."[135] Even a friend and former colleague from Manchester saw in Burke's book a new but ultimately unhelpful twist: "While defending his radiobes from the imputation of being dead bodies, [Burke] turns the difficulty by asserting that the radium from which they sprang was itself alive. Put in this way the whole matter resolves itself into a question of words, which is of no interest to the general reader." Such play with words could only lead down a thorny path of "merely dialectic exercises." In a critique similar to those made of Soddy at the time, this reviewer concluded that Burke "is sometimes apt to be carried away by a flow of language which suggests rather than conveys his meaning."[136]

Burke's reputation took a beating even in contemporary literature. His experiments were depicted not only in Arthur B. Reeve's *The Poisoned Pen* (1913), but more extensively in W. H. Mallock's novel *An Immortal Soul* (1908) as the odd doings of a scientifically inclined boy named Mr. Hugo. Pointing to some vials, Mr. Hugo tells an elder at one point, "Those . . . contain sterilized gelatine. As soon as I can get a little radium I am going to produce life." Later in the book one of the characters recounts a conversation with Mr. Hugo that explicitly linked the new atom with the creation of life: "He's been telling me all sorts of things about the sun and the earth's shadow; and he's going to reform humanity by manufacturing a new Adam; what is it out of, Mr. Hugo—a mixture of glue and radium?" "'Well,' said Dr. Thistlewood, taking Mr. Hugo's hand, 'I suppose she is thinking of radiobes.'"[137] Other passages in the book include a description of Mr. Hugo thinking "that human beings can be made out of beef-tea"; of his creations as "something like the radiobes, which I hope I may be able to show you in my bottle"; his statement to another character: "I'll show you something to-morrow. I am actually producing life with radium in a closed glass vessel"; and of his response to the offer of "a good rat-hunt" at a nearby lord's estate: "'Would you,' asked Mr. Hugo, aghast at this bold proposal, 'like that better than looking at my radium and the beginnings of life in my bottle?'"[138]

The radiobes in Mallock's novel, as the putative origin of life in a bottle, are a laughingstock, a gag line even as they also represent the sublimated essence of human nature. As one of the main characters of the story is said to wonder (as if echoing H. G. Wells's *Tono-Bungay* of the same year): "Was she merely an iridescence, a phosphorescence, on the quagmire of organic matter?" The only proper response to such materialistic metaphysical musings is apparently action, as the book ends on a skeptical note about the power of mere metaphor: "'It's idle to talk,' he said, 'if we are to canter off on a metaphor.'"[139] Full of activity and conversation but strangely without a real sense of depth, Mallock's novels were intriguingly described by one reviewer in terms that seem reminiscent of Burke's own experiments: they were said to have "the semblance of life—of fine-spun energizing life—without the colour of it."[140]

A Defense

As his role shifted from provocateur to disillusioned bystander, Burke rapidly tired of the limelight—or rather, the misunderstandings and misrepresentations of his work that being in the limelight involved. As he noted in a weary swan song published in *The World's Work* in September 1907, over a year after the initial bout of publicity, his experiments "have been, in some instances at any rate, somewhat exaggerated in other respects, perhaps unduly misconstrued or misunderstood." Reprimanding those most responsible, he portrayed the turn of events as "a less excusable misrepresentation on the part of some of those who, as critics, should have been better acquainted with the subject under discussion."[141]

Burke mounted a strong counterattack against Rudge, retaliating one final time in the press before Rudge's interpretations carried the day. He noted that Rudge's claims that radium had no effect—that barium alone was responsible for the formation of the radiobes/precipitates—seemed especially "bold" given that the two elements had similar *chemical* properties while differing in their *physical* properties. More dramatically, Burke snidely drew attention to Rudge's observations of N-rays as being of "rare interest, as he was the only man in England who could see anything with them." Such a statement in 1907—well after the decline of N-rays—was critical indeed. Long opposed to N-rays, Burke must have felt insulted to have the validity of his own work impugned by a man who still believed in them (and who then worked—far from the Cavendish—as a science instructor at a local grammar school in Suffolk). Burke let the vitriol flow: "The schoolmaster above referred to has made some experiments with gelatin, agar, starch, and isinglass, but none of these substances contains albumin. And the results have been, as they might well have been expected to be, negative. In fact, nothing is easier than to obtain negative results. We have merely not to do the right thing and there it remains undone."

Burke portrayed Rudge as an incompetent investigator who had mistakenly used commercial *gelatin*, containing "sulphuric acid and other common impurities," rather than the *gelatine* Burke had employed. (As Burke made a point of noting, "This is generally spelt *gelatine* by chemists, to distinguish it from the commercial product." Even the novelist Mallock had managed to spell the word correctly—although Burke's own letter to *Nature* had referred to "gelatin.") As radium would have had no effect on glycerin or gelatin but would have coagulated the albuminoids present in bouillon—which Rudge had neglected to include—Burke argued that Rudge hadn't even properly approximated his experimental technique. Burke pulled chemical rank on the schoolmaster Rudge: "Gelatin, as every chemist knows, does not contain albumin, and the radium effect on it is *nil*."

Burke reiterated his discoveries: The radiobes looked "like a diplococcus" and, *pace* Rudge, were not produced by barium, strontium, or lead. They grew, subdivided, and multiplied, "but unlike bacteria, they possessed a nucleus." If the secret of life resided somewhere in the cell nucleus, and if radium truly bridged the inorganic and organic worlds, then it stood to reason—as Burke found to be the case in his experiment—that "this nucleus seemed to be in some way associated with the radium emanation."[142]

Burke acknowledged the difficulties he faced in obtaining quality photographs in his earlier work—and the poor quality of the ones that

Rudge had characterized as mere drawings—but remarked that "there are good ones given in my recently published book." Burke also acknowledged the incredible rarity and expense of radium as one reason for the slow progress of his work—and as a possible reason why Rudge may not have carried out his experiments in the same way as Burke did. Indeed, on the day of the public announcement of his results in June 1905, Burke had remarked that his experiments were "necessarily expensive" and that as he was "working privately and without the support of any public body they are rather hampered by the lack of funds."[143] The situation had not changed a year later:

> If progress has not been as rapid as might have been expected, it is, to some extent at least, due to the enormous expense involved. For radium now is scarcely procurable and almost priceless. One therefore feels it half a sin to put a pinch of this rare substance in gelatin and bouillon, where it gradually spreads, never perhaps again to be separated out in its entirety.

Some of the radium, in fact, appeared to disappear soon after it was added—a phenomenon that Burke noted "has puzzled a good many observers; and they are therefore rather chary of trying the experiment."[144] This was not, he emphasized, reason to substitute other metals, such as barium, strontium, or lead, that would give only negative results—as Rudge had done. Radium alone was capable of producing radiobes.

Burke's Swan Song

In setting forth his case one last time, Burke scaled back the nature of his claims. He had not "solved" the "great enigma of life's origin"; he had merely found a provocative clue. Commenting on the long and not-so-distinguished tradition of artificial cells and other forms mimicking life, which would soon be labeled "synthetic biology" by Stéphane Leduc, Burke declared, "We should dismiss from our minds the illusion that we may find the final solution of this enigma in the laboratory, in bottles, or in test-tubes." We should never expect to be able to produce living forms identical to those extant today—"it is not likely, nor even to be expected, that we should obtain by such means life: such life as that which we see existing naturally around us"—because these forms are all the result of a long evolutionary history. But we might be able to produce what he thought were "simpler imitations" of them.[145]

Indeed, in discussing his 1906 book *The Origin of Life* and the "violent opposition" it encountered in some quarters, Burke acknowledged that he might have "more appropriately" titled it *The Origin of Cells and the Physical Aspect of Life*. Burke also acknowledged other claimants to the throne of "artificial life," such as Leduc's artificial cells, and diplomatically declared that Leduc's inorganic morphological mimics of living things (like plants and mushrooms) "belong most probably to the same category of microscopic forms."[146] Leduc, who in a few years was to publish both *Théorie physico-chimique de la vie et générations spontanées* (1910) and *La biologie synthétique* (1912), had claimed much for his forms.[147] For Burke, however, the point was to get at something more than mere mimics: to get at the nature of life itself. Leduc's forms, though they may look like "blades of grass, leaves or ferns . . . *have not the inherent and characteristic directive power of the living organism* . . . that depends on the physical and chemical properties of the nucleus, wherein the mystery of life and of life's origin now rests."[148]

Burke was also well aware of prior attempts to create something approaching "artificial life," such as various attempts by Sachs and Lehmann, though he claimed to be unaware of M. Raphael Dubois's production of so-called "eobes," despite their eerily similar name; there was a simmering priority dispute between Burke and Dubois.[149] Burke had also faced a priority dispute with Martin Kuckuck of Saint Petersburg, whose *Die Lösung des Problems der Urzeugung* (1907) described similar experiments with radium and gelatine undertaken in February and March of 1905. (Kuckuck in his work argued that ionization led to organization, from "inorganic stuff" to "organic substance" and from thence to "organized substance" and "organisms.") Burke was thus one of a diverse set of theorists and experimentalists actively trying to move the conceptualization of life to a new basis, and he generally readily acknowledged and even referred to others' earlier attempts to create artificial cells, cells that incorporated foreign material, and cells that appeared to grow.[150] It had "long since been discovered," he noted, "that the action of potassium ferrous cyanide upon gelatine produced cells which were capable of absorbing water, and apparently 'growing,'" but these earlier attempts did not show the phenomena of subdivision or reproduction. Burke thought these forms to be more like vacuolides and held that his own growths were something else altogether—that the sheer number of life-related phenomena they exhibited far surpassed earlier attempts to mimic life.

Fully aware of the history of the critique of analogies experienced by Otto Bütschli and others, Burke nevertheless held that his efforts were something closer to getting at the nature of life than a mere model.

Burke didn't want to just mimic life—he wanted to get at its underlying features. Convinced that he had produced something lifelike, or approaching the nature of life even if not quite living, Burke labeled his results "artificial life" in order to adequately distinguish them from the various forms of real life present in the world. And Burke's most powerful argument for the validity of his radiobes as primitive forms of life was that they were distinctly *not* life as we know it. The very features that called them into question as living things—they were demonstrably not bacteria, and they were curiously soluble in water—were, for Burke, proof that he was onto something that was different and that may once have existed, even as these same characteristics were fodder for his opponents' criticisms. Burke even proposed that perhaps the insolubility of the cells we know today was the result of natural selection from an earlier and different state.

In trying to save the phenomena with his theory, Burke thus played the last and most powerful card in biology that he could: the name of Darwin. He argued that his artificial cells followed the "same principles" outlined in Darwin's *Origin of Species*, and that he was applying the doctrine of evolution to the evolution of life itself, wherein "the problem of life thus becomes resolved into a problem of physics, wherein the individual atoms themselves by natural selection in forming suitable aggregates play their part in the struggle for existence by the survival of such of them as may be best fitted to live."[151] The reason why radiobes didn't exist in the present as a stepping-stone from nonlife to life, Burke said, was the same reason that there wasn't as much radium around as there once must have been: natural selection. If his radiobes were truly simpler forms of life, Burke noted, they "may not possess all the properties of bacteria," such as insolubility, or alternatively, "radium may convert insoluble proteids into soluble peptones under the action of water. The point being that there is no *a priori* reason for supposing that any primitive form of life hitherto undiscovered should be insoluble in water."[152] Natural selection thus operated not only in organic evolution (as Darwin had shown), and not only in cosmic evolution (as Soddy had argued), but in the very singularity where the two came together: in the origin of life.

Indeed, as one commentator noted, Burke never claimed that his radiobes were the "actual ancestors of living things, but rather [were] early forms which were so inefficient as to be crushed out in the struggle for existence by their more vigorous rivals from which life as it exists has been derived. And these true progenitors remain yet to be discovered."[153] Long before Aleksandr Oparin's theory of gradual chemical evolution, and even before Benjamin Moore's 1913 coining of the term "chemical evolution," Burke thus proffered a theory that transferred natural selection from the biological realm to the pre-biological:

> There is in this so-called dead, inert, inactive, inorganic matter a process not unlike that of natural selection or survival of the best adapted types, which in the long run find their level in the adjustment or evolution of inorganic as well as organic matter.

Living matter, as we know it, is but a species of matter which has been sifted out as the fittest to survive. In the infinite gradation from the most complex to the most simple we may perceive the same process in an ever simplifying degree. The fact of self-reproduction was an accident, and a happy accident in a particular type.[154]

For Burke, the transfer of a property from a group of living things to nonliving things—namely, natural selection as the mechanism of evolution—meant that there was no line to be drawn between the physical, the chemical, and the living: we can "deduce that the atoms and molecules of the chemist and physicist are of the nature of living things." Because natural selection took place in both, Burke thought himself justified in saying that "in truth, life exists as much in one as in the other and the difference is only a question of degree." He was surprised to have found himself the first (or at least he thought so) to have proposed a theory calling for such an overlap between the physical and biological realms, but went so far—despite all criticism—as to predict that "molecular physics will doubtless yet become a branch of biology."[155]

Burke's position was a precarious one, establishing the realm of the half-living by claiming lifelike characteristics for assuredly nonliving things, and his explanations routinely stepped into philosophical territory. His experiments got at the processes of natural selection involved in the emergence of life, and yet *he was not claiming to have discovered the means by which life first originated*. He was out to investigate the *physical conditions* for the origin of life, rather than attempting to answer questions about its *actual*, unique origin.[156] That life "belongs to the evolutionary series is true of such life as has survived," Burke noted, "but what of that which has been eliminated, which we are trying to produce in the laboratory?"[157]

Accordingly, Burke's work did more to establish the conditions of possibility for later research into the origins of life than to provide any firm findings regarding its actual historical emergence. From Burke's point of view, his investigations were intended to clear up the statement of the *problem* of the origin of life, not to have "accomplished its solution."[158] But, Burke concluded, "whether biologists will yet accept my view is not for me to say. If it is admitted to be a new view, it is no argument against it to say that it is not the accepted view at the present time."[159]

The Aftermath

Burke's work on the origin of life failed to gain the acceptance of those scientific experts with whom he had been on intimate terms. Having worked with J. J. Thomson, communicated with Soddy, interacted with Ramsay, and worked at the Cavendish Laboratory, he was viewed within the laboratory as having "caused a little amusement."[160] In a statement that reflects the internal politics at the Cavendish during his time there, Burke defended his work and his interpretations:

> The study of these questions has occupied my attention for many years, and the ground of tread is on the whole a pretty sure and sober one. A number of distinguished precedents have been quoted by my friends, and relieve me at least of any feeling of depression from the discouragement which so invidious a position as that which I have adopted may appear to have evoked. The more especially must my attitude seem novel in so conservative a place as that in which I find myself. Yet I maintain that there is no reason why men should not work independently of each other here and still be friends.[161]

Nevertheless, a photograph of Burke with the rest of the group at the Cavendish shows him looking distinctly uncomfortable, seated with legs and arms crossed, and seemingly out of place with the confidence exuded by many of the other members of the group. Once an up-and-coming young scientist with publications on fluorescence and phosphorescence, published in *Nature* and other respected journals, within a few years Burke seemed impelled to flee the centers of scientific orthodoxy, leaving the academic spires behind to better publicize his work:

> So invidious, indeed, did the course I had decided to take appear that certain dons of unspeakable nervousness were said to have got into hysterics like militant suffragettes, and their tarantic behaviour equaled only that of corybantic Christians of the Salvationist School; nor have they since ceased to hurl their

> boomerangs of unseemly epithets against me on every conceivable occasion.[162]

Burke left the Cavendish in 1906. While most others who had passed through the laboratory went on to academic careers in one form or another by 1910, Burke's entry in *A History of the Cavendish Laboratory* indicated that in the time since his departure in 1906, he had been "engaged in literary and scientific pursuits."[163]

The level of excitement at the Cavendish went down a notch with the start of the second decade of the twentieth century, as Thomson continued to hold to the vortex model of the atom, and especially as other laboratories in Paris (Curie) and Manchester (Rutherford) made more significant advances in radioactivity. Although the Cavendish "still carried out some important work between 1910 and 1914," one historian of the laboratory has noted, "the Cavendish was losing vitality."[164] Soddy never publicly proclaimed his support for Burke's findings, making only passing reference to Burke in his *Annual Progress Report to the Chemical Society for 1906*.[165] Soddy came to support Rudge instead, and in time Soddy disavowed any close link with Burke's experiments. Burke's departure and the loss of vitality in radioactivity research at the Cavendish went hand in hand.

Burke's departure in a minor key of ignominy was perhaps less the result of steady misrepresentation of his experiments and his claims—it is worth recalling Burke's scathing refutations of Rudge's failed attempts at replication—than of the sheer difficulty of arguing a complicated philosophical position on the nature of life as it may be or may once have been. This position proved to be too much for his contemporaries to handle, although they were able to appreciate nascent attempts at "synthetic biology" so long as these attempts stayed within the realm of mimicry and models. Such models might be useful for understanding the nature of growth or development, but were generally not useful for understanding metabolism and heredity (as Loeb had pointed out). To call any newly produced forms such as these "living" in any expanded sense was to pass the bounds not only of credibility, but even of pragmatic utility.

Conducting his experiments in the context of overwrought spontaneous generation debates, Burke was in the unenviable position of wanting to produce life but being unable to, and instead producing something that was neither fish nor fowl. Occupying an inherently unstable in-between space between physics and biology—a position he justified by an appeal to history and to the effacing effect of natural selection operating in both inorganic and organic evolution—Burke's radiobes embodied a sophisticated claim. To his colleagues, however, radiobes were either physical phenomena or biological phenomena. As they were not the latter, they clearly had to be the former—although if the radiobes were to have *anything* to do with the nature and origin of life, they obviously had to be something more than *just* physical phenomena.[166] Although he carefully positioned himself in a sort of limbo so as not to collapse the radiobes solely into the realm of either physics or biology, Burke's efforts at historical nuance were lost on both physicists and biologists. Physicists were more than happy to make the radiobes into physical phenomena, while biologists—looking deeply askance at Burke's ignorance of basic biological details—were all too happy to let them do so. Burke's standing as a physicist and his stated intent to rescue biology from the ailing hands of biologists did little to help establish his claims among those entrusted with policing the meaning of "life." Even with all the evolutionary discourse surrounding cosmic and organic evolution, living things, and radioactive phenomena, neither physicists nor biologists seemed keen on Burke's claims.

Intriguingly, it was the popular science writer Saleeby who perhaps best realized the predicament of any firm response to Burke: "We must define life, and since no one need accept any one else's definition of life, nor need adhere to his own any longer than he pleases, we are likely never to reach any possibility of returning a definite answer to the particular question concerning Mr. Burke's radiobes."[167] The *New York Times* concurred:

> If it were shown that what has hitherto been regarded as the creative miracle of the vitalization of matter is possible of performance as the result of such conjunction of materials and forces as may be brought together in the laboratory, its significance would probably be found to depend a great deal upon one's concept of the nature of life.[168]

Claims residing on this knife's edge of "life as we have never known it" are destined to be rapidly designated as redundant and simply a part of physics, or as pseudoscience; benignly forgotten; or hailed as pathbreaking experiments that will become the foundation for a new field (but then fall into the physical or biological camp in short order). "Life as we have never known it" is an inherently unstable place to rest one's research.[169] Burke's claims simply *could not* be allowed to remain problematic, because to do so would be to necessarily recognize the problematic character of the category of life itself. Better to forget that there was a problem. Better to forget about Burke. Better to forget about the radiobes.

A few years after Burke's work, Sir Edward Schäfer delivered his presidential address to the British Association for the Advancement of Science, where he raised the issue of the status of research into the origin of life. Though Schäfer called for further investigation into lifelike phenomena, Burke's work was not mentioned. Burke's role in inaugurating a new experimental approach to the historical origin of life was effectively forgotten only a few short years after his name had resounded across the world. As the *New York Times* reported on the meeting, "Many differences of opinion were revealed in the debate, but on one point there was complete agreement—that we are no nearer a solution of the problem than we were a thousand years ago."[170]

Nevertheless, Burke's legacy of theorizing and experimentally producing "precursors" of living things remained. Some "first steps" were soon announced, including Benjamin Moore's synthesis of organic compounds from inorganic starting ingredients, as well as the discovery that a mixture of colloids, water, and carbon dioxide "in the presence of uranium salt" would produce formaldehyde, "the simplest organic structure" and "the first step in the evolution of life." Without claims to have produced something half-living, but certainly having produced something organic from a radioactive element (and in circumstances distinctly different from those of Wöhler's synthesis of urea in 1828), the experimental search for precursors to the first living thing had begun in earnest.[171]

Indeed, the "precursor" approach formed the heart of origin-of-life studies for decades to come. Burke's inaugural experiments thus undoubtedly place him as one of the pioneers, if not *the* pioneer, of an experimental approach to the question of the historical origin of life on earth. Whatever the accuracy or longevity of Burke's particular theories about radium and life, his work was undoubtedly a powerful stimulus to the experimental study of the origin of life, and his experiments opened the door for the later and perhaps more familiar origin-of-life theories and experiments of figures such as Aleksandr Oparin and J. B. S. Haldane.[172]

Off the Deep End

Burke's private fortune ensured that he was able to move on to other activities, to the point that his youthful indiscretions with radium were conveniently forgotten. Something of a self-made polymath later in life, Burke lived the good life in London, at 63 St. James (just around the corner from Christie's), and in northern Italy, at his villa in Merano. He was reported to have spoken eight languages and came to be widely known as a "physicist, inventor, and scientific author." By 1924 he had become best known for his work on automatic typewriters, new methods of typesetting, and automatic printing of telephone messages. By and large, he had left "life" behind, with the sole exception of a curiously impenetrable book entitled *The Emergence of Life: Being a Treatise on Mathematical Philosophy and Symbolic Logic by Which a New Theory of Space and Time Is Evolved*, published alongside a popularized version entitled *Mystery of Life* in 1931.[173]

For Burke, the origin of life had been equivalent to the fundamental mystery of the origin of matter: "The mystery of both still remains where it was, the inconceivable, impenetrable, source and nucleus of our being, which lies hidden for ever from us. I can find in that remote immutable and distant origin which loses itself in infinity of space as well as of time the only origin not merely of life but of mind."[174] This equation of life with nonlife with mind—which "implies and even demands that atoms and molecules are thinking and alive"[175]—took Burke down increasingly bizarre roads. His *The Emergence of Life* was described by one reviewer in the history of science as little more than a "curious mixture of the metaphysics of monadology and the mathematical methods of symbolic logic," that yet somehow managed to incorporate "the philosophies of Kant, Schelling, and Hegel into this symbolic language"—a striking example of a physicist stark mad in metaphysics if there ever was one.[176] Burke's obituary was considerably kinder, generously calling these latter exercises "richly eclectic, openly professing a synthesis of the Platonic theory of ideas with Leibnitz's monadology and with the mathematics of relativity and modern theory of numbers. . . . The greatest value of the book lay, perhaps, in its demonstration of the heuristic value of mathematics in philosophical investigation."[177]

Burke had gone right off the deep end, even claiming at one point in the book that "it can be shown that the phenomena of karyokinesis or sub-division of the nucleus can be explained by the theory of relativity."[178] But by this time Burke had company. Among others in the 1920s, the Russian-born naturalized Frenchman Georges Lakhovsky had compared the nucleus of a cell to an electrical oscillating circuit, calling the interaction between a living thing and microbes a "war of radiations" and characterizing health as an "oscillatory equilibrium." (According to Lakhovsky's translator, "The foundations of Lakhovsky's theories rest on the principle that life is created by radiation and maintained by radiation.") The American surgeon George Crile, on the other hand, "whose great work on surgical shock has earned him an international reputation," was reported to have argued that "man is a radio-electrical mechanism and stresses the significant fact that when life ends, radiation ends."[179] Both espoused theories of radiation produced by and emanating from living things—*radiogen* for Crile and *biomagnomobile* for Lakhovsky. While in his *Secret of Life*, Lakhovsky held that life was created and maintained by radiation and "destroyed by oscillatory disequilibrium," Crile's two books—*A Bipolar Theory of Life* (1926) and *The Phenomena of Life: A Radio-Electric Interpretation* (1936)—endeavored to rework contemporary notions of the proper reach and course of biophysics. Meanwhile, a Becquerel of another time and place drew on the discursive storehouse of radium to propose a theory of life's origins all his own in 1925: "Did the radiations from radium minerals, which have either stimulant or deadening power on vital processes . . . once act in just such proportion and under just such circumstances that the chemical atoms combined into living protoplasm?"[180] *Plus ça change . . .*

Burke made one last attempt in his final works to shore up his reputation in origin-of-life studies—one final attempt in 1931 to clarify just what he had tried to do:

> Evolution has been continuous, and life in its primitive states must have been different from anything now observed in Nature. The products of spontaneous generation, if such were possible to-day, would be quite different from anything in the evolutionary series. I have emphasized this again and again; and the ever-recurring criticism, with almost obstinate persistency, has been put forward that the bodies I had obtained were not, and on my own admission, could not have been, bacteria at all! That was just my point. I drew a distinction between natural and artificial life.[181]

For Burke, the creation of artificial life in the laboratory—by which he meant precursors to living things or to things that might not traditionally have been considered living, and not the immediate production of a living thing itself—was one and the same with an attempt to investigate the historical origin of life in the laboratory.

But Burke, perhaps aware that his reputation in the matter was beyond salvage, finally gave up on thinking that either his experiments or his thought experiments could help him get at both the nature and the origin of life. In fact, he argued, a distinction needed to be made between the two:

> If we were satisfied as to its nature, we might or might not know anything about its origin: and conversely even if we were acquainted with its origin, that would not necessarily satisfy us as to its nature.
>
> The material phenomena with which life becomes manifest, or is enveloped, would seem to throw little or no light on its origin or its nature: except perhaps as an intermediate step or stage of its history on the one hand, and its behaviour on the other. This is as regards the purely scientific aspect of the question. It obviously deals neither with the Riddle, the Enigma, nor in other words the Mystery of Life. The discussion must needs close with the admission of mere nescience.[182]

Burke himself had reached a point of mere nescience on the matter, or, as he declared elsewhere in the book, "Life is what IS."

Burke died shortly after publishing these final remarks. Not even the vitalizing power of radium could save him. His obituary made no mention of the radiobes that had made him famous in his youth.[183]

: : :

Sitting at the intersection of a discourse of living atoms and atoms of life, reworking preexisting traditions ranging from *Naturphilosophie* to crystal analogies, deeply embedded in studies of the phenomena of phosphorescence, bioluminescence, and radioactivity, and weaving radium into the history of life on the early earth, Burke's work explicitly linked the previously separate discourses of cosmic and organic evolution for the first time with concrete experiment. His work not only proved pivotal in the redefining of "spontaneous generation" in the Anglophone context, but also served as a founding moment in the history of experimental research into the origin of life. Revealed at the height of the radium craze, his findings also demonstrate the rapid sedimentation of vitalistic metaphors of radium into a novel and provocative experimental system that relied as much on metaphor, metaphysics, and careful philosophy as on petri dishes and test tubes. These connections between radium and life proved more than merely metaphorical and more than airily metaphysical. Not just reminiscent of life, radium reached its apotheosis in experimentally *vitalizing* matter.

Moreover, despite Burke's failure, his work had pushed the realm of biological possibility for radium to its limits. The half-life of these connections between radium and life would play out in ever more concrete ways over the succeeding decades. New experimental systems emerged out of the same generative metaphorical and metaphysical hot dilute soup that had spawned the radiobes, each with its own life history and each interacting in its own ways with the ongoing conceptual, technical, and technological changes that were driving the transmutation of the associations between radium and life still further.

One prominent botanical investigator working early in the twentieth century roundly criticized Burke's work as ridiculous, but felt compelled to ask, if radium could not be used to *effect* life, could it nevertheless *affect* life? Radium's powers were soon to be tapped in the quest to gain control over the very processes of evolution itself.

3

Radium and the Mutation Theory

The demands of the biologists and the results of the physicists are harmonized on the ground of the theory of mutation.
—Hugo de Vries, *Species and Varieties*

No sooner had Burke's results begun to come into question than the *New York Times* announced (on Christmas Eve of 1905) yet another sensational new finding bearing on the nature of life:

> A scientist in New York, Dr. Daniel Trembly MacDougal, pursuing in the domain of botany investigations into the origin of species, has, by injecting into the ovary strong osmotic reagents and weak solutions of stimulating mineral salts, succeeded in causing changes in the egg cells of a plant before fertilization so that the altered eggs give rise to a new form or species.

Burke's radical and hotly contested claim to have created life, or at least something like life, de novo now faced a new competitor. Moreover, MacDougal's findings seemed like the stuff of real science, with defensible claims. If scientists could not hope to produce life itself, and if Burke's results were increasingly called into question—explained away as mere physical epiphenomena—and his attempts to extend the concept of life into the inorganic realm rejected as unworkable or excessively metaphysical, MacDougal's work showed that scientists could at least gain some sort of mastery over the production of new species. Radium moved from the heart of debates over the origin of life to experimental investigations into the origin of species.

Moreover, as biologists in the early twentieth century developed new theories of heredity and evolution, they drew on a widespread sense of a fundamental homology between physical and biological transmutation to suggest novel and often surprising ways in which the new findings and tools of physics might be deployed in biological experiments. In short, an elementalist theory of heredity that had atoms of life all its own met up with a novel and provocative radioactive account of speciation. This, then, is the story of how radium came to life in a third way: by entering the rich realm of research into the nature of biological transmutation. After a birth steeped in metaphor, and a baptism by fire in the production of radiobes, by the first decade of the twentieth century radium had everything to do with experimental attempts to get at the question of the origin of species.

By 1899 Jacques Loeb was already famous for his induction of "artificial parthenogenesis": the artificial reproduction of sea urchins from unfertilized eggs. By 1903 others had used radium to achieve the same effects.[1] MacDougal's work at the New York Botanical Garden extended this reproductive promise from the generational to the species level. As the *Times* noted:

> This achievement, which is expected to cause a revision of long-adopted theories as to the progress and processes of organic evolution, is herewith announced to the general public for the first time. . . . This is believed to be the first conclusive proof yet obtained that agencies external to the cell may induce mutations, and consequently exert a profound influence upon heredity.[2]

While sulfates had proved to be the undoing of Burke's reputation, for MacDougal they were a primary means of success. First imitating Charles Darwin's experiments on leaves with some "equally crude but successful attempts to modify egg cells by injecting zinc salts into pistils in 1905," MacDougal proceeded to inject plant capsules with solutions of zinc and copper sulfate as well as magnesium chloride, sugar, and calcium nitrate, among other solutions, timing his injections to occur "in the forenoon of the day at the close of which pollination would occur."[3]

The result was that MacDougal discovered "a possible new method of forcing variations"—and heritable variations at that.

MacDougal first reported on his success in inducing mutants by means of "chemically and osmotically active stimuli" in a lecture to the Barnard Botanical Club on December 18, 1905.[4] He noted that his experiments produced specimens "of the normal, parent forms, and aberrant mutants," and that he was soon to be in a position to offer

> conclusive proof that agencies external to the cell may induce mutations, and consequently exert a profound influence on heredity. It would not be well to exaggerate the importance of this result, yet it is evident that the establishment of this fact marks a long step forward in the experimental study of inheritance and the origin of species.[5]

MacDougal published various articles on his results, and he was widely cited as having produced "definite germinal mutations." Even Wilhelm Johannsen, in the very piece in which he first introduced the word "gene," referred to MacDougal's experiments as "highly suggestive."[6] MacDougal's "artificial production of mutation," as it was called in 1905, indeed inaugurated a new realm of research into experimental evolution and served as a significant first and now largely forgotten step on the longer road toward H. J. Muller's memorable "artificial transmutation of the gene" in 1927.

The Mutation Theory

While Burke had used radium to attempt to produce artificial cells in an experiment in what some viewed as "heterogenesis" (the origin of life from nonliving but once living parts), MacDougal was soon to turn to radium himself in an effort to induce "heterogenesis" of a different order: mutation. Straddling meanings in two realms at once, the word "heterogenesis" in the Anglophone context of the time meant *both* a subset of issues in ongoing debates over spontaneous generation and the origin of life (as James Strick has shown) *and* a set of issues in the context of evolutionary theory.[7] MacDougal himself had noted as early as 1902 that "heterogenesis" meant the origin of species by mutation (the word "heterogenesis" was earlier used in this way by Sergy Ivanovich Korschinsky, who had promulgated his own theory of the discontinuous origin of species).[8] And already by 1907, Vernon Kellogg had noted that "under the name heterogenesis we have to consider a theory of species-forming which is more popularly and widely known under another name, viz., the mutations theory."[9]

The idea of a mutation—indeed, the very application of the term itself to large-scale variations in biological phenomena—was relatively new to the life sciences, only having been properly introduced by the Dutch botanist Hugo de Vries in the two volumes of his provocative landmark text *Die Mutationstheorie* (published in 1901 and 1903).[10] MacDougal regularly and prominently acknowledged de Vries as an important source of inspiration for much of his work, and he was the first American to find experimental confirmation of de Vries's theories.[11]

At a time when the basic mechanics of Darwinism were being seriously challenged within the ranks of practicing biologists, the mutation theory—all the rage in biological circles at the time—served as a way out of the turn-of-the-century hardening of Darwinism into an *Allmacht* selectionist dogma.[12] Ironically, this intense focus on Darwinism-as-selection had brought to light a distinct problem: How could natural selection explain all the phenomena of evolution? More to the point, as one critic harped, natural selection could explain the survival of the fittest, but what about the *arrival* of the fittest—what was the origin of that variation on which selection acted?[13] "Darwinism" faced several other issues at the turn of the century, including most notably the age of the earth (from his calculations, Lord Kelvin had concluded that the globe was too young for the time required by Darwin's account) and the lack of a well-developed theory of heredity.

Developed at precisely the same time that descriptions of radium were resonating with both discontinuous and vitalistic overtones, de Vries's mutation theory—the result of more than a decade of research and theorizing—was the first attempt of its kind to synthesize the Darwinian natural selection of varieties with a new kind of account that explained the origin of species in terms of large-scale, *internally* derived abrupt "mutations." According to de Vries, new species could emerge in the space of a single generation as the result of large-scale discontinuous variational jumps between a parent and its offspring. De Vries thought one could learn much about the nature of evolution by studying an organism that periodically experienced "mutating periods"—as he believed his own favored specimen, the evening primrose *Oenothera lamarckiana*, did. Of the newly born species, de Vries wrote:

> They came into existence at once, fully equipped, without preparation or intermediate steps. No series of generations, no selection, no struggle for existence was needed. It was a sudden leap

> into another type, a sport in the best acceptation of the word. It fulfilled my hopes, and at once gave proof of the possibility of the direct observation of the origin of species, and of the experimental control thereof.[14]

The single greatest strength of his theory, de Vries thought, was that it "answer[ed] in an unexpected and decisive way the numerous and in part very grave objections which have been brought forward against the theory of Darwin." Above all else, he thought the mutation theory "release[d] the theory of evolution from the serious difficulties which its adversaries have never ceased to urge against it."[15] The mutation theory thus provided a ready alternative to the apparently lengthy time requirements of Darwinian natural selection even as it offered a novel way of explaining the *origin* of new traits and species (and not merely their selection). It also, or so de Vries was reported to have thought, accounted for the evolution of the "lords of creation" from the very origins of life itself in a "primordial protoplasmic atomic globule."[16]

De Vries's *Die Mutationstheorie* was a tour de force. Far less controversial and with much greater staying power than Burke's *Origin of Life*, its publication marked "an epoch, not only in the history of botany, but of all biological science," according to one prominent reviewer—"and the mutation-theory itself is, in all probability, the most important contribution to evolutionary thought since the publication of Darwin's '*Origin*.'"[17] As another contemporary noted, "Whether de Vries' theories are correct or not, wholly or partly, is of far less importance to agriculture than the stimulus he has given to the experimental study of plant variation."[18] According to historian of science Bert Theunissen, "The response to the theory was overwhelming. . . . [I]t enjoyed a tremendous popularity in the first decade of the century."[19]

De Vries delivered a series of lectures propounding the mutation theory at the University of California, Berkeley, in the summer of 1904, later published as *Species and Varieties: Their Origin by Mutation* (1906). Synthesizing Mendelism and Darwinism with his own peculiar theory of intracellular pangenesis, de Vries "aimed at nothing less than a complete explanation of variation, heredity, hybridization, speciation and evolution," according to Theunissen. "Moreover, the end result was not merely a theoretical construct, since de Vries also succeeded in lining up the results of his ten years of wide-ranging experimentation in support of the theory."[20]

Indeed, experiment was the password of the day. De Vries eventually succeeded in observing the emergence of numerous aberrant forms, and he found that these forms bred true when propagated by seed for what ended up being over twenty-five years—that is, they were "true mutations."[21] As he remarked of his discovery, "That I really had hit upon a plant in a mutable period became evident from the discovery, which I made a year later, of two perfectly definite forms which were immediately recognizable as two new elementary species."[22] De Vries's contemporaries understood the value of his work not only in elaborating "the theory of saltation as an adequate method of the origination of new forms in the organic world, but (and more especially) in *removing the entire question forever from the realm of ineffectual debate, and establishing it upon the firm basis of experimentation.*"[23] With mutation rapidly taken to be the solution to the perplexing problems of evolution, agriculture, and breeding alike, the new theory provided a widely accepted new framework for the experimentalization of Darwinism.

In the mind of Charles Stuart Gager, a colleague of MacDougal's at the New York Botanical Garden and one of de Vries's American translators, there was little doubt that de Vries's theory had been established by experiment:

> The deciding test as to whether a given new form, arising without crossing from a form that has bred true for at least two generations, is really a mutant or merely a fluctuating variant, is to see if it breeds true to seed for the new character or characters. If it does it is a mutant; otherwise it is not. It is clear, therefore, that the only way the problem can be followed out is by experiment—hence the term *experimental evolution*.[24]

The mutation theory thus proved fertile soil for biologists seeking to put the new half-living element to work in experimental evolution.

MacDougal and De Vries

Forty years old at the time of his experiments, MacDougal was described as "modest, unassuming, sympathetic, and of unfailing courtesy," and he was "much liked by his associates" and students alike. The author of both elementary and advanced textbooks on botany, MacDougal carried out his mutagenic experiments in a "special greenhouse, of which he alone has the key" in the New York Botanical Garden, where he was assistant director.[25] MacDougal first joined the botanical garden in 1899, five years after its organization, having come from a position in plant physiology at the University of Minnesota, where he had been for six years. He was later to serve as the director of the Department of Botanical Research at the Carnegie Institution.

As Jane Maienschein has noted, American biology underwent several significant transformations in the decades around the turn of the century, becoming increasingly professionalized, interventionist, and experimentalized and making use of new technologies, equipment, and funding sources. Although a definitive biography of MacDougal remains to be written, he is a classic exemplar of these changes.[26] As early as 1902, he had made his allegiances clear, allying his research program with that of the new experimental biology then taking the American scene by storm:

> Within the last decade the conviction has been growing among both botanists and zoölogists that polemics, the array of recapitulative facts offered by the organism in its younger stages, or the fact of comparative anatomy might not offer any convincing evidence of the manner by which the different species actually have arisen, although the results of these studies have been of enormous value in relation to other problems of biology.[27]

MacDougal firmly linked problems of organic evolution with the physiology of heredity, arguing that the mechanism of heredity, and of "saltatory inheritance" more specifically, could be elucidated only with "accurate observations and experimental tests with active or living material."[28]

In speaking to curious reporters, MacDougal recalled that he had first been led to his researches "by a note by Charles Darwin as to some 'fool experiments' he had made in injecting chemicals into leaves with the idea of bringing about morphological alterations."[29] It was early in 1905 that MacDougal first came up with the idea that his methods "might secure some evidence of value in its bearing upon the influence of environic factors upon germ and soma and their inheritance."[30] MacDougal's "original purpose" was "to test the matter of localization of the supposed alterations which by discontinuous variations occur in hereditary lines," something to which he acknowledged de Vries had first directed experimental attention with his speculations that mutations were due to changes in the germ-plasm prior to the reduction divisions. "If such localization were established," MacDougal thought, "it was hoped that new mutations might be induced experimentally by controlled conditions or reagents."[31] Such an approach would enable him to see not only "the factors which operate as stimuli" but also how changes in chromosomes might be related to "saltations in inheritance."[32]

Although MacDougal did not believe that all species needed to have arisen "in the same manner" in the course of natural evolution, he saw definite promise in de Vries's approach.[33] He was much impressed by de Vries's use of pedigree cultures, calling it "one of the most efficient forms of research yet used by the biologist, and its usefulness is hardly beginning to be realized."[34] Turning to pedigree cultures himself, MacDougal soon came up with the idea of injecting chemical agents into the embryo sac or the pollen mother cells early enough to cause mutations.[35] MacDougal's own experiments thus revolved around ovarial injections, rather than injecting chemicals into leaves, as Darwin had done. After first attempting to induce mutations in *Begonia rotundifolia*, various species of *Cleome*, and *Abutilon abutilon* with less than positive results, MacDougal turned to Lamarck's evening primrose, *Oenothera lamarckiana* (also known at the time as *Onagra biennis*), de Vries's own plant of choice for illustration of the mutation theory.[36] De Vries had claimed that *O. lamarckiana* was in the middle of a "mutating period" and that it was spinning off new mutants of its own accord in every generation; some nine descendant mutant species had been identified by 1902.

MacDougal began his first pedigree cultures in 1902. He noted that of seven generations totaling fifty thousand seedlings studied by de Vries up to that point, some eight hundred seedlings (about 1.5 percent) "were mutants or forms sufficiently divergent from the normal to be designated as new species."[37] In August 1904 he sowed a packet of seeds that had been harvested directly from de Vries's cultures in 1901. (De Vries also supplied the first round of paraffined paper bags necessary for the fieldwork.) In late September 1904 de Vries helped MacDougal inspect the small rosettes that had grown and "kindly assisted in the identification of a few of the mutants included." As early as October 1, 1904, MacDougal was able to report "seven known mutants which had been seen to originate previously in Amsterdam," and that "seven other forms could be distinguished which could not be identified with any forms heretofore observed by Professor De Vries or the authors."[38] MacDougal noted, however, that

> some confusion in the record makes it impossible to give the exact census of the culture, but it comprised between 500 and 600 seedlings, among which 26 mutant derivatives were identifiable, and, so far as possible two representatives of each type were

> transplanted to the experimental garden in May, 1905, coming into bloom about 60 days later.[39]

As a result of this confusion in the record, MacDougal went on to plant "a lot of purely fertilized seeds derived from one individual," harvested in-state in 1903, sowing these in his greenhouse in August 1904. The basal mutation rate in this group appeared to be about 6 percent, but yet again confusion in the record led to a mistaken claim from inconclusive evidence that he had "succeeded in modifying the coefficient of mutability."[40] Nevertheless, it was clear that MacDougal had again succeeded in producing mutants, of some indeterminate number.

Cultivated in pure pedigreed strains in his experimental garden,[41] abundant and readily available otherwise, and blessed with a large number of ovules in one ovarial cavity, the well-studied evening primrose was especially suitable as a test organism.[42] MacDougal noted that "thousands of individuals of many generations . . . had been cultivated, and in no single instance has anything beyond the well-known forms of fluctuating variability been shown, except when diseased plants were encountered. Better authenticated material would be difficult to procure."[43] And as he noted elsewhere, "Perhaps no plant is known in which the purity of the strain has been so critically examined as Lamarck's evening-primrose."[44] Accordingly, the behavior of some of the mutant primroses was downright compelling: the *O. scintillans* mutant had reappeared some fourteen times in de Vries's own cultures and four times at the New York Botanical Garden, irregularly but continually throwing off various daughter species.

"The oenotheras have furnished so much evidence of importance in connection with saltatory action in heredity," MacDougal wrote, that he was rightly concerned that it was possible, as de Vries had theorized, that "within the next few years . . . the botanist [might] actually witness the closing of the mutative period in this plant which has furnished material so rich in practical and theoretical results."[45] Or, as he concluded elsewhere, "it seems very probable that no plant will exhibit the tendency to produce mutants in greater degree than the one which has been selected for these notable experiments."[46]

MacDougal had found that the new mutants were often weaker in "strength and general virility" when compared with the parent type, and concluded that

> the few individuals representing some of the new species in any community would have but little chance of survival in the strug-

> gle for existence with the thousands of their fellows of the parent type. When isolated, however, and relieved from the fiercer competition met under natural conditions, the majority were independent constant types.[47]

The issue of "natural" vs. "laboratory" conditions was a crucial point MacDougal had to navigate before he could successfully claim to have "induced mutation." It was this structurally identical passage point that Burke had had difficulties maneuvering through with his experiments in the origin of life. If an experimenter must deliberately create an *artificial* situation in which the forms of interest can be observed—and both Burke and MacDougal did so—then the issue becomes one of how to establish to the satisfaction of others that the experimental setup is sufficiently representative of "natural conditions" that the phenomenon studied and explained can rightfully be viewed as contributing to knowledge of processes in the natural world.

MacDougal was able to convince others that he had produced new mutants: they were plants of the same genus that were clearly different from the parent type, even to the point of being morphologically identifiable as new species, and yet they had clearly come from a preceding generation of plants. Burke had no such "reality effect" to fall back on with his radiobes, brought into existence by an element that was only putatively living in the terms of popular discourse, or half-living at best. And while Burke had to deal with the origin of life, a particularly messy singularity at the confluence of several fields and discourses, MacDougal had only to convince trained observers that he had produced in the space of one generation new and sufficiently different plants as a result of his treatments.

The nature of the stimulus that caused MacDougal's mutations is worthy of further analysis. He claimed that the "saltations arising from the non-uniform action of the chromosomes, must take place in response to some stimulus outside of the protoplast in which it actually occurs." This stimulus, he noted, need not be environmental—it could be enzymatic, or of some other nature. It could even be, as MacDougal suspected, an effect of concentration: "It has been possible to induce new mutants by the use of solutions of strong osmotic activity and by highly dilute preparations of mineral salts, some of which are poisonous to plants in high concentrations and stimulative in low concentrations."[48] Regardless, the fact that there *was* such a stimulus implied that "we may hope to be able to duplicate the process in our cultures and call out a proportion of mutants at our will."[49] This was not yet the ability to produce *specific* mutants, but it was at least some form of control over evolution.

: : :

According to *Science*, MacDougal was de Vries's "foremost champion" in America: "He has largely contributed to the popularity of this theory," the journal reported.[50] A committed de Vriesian, MacDougal is reported to have said that "there was no more profitable subject for research in all of natural history than the causes that produced new species."[51] Elsewhere, he remarked:

> The conceptions of de Vries as to the origin of species may be rightly understood only when his analysis of the character of consistency of a species is borne in mind. His interpretations of the facts lead him to the conclusion that the characters of an organism are made up of well-defined and separate units, or elements, and that these elements are associated in groups; the same elements or groups of elements may, and supposedly do, recur in related species. The origin of a species by mutation would imply the substitution of a new elementary character, or quality, in the combinations, or groups, much after the same manner in which changes in the constitution of chemical bodies are effected.[52]

MacDougal's work was among the first to take up and experimentalize de Vries's theory, and his success and the widespread reports of his experiments undoubtedly contributed to the popularity of the mutation theory at the dawn of the twentieth century.[53] De Vries himself later publicly thanked MacDougal for his efforts.[54]

Physics, Biology, and the Mutation Theory

Frederick Soddy, with his "ancestral prowess" in chemical endeavors, had been described by one of his biographers in the terms of the new heredity: "Perhaps his was just one of those strange mutations in which the geneticists rejoice."[55] Soddy had been fully aware that there was grist for disanalogy between the "life" of radium and its actual behavior—that one atom could spontaneously disintegrate while a neighbor might not, with no relationship to their respective ages. Lord Kelvin had similarly wondered to J. J. Thomson, drawing on a biological analogy, "What would be the difference, between radium atoms in a piece of radium bromide, of the atoms which are nearly ripe for explosion, and those which have the prospect of several thousand years of stable diminishing motions before explosion?"[56] Similar questions occurred to the early mutationists: *Why* was mutational change happening at the moment that it did? While historian of physics Abraham Pais has noted that "the lifetime paradox simply did not lend itself to the statement of new hypotheses subject to test," and that "the problem was so difficult that it was hard even to get a wrong idea about it," assessing just why *biological mutations* happened when they did was a problem that became increasingly tractable.[57]

The problems in physics and biology were analogous. Deborah Coen described the situation in physics:

> Marie Curie laid out this dilemma . . . why a certain atom would decay at a certain moment. Since experiments had failed to detect any influence of external conditions on the rate of decay, physicists faced a quandary: "If we renounce making external causes intervene, it becomes difficult to conceive of a mechanism leading to the exponential law." She and her collaborator André Debierne had considered two possibilities: a disordering surrounding the atom or internal to it.[58]

A similar problem existed in biology: What was the cause and source of the variation in living things? The mechanism was clearly in the hereditary material somewhere, but were the mutatory changes due to external influence or to internal factors? Or, if such mutations happened "randomly" (whatever that meant) in biology, as in physics, then perhaps at least the laws of this randomness could be established.

Coen notes that physicists "repeatedly shifted between an applied and a theoretical context. This flexibility allowed them to assert that alpha decay was 'random' and yet to remain agnostic about the metaphysics of that statement. The decay was random for the purposes of interpreting the unpredictable behavior of their new counting methods."[59] Biologists similarly shifted between applied (agriculture, breeding) and theoretical contexts (the study of the hereditary elements), but remained pluralistic in their own understandings of mutation (ordinary fluctuation and variability as well as sudden changes counted as forms of mutation) and its causes (perhaps internally derived, perhaps effects of the external environment). Nevertheless, they characterized mutation as an essentially random phenomenon. This growing attention to probabilistic reasoning in the study of heredity and evolution mirrors that in many other scientific fields at the end of the nineteenth and early twentieth centuries, a period that has been called "the probabilistic revolution."[60] And yet, reaching further than the physicists, they quested after the pragmatic goal of trying to *induce* mutation—a goal they achieved more than a decade before the physicists were to induce artificial transmutation in the elements.

This experimentalization of the heredity of mutation, and its attendant goal of control, was de Vries's greatest accomplishment. According to Sharon Kingsland, his contribution was "not so much in his new ideas of the origin of species but in the way it made the origin of species into an experimental science." It was de Vries's "dissolution of the distinction between artificial and natural creations," she argues, itself reliant on "scientific advances in agricultural research since Darwin's time," that had at last "enabled the analogy between artificial and natural selection to be properly grasped."[61] The experimental angle that had failed Burke—the provocative elision of the difference between the natural and the artificial, which his critics were unwilling to grant—was proving wildly successful for de Vries and his followers.

More directly, a fundamental homology perceived to exist between physical and biological transmutation contributed directly to the further importation of radium into the biological sciences and into the study of experimental evolution in particular. De Vries had unabashedly put forward his mutation theory at precisely the same time that Rutherford and Soddy "unabashedly put forward the idea that some atomic species are subject to spontaneous transmutation."[62] Johannsen held that "*Natura facit saltus*" in the much the same way that Soddy had earlier described radium's radioactive changes as taking place "*per saltum*."[63] And de Vries's theory was routinely described as an "elemental" theory of heredity. As he himself had noted:

> The characters of the organism are made up of elements that are sharply separated from each other. These elements can be combined in groups, and in related species the same combinations of elements recur. Transitional forms like those that are so common in the external features of animals and plants do not exist between the elements themselves, any more than they do between the elements of the chemist.[64]

By 1912 *Science* had similarly likened the unit characters of inheritance to the chemical elements: "We may think of these unit characters as organic elements similar to chemical elements, that by their recombination through hybridization, form new compounds—new plants—of distinctly different appearance, but which in turn do not affect the unit characters, which may again be separated and led to form other compounds, again resulting in distinct organisms."[65] Moreover, these "new elementary species," de Vries noted,

> arise suddenly without transitional links; for the most part they are quite constant; within the limits of their essential constancy they exhibit similar minor fluctuations; they are usually represented by numerous individuals within the same period of time; the observed changes affect many organs and parts, and in no definite direction; and the mutability seems to be periodic, not continuous.[66]

Periodic mutability, essential constancy, minor fluctuations, elements recombining in groups, and the absence of transitional forms—the evolution of mutant forms was discursively close to the evolution of the radioelements. These two forms of evolution were soon to be drawn even closer by major scientists in this period.

Although de Vries was far from the first to propose the discontinuous nature of variation in the living world, his mutation theory was a major contribution to an ongoing series of debates at the end of the nineteenth century over whether organic variation was continuous or discontinuous. One of the chief proponents of the discontinuous camp, William Bateson, had remarked in 1894, "Species are discontinuous; may not the Variation by which Species are produced be discontinuous too?"[67] Bateson had suggested that it was change from within that was responsible for evolutionary mutation. Talk of intrinsic, internal sources of discontinuity became inherently more suggestive following the discovery of radium, with its own internal source of discontinuous change. Proposed only a year after Planck's theory of the discontinuous quantum of energy, de Vries's mutation theory—with its focus on both internally derived and possibly externally induced mutations—resonated strongly with these ideas of atomic discontinuity.[68]

: : :

It was the ready transfer of metaphors of life to the radioactive elements that had led Burke to question the existence of any firm line between the physicochemical and the living. Burke was far from the only one calling for evolutionary overlap between the physical and biological realms, however. Though he was the first to do so in the context of experimental studies on the historical origin of life, the sense that organic and inorganic evolution were inherently related was more widespread among those concerned with the roots and mechanisms of biological evolution in his day. Indeed, radium was characterized almost immediately after its discovery not only with the living metaphors that Burke had made such productive use of, but also with explicitly evolutionary metaphors.

"There exists a resemblance between the two realms of nature," that of life and that of matter, Sir George Darwin noted in 1905, "which is not merely fanciful."[69] The second son of Charles Darwin and an accomplished astronomer in his own right, Darwin delved deeper into this radioactive theme in his presidential address to the British Association for the Advancement of Science at Cape Town in 1905. He stated that the "struggle for life" occurred in living and nonliving worlds alike, and famously proposed a relationship between the transmutation of the radioelements and the transmutation of species. Punctuating the equilibria of the stable elements, the radioelements (and radium in particular) were for the younger Darwin the model for understanding the nature of evolution, inorganic and organic.[70] Such new discoveries, Darwin held, led him

> to express a doubt whether biologists have been correct in looking for continuous transformation of species. Judging by analogy [to radium] we should rather expect to find slight continuous changes occurring during a long period of time, followed by a somewhat sudden transformation into a new species, or by rapid extinction.[71]

Darwin rapidly moved from mere talk of evolution taking place in the physical and biological worlds alike to an immensely more interesting position—the equation of unstable chemical elements with mutant biological species themselves:

> In the world of life the naturalist describes those forms which persist as species; similarly the physicist speaks of stable configurations or modes of motion of matter; and the politician speaks of States. The idea at the base of all these conceptions is that of stability, or the power of resisting disintegration.[72]

Just as the struggle for life in the biological world "is held to explain the transmutation of species," Darwin said, so, too,

> although a different phraseology is used when we speak of the physical world, yet the idea is essentially the same. Theories of physical evolution involve the discovery of modes of motion or configurations of matter which are capable of persistence. The physicist describes such types as stable; the biologist calls them species.[73]

Darwin readily interchanged metaphors of stability, selection, and mutation, conflating the biological and physical worlds. Moreover, the shared metaphoricity of evolution and radioactivity here was clearly meant to provide fuel for the mutation theory: stable elements in physics were analogous to species in biology. By implication, unstable elements, such as radium, could be understood as analogous to disruptions in those species boundaries—that is, mutants. Radium was a mutant element, and some had described radium as "monstrous" during the same period. The production of mutants, understood as momentary instabilities in the process of originating new species, was linked by the younger Darwin not in a "merely" metaphorical mode but, more significantly, with a sort of *predictive* capacity about the nature of the natural world and how science could understand it. If both chemical and biological species can transmute, and if they are so closely related not only metaphorically but also in their actual nature, then why not use a transmuting element to transmute species, a mutant element to produce biological mutants? Perhaps radium held a clue even to the origin of species.

Numerous thinkers cast around for additional inorganic analogies to the process of speciation. Analogizing speciation to radioactive decay was common, but by no means exclusive of other analogies. Echoing de Vries's "elemental theory of heredity" and his talk of "elementary species," W. Hanna Thomson had already made the following analogy by 1909:

> Now the species of animals and plants are very like chemical elements. . . . Perhaps the change from one organic type to another, a brusque change which we call a mutation, is comparable to the [radioactive] change from uranium to lead. Perhaps the change from one variety to another, which we call a fluctuation or minor variation, is comparable to the change from one ethyl-compound to another.[74]

Others were keen to make the powerful link with radioactivity even more explicit. E. G. Conklin thought it possible "that germinal variations, and new hereditary characters, may result from intrinsic changes in the germplasm, comparable to the spontaneous changes which occur in radium." Conklin even discussed "the precise manner in which the structures of the germ become *transmuted* into the structures of the adult."[75] Even as late as 1920, Conklin was still comparing the mutation theory with radium:

> Overemphasis upon the intrinsic causes of evolution and neglect of the extrinsic causes has led to the extreme view that elementary species, pure lines, unit characters or inheritance factors are immutable, except that in some instances they may undergo digressive changes like those of the radium atom, which changes are wholly independent of environment.[76]

This idea was necessarily related to the genetic presence-absence theory of Bateson—the idea that, in Charles Davenport's words, "the foundation of the organic world was laid when a tremendously complex molecule capable of splitting up into a vast number of simpler vital molecules was evolved."[77] This view of evolution as "the unpacking of an original complex," itself a development of Bateson's views, meant for Conklin that all evolution would be merely "a process of devolution, or simplification. According to this bizarre view, man would be little more than, as [W. E.] Castle has said, 'a simplified ameba.'"[78] (H. J. Muller would later ridicule the presence-absence theory as the "perverse view, championed by a few mystics, that modern organisms, in general, are degeneration products from some golden age of living matter."[79]) According to this view, the evolution of genes themselves over the eons followed a decay process from "more" to "less" that was more or less analogous to radioactive decay, throwing off new elementary species along the way. In short, Bateson's presence-absence theory of genetics was read by some as an evolutionary analog of radioactive decay.

That it was a common metaphysics, rather than merely felicitous modes of description, that underlay these comparisons is readily apparent: Davenport bought into Bateson's presence-absence theory because he saw it as a necessary consequence of an evolutionary connection across the organic and inorganic realms. In his 1916 paper "The Form of Evolutionary Theory that Modern Genetical Research Seems to Favor," in a section revealingly entitled "Evidence from Evolutionary Changes in the Inorganic World: Radiation Studies," Davenport noted that "the view that evolution is primarily by internal changes receives unexpected support from the recent discoveries concerning the evolution of the elements." He later noted, in another section entitled "Certain Consequences of the Theory," that the acceptance of Rutherford and Soddy's theory of radioactive decay "requires a special *explanation to account for adaptation*."[80] Davenport concluded that "a theory of evolution that assumes internal changes chiefly independent of external condition, *i.e.*, spontaneously arising, and which proceeds chiefly by a splitting up of and loss of genes from a primitively complex molecular condition of the germ plasm seems best to meet the present state of our knowledge."[81]

That this theory of evolution, mirroring the earlier presence-absence theory of genetics, also curiously mirrored the evolutionary history of an atom of radium was for Davenport and others evidence of its *likelihood* rather than evidence of a strained analogy. Indeed, Davenport held that "such a theory receives support from various fields," including ontogeny, paleontology, experimental breeding, and "from analogy, with evolution in the inorganic world, so far as may be inferred from the studies on the 'rare earths.'" He concluded that "such a theory makes clear that success in 'selection' depends on rate and amplitude of internal change and ability to judge of germinal from somatic conditions."[82] The popularity of Bateson's own presence-absence theory of hereditary elements in some quarters in the earliest years of the twentieth century seems integrally related to understandings of radium as an element with a heredity, metabolism, and evolutionary history all its own.

MacDougal was drawing on these same reservoirs of inspiration. He noted that just "at this time when the physicist is successfully concerned with the resolution of the elements into constituent forms of energy or matter . . . the physiologist is taking the living unit apart and variously manipulating its chromosomal particles."[83] De Vries's elemental theory of heredity and evolution thus emerged and gained ground not merely for the program of experimental evolutionary research it promised, but also for its readily apparent analogies to the world of the radioactive. The idea that mutations happened, and happened abruptly, led many to wonder whether such mutations could be induced, and if so, by what means. Radium was soon to figure in these plans. Indeed, one of the first to suggestively connect radium and evolution had been Becquerel, who in his Nobel lecture of December 1903 noted that "radium rays . . . seem to act with particular intensity on living tissues in the process of evolution."

: : :

From the earliest days of the century, it had been noticed that "strong exposure to radioactivity is always injurious to tissues." As *Science* reported:

> The literature on radioactivity and its biological effects is voluminous, but there are only a small number of papers dealing with the question from a biological point of view. . . . Most of the work done [from 1903 to 1905], and indeed the majority of all work on the problem, has sought to use radioactivity for the study and solution of questions which were purely medical . . . with very little work on the biological phases of the problem.[84]

Experimental evolution became the bridge that linked the transmutations of the radioactive elements with the transmutations of species and turned the biological metaphors of radioactivity to productive experimental use in a way that the production of radiobes had not. It is perhaps not surprising that American textbooks were among the first to teach atomic transmutation, or that the mutation theory attained its greatest popularity in the United States.[85]

In retrospect, then, the application of radium to questions of the nature of organic evolution seems not only understandable, but overdetermined, following as it did upon earlier developments: Rutherford's and Soddy's roles as midwives to the birth of living radium; a series of early and well-known experiments investigating the effects of radiation on living cells; Becquerel's early remarks that those tissues in the process of evolution seemed most susceptible to the effects of radiation; the widespread observation that the discontinuous nature of radioactive decay paralleled the discontinuous nature of speciation (as proposed in the writings of Bateson and others); the establishment of Planck's quantum and its resonance with talk of discontinuity in the realm of the organic; the publication of de Vries's mutation theory, which held that mutation happens spontaneously from within, and without any necessary external force or effect (just as was the case with quantum changes); and the recognition by Conklin and others that the germ-plasm varied intrinsically, much like radium. Suddenly it seemed worth investigating radium's effects on the processes of evolution. Metaphor melded with metaphysics and experimental practice not in that most contested of realms where the inorganic and the organic came together—the origin of life—but in a realm seemingly more accessible to proper experimental investigation: the origin of species.

It all came to a head at the dedication of the Department of Genetics of the Carnegie Institution at Cold Spring Harbor, New York, on June 11, 1904 (an institution soon to be commonly referred to as a "nucleus" of research). While physicists knew that radium could not be induced to transmute at will, de Vries wondered aloud before the crowd of celebrants whether *organisms* could. In an address entitled "The Aim of Experimental Evolution," de Vries thrilled his audience by suggesting that "the rays of Roentgen and Curie" might be successfully used to induce mutation, grant humanity control over evolution, and lead to the production of new and useful varieties—all extraordinarily popular topics at the turn of the century.[86]

De Vries was just as entranced as his contemporaries with the suggestive possibilities of radium, and in his own way, he was part of the popular "radium circuit" touring the North American continent. Lecturing widely on his theory and visiting colleagues from sea to shining sea, de Vries in 1906 even carried a sample of Rutherford's radium from Canada to California, where Jacques Loeb was soon to begin his own experiments on the effects of radium on flies and other organisms. De Vries kept in regular contact over the years with various investigators studying the effects of radiation on living things.

De Vries's interest in using radium was a natural outgrowth of his interest in experimental evolution. As MacDougal's colleague, Charles Stuart Gager, later recalled:

> In one of his early papers on mutation, de Vries noted that, if the chromatin in reproductive cells could be altered by some external agent, artificial mutation might be produced. He suggested that, by skillful manipulation, this might be accomplished by bringing the sun's heat to a focus on a nucleus by means of a burning glass. That experiment appears never to have been successfully carried out, but the discovery of such penetrating radiation as that given off by radium placed a convenient device in the hands of experimental biologists.[87]

The connections between the instability of radium and what famed cytologist E. B. Wilson had called the "instability of idioplasm" (an earlier term for the hereditary elements, derived from Weismann) had become too provocative to ignore.[88] Like the unstable radioelements whose action they were induced by and whose half-lives in some deeper sense seemed to parallel their own, "new-born species" were described as "wobbly and variable" in their "organs and character," but each would become "more steady, more constant, or more true to its type" as the generations passed.[89] The cytologist and chromosome expert Reginald Ruggles Gates had reported in 1911, for example, that some particularly mutatory crossings were "accompanied by a disturbance of the germ plasm," which then manifested itself "in the occasional production of various aberrant types displaying whole series of new characters." He concluded that "mutation appears, therefore, to be not a simple unitary process of splitting, but to be the result of a condition of instability in the germinal material, which is again probably a result of previous crossing, and which leads to various types of departure from the parental race."[90] In 1920 Gates remarked that

> a great deal of ink would have been spared if it had been recognised that for plants as for animals, for *Œnothera* as for *Drosophila*, mutation is a process *sui generis*, a "spontaneous" disintegration or alteration of elements in the germ plasm which finds certain physical parallels or analogies in the behaviour of the atom of radium and other radio-active substances.[91]

The nucleus of the cell, with its connections to the phenomena of heredity, was thus perceived in terms much like those applied to the nucleus of the atom: radium "threw off" new daughter products of decay just as the unstable nucleus of an unstable species "threw off" varieties, or perhaps even de Vriesian mutants. (The fact that Darwin had referred to his pangenetic "gemmules" being "thrown off" helped to reinforce the link at the ultramicroscopic level.) While Darwin had regularly referred to the "transmutation" of species, the de Vriesian understanding of species added a new wrinkle to the story: in de Vries's account, species transmuted because of a mutation in the particular configuration of pangenes that made up a species. This concept lay at the core of de Vries's idea of mutation. (It was also something that H. J. Muller would later, and probably unknowingly, resurrect in his idea of the gene as "the basis of life" and his identification of mutation with a change in the configuration of a gene—see chap. 5.)

Well aware of the provocative connections between radium and life in his day, de Vries even chose to name the most unstable of his new evening primrose mutants, "constant in its ever-sporting character," *Oenothera scintillans*, in apparent honor of the scintillating element. It was called *scintillans*, he said, "or the shiny evening-primrose because its leaves are of a deep green color with smooth surfaces, glistening in the sunshine," and because "the progeny of the *scintillans* appears to be mutable in a large degree, exceeding even the *lamarckiana*." Moreover, he remarked, "the instability seems to be a constant quality, although the words themselves are at first sight, contradictory."[92]

If elements and species alike transmute, why not use one to induce transmutation in the other? More specifically, if species are what they are because of their hereditary "elements," then why not attempt to transmute the atoms of life with the transmuting living atom? Sedimentation of such metaphors proceeded apace, and the link between the transmutation of radium (taking place from within and without any external force) and the transmutation of species in de Vries's mutation theory (by means of large-scale, internally directed mutations) became ever more convincing. If both chemical and biological species can transmute, and if they are so closely related not only metaphorically but *metaphysically*—as the physicist George Darwin had implied—then *why not use a transmuting element to transmute species*? As with Burke, the analogies were to prove as productive as they were provocative. Just as prominent physicists were engaged in the search for the environmental causes of radium's untold energy, many early proponents of the mutation theory were looking for the environmental causes of mutation. It seemed only a small leap of logic, if even that, to use radium as a mutagen, and a number of experiments testing the effects of radium on plants and animals were undertaken during precisely this period.

MacDougal Turns to Radium

An elementalist theory of heredity had at long last met up with a radioactive account of speciation. The idea of mutations was enticing enough: it made sense of a number of phenomena apparent to breeders and hybridists; it fit well with de Vries's own particular (and particulate) theory of heredity, first outlined in his *Intracellular Pangenesis* (1889); and it solved the problem of time facing Darwinism. Dovetailing with de Vries's own belief that all mutations arise in the process of the formation of germ cells, the effect of radium on life in those cells most intimately related to the nature of heredity and the propagation of life was too highly suggestive to let pass unexplored.

Enter MacDougal. Searching for the experimental element that would enable him to investigate the nature of the hereditary elements themselves, having tried every other sort of strongly osmotic solution and highly dilute preparation of mineral salts under the sun, and eager to extend his research in the direction that both Darwin and de Vries had pointed out, MacDougal turned to radium:

> It is of the greatest interest to note that in the effort to correlate the larger generalizations in the various departments of science in the concept of mutation we have hit upon a principle strongly favored by a modern system of mathematics, well exemplified by the spontaneous breaking-up and rearrangement of the complex atoms in radium, uranium, and allied metals, and which has been recognized by Prof. George Darwin. . . .
>
> In the long-continued narrowing of the range of fluctuation in the various organs, coming to saltations, or direct origination of new forms, as the plant passes from generation to generation, we have as perfect a fulfillment of this motion as might be expected when an attempt is made to interpret the action of the living by the properties of the non-living.[93]

While "compatibility with physics lent the mutation theory an aura of correctness," as Sharon Kingsland has suggested, there was much more than "physics envy" going on. Rather, the move to view organic evolutionary phenomena in radioactive terms and to design corresponding experiments reveals the underlying persistent link between radium and life in the first half of the twentieth century and, in fact, closes the circle: from life to radium to life once again. Kingsland notes that MacDougal saw the mutation theory as a way to bring "botanical research to the forefront of theoretical debate in evolutionary biology." There was simply no better way to be on the theoretical forefront than to produce grand narratives of evolution that spanned the physical and the biological, and that explained a common "evolutionary pattern of long periods of gradual linear change punctuated by rapid speciation and divergence."[94]

Burke had linked evolution across the inorganic and organic realms by providing an experimentally accessible scenario for the origin of life. MacDougal linked the very transmuting processes to real-time mutational change: what he produced was not an intimation of some possible result or some putative "ancestor" of a result, but a brand-new species. Newspapers began to shift their sensational focus from the troubled realm of Burke's experiments on the origin of life to comparatively firm experiments in plant physiology. The persistent connections between radium and life entered a new stage. While many investigators in the first decade of the twentieth century were actively investigating the biological effects of radiation, MacDougal was arguably the first to make the evolutionary link between the transmutation of radium and the transmutation of a species *experimentally* explicit.

Preliminary reports were already indicating that radium and X-rays had especially injurious effects on germ cells, especially near the time of fertilization. The first cytological studies available on the effects of radiation damage had led early researchers to conclude that the effect of radiation "is a direct one on the chromatin of the radiated cells, not an indirect one . . . and further, that the seat of the injury if not exclusively in the chromatin is certainly chiefly there."[95] Even Thomas Hunt Morgan remarked that "how the action takes place is not definitely known, but X-rays and radium emanations appear to be almost specific agents for sperm cells; at least they are more quickly injured than the other cells of the animal."[96]

Although MacDougal was not the first to have used radium to induce heritable changes—that distinction fell to E. Aschkinass and W. Caspari, who had induced heritable changes in a bacterium in 1901—most of the literature on the biological (nonmedical) effects of radium date from *after* MacDougal's earliest experiments. MacDougal was therefore among the very first to investigate the mutagenic effects of radium in the context of de Vries's theory, with all its radioactive overtones.

In characterizing the "direct influence of the environment," MacDougal held that "radiant energy in its various phases" was among the "more important external, direct, or physical factors, the influence of which induces adjustments and engages the activities of protoplasm." Along with the "chemical structure of the medium, substratum or substance coming into contact with the living matter and included with its intake and output," MacDougal argued:

> These agents interlock intimately with the parts of the self-generating protoplasmic machine, furnishing building material, energy in various forms, catalysts, and control reactions in a manner so intimate that it is impossible to think of living matter free from its environic setting.

MacDougal was keenly interested in the effects of radiation on the germ-plasm, noting that "in general it may be said that such forms of energy retard growth and compel an incomplete differentiation of tissues when applied to an individual before maturity, producing serious deleterious effects afterward."[97] Or as he wrote in the *American Naturalist* in 1911: "The briefest comprehensive view of the physical sciences will show that here also the chief advance lies along the way of the study of energetics, and that the fundamental problems are those lying about the mode and means of transformations of energy."[98] Radium was the right element for the organism[99]:

> If we seek a similar possible intervention of external forces which might act upon the plant unaided by man, we might find such influence coming from radio-active substances, such as spring- and rain-water, or from the effects of sulphurous and other gases which are being set free in numberless localities.[100]

MacDougal did not move to using radium as a mutagen until 1904, the same year he brought more than a "dozen of the various forms" he had produced to maturity. (Prior to that year—the same year that saw the dawn of the popular radium craze—"no attention was given to the possible occurrence of mutants among the seedlings, although many might have been present.")[101] While the results of Burke's experiments resulted in a move from crediting radium to crediting sulfates for the odd forms produced, MacDougal moved from using nitrates and sulfates toward using radium as the mutagen of choice, exposing plant ovaries to its radioactive power.

With no internationally established unit of radioactivity yet decided on (the curie was not established until 1910), and with no "precise quantitative method" for radioactive strength, "certain crudities were inevitable" in the methods of these experiments up to around 1908.[102] Nevertheless, the results were striking: "It was evident that a mutation had appeared following the injections and nowhere else, and thus [had] some direct relation to the operation," MacDougal reported. Moreover, the plant that de Vries had found most promising for investigating and demonstrating his mutation theory was, coincidentally, also the plant most susceptible to the transmuting power of radium: MacDougal's results in *Oenothera* were so much more convincing than those in *Raimannia*, another test organism he had found suitable, that they left "but little doubt as to the nature and character of the changes induced."[103]

The stimulating, accelerating effects of the radium were often readily apparent.[104] One mutant, for example, was described as being "characterized by a much deeper green color than the parental form and the leaves are slightly curled and twisted, owing to inequalities of growth, and it reaches maturity quite early in the season." Among other mutants, "the corollas of many were so retarded that they failed to open and fell off prematurely. At greater distances [from the radium] development of the ovary proceeded but slowly and normal size was not reached." MacDougal hypothesized that "the treatment has simply thrown certain parental characters into a state of latency and awakened others with which the parental characters are mutually exclusive as to external manifestation." He concluded from these results that "a variety of agents act in inducing discontinuous variation in the progeny" with effects that sometimes last "to the third generation." These "atypic forms," he wrote, could thus "transmit their qualities perfectly from generation to generation, and the third generation now in hand are like the first from which they came originally."[105] MacDougal gave an even more complete characterization of "the technique of pedigree-cultures and of the methods employed in the stimulation of ovaries" in a lecture he delivered at the Woods Hole laboratory in July 1906. It was with this lecture, "Discontinuous Variation in Pedigree-Cultures," that his mutagenic success with radium became more widely known.

MacDougal and the Meaning of Mutation

MacDougal concluded that "saltatory inheritance has been induced by the action of external agents upon the ovules of two species of seed-plants," and that he had produced true mutations in the "de Vriesian sense": his mutants were "real and actual departure[s] from the course of the hereditary strain." The concept of a biological "mutant" was a relatively new one, recently introduced by de Vries, so establishing whether one had found a true mutant in the field or had produced one in the experimental garden was far from simple—a fact that MacDougal himself acknowledged, with reasons ranging from "the inexperience of the experimenter" to the related "plain mechanical fact that the selection of various forms is generally done in the seed-pans in which germination occurred in order to save the labor necessary in transplanting them to small pots."[106]

There were several other challenges to MacDougal's claim to have produced mutants. Mutants were clearly different from mere hybrids, and any suspicion of hybridity had to be ruled out (MacDougal endeavored to clearly distinguish the two categories in his *Mutants and Hybrids*). Using a well-studied organism whose range of variation was already well known, such as *Oenothera*, was especially important in this regard. While breeders at the turn of the century had achieved remarkable success in producing all sorts of novel hybrid forms, and while many newly discovered forms apparently intermediate between two putative parental types were often referred to as "hybrids"—a fact about which later investigators were to complain—MacDougal worked hard to establish that his forms were true mutants.[107]

Just as earlier investigators had marveled at the ability of radium to maintain its properties regardless of its external environment, one of the true tests of a mutant was the reproductive stability of its type amid a statistical cloud of variants: putative mutants had to demonstrate a persistence of type and breed true for several generations. MacDougal reported in *Mutants and Hybrids of the Oenotheras* (1905) that the effects of radium exposure produced plant forms that remained constant into the second and third generations, consistently "com[ing] true to their newly assumed characters."[108] He continued the lineage for five generations, with the same results.[109] This reliance on a characteristic test of the purity of a species was widely employed by botanists of the period and was the same criterion de Vries relied on to demonstrate the mutants he found in *Oenothera*.

Establishing a non-mutatory baseline for the species was vital. The more study had been done on the non-mutating members of the species, the more clearly one could establish that one had in fact produced a new mutant. It was for this reason that MacDougal had made it his "chief purpose . . . to make comparative studies of the parent-form with its mutant derivatives, and also to test the stability of all of the types concerned when cultivated under climatic conditions widely different from those under which the mutants arose."[110]

MacDougal was well aware early on, however, of the "prolific source of confusion" that existed in the "widely different conceptions as to the nature of the taxonomic units used in zoological and botanical writings . . . as a consequence of which we have some zoologists calling attention to the supposed fact that certain botanists of differing views have no real conception as to the nature of 'species' and 'varieties.'"[111] This confusion as to just what a "mutant" was and how it was to be defined—the shifting meaning of mutation—would in time contribute to the eventual near excision of MacDougal's pathbreaking work on the induction of mutation with radium and its significance from the historical record (see chaps. 4 and 5).[112]

MacDougal used a variety of words to describe the new forms he and de Vries had found: they were "alterations," "divergent types," "derivatives" and "mutant derivatives," even "breaks in heredity." What was clear to MacDougal was that the new probabilistic reasoning could assist in the assessment of mutation just as it assisted with the question of radioactive decay:

> If now the individuals of the mutant progeny are placed in a series with respect to any given quality, statistical observations may show whether it is included within the range of fluctuating variability of the parental type. The question therefore as to whether a plant is a continuous or discontinuous variant is one of simple measurement and estimation of qualities, not a matter of opinion.[113]

The relationship between mutants breeding true and other "inconstant forms" of the species remained something of a vexed concept, however. De Vries himself had identified several "inconstant forms" of *Oenothera* that he considered still legitimate mutants—*O. scintillans*, *O. sublinearis*, and *O. elliptica*—and on top of these specific cases, de Vries also recognized three main *types* of mutants: fluctuations, mutations, and ever-sporting plants.[114]

MacDougal showed a modest concern at this time—presaging the many more debates that were later to emerge when de Vries's mutation theory was cast into doubt—as to whether the mutant forms he had produced were themselves *true* mutants, a theme that would recur time and again over the subsequent decades as the study of radiation-induced mutation developed. A good portion of MacDougal's reports is devoted to justifying his claims to have produced mutants. These mutant forms were "not only physiologically differentiated" but also "easily separable from one another and from the parental type when tested by accepted taxonomic criteria, and by an examination of the features of their life-histories." The actual *characteristics* that defined the mutant were at issue, as was the *relative proportion* of these characteristics that a putative mutant had to have.[115]

Bud mutations—changes in one part of the plant that did not affect the overall appearance of the plant—were another challenging form of mutation that had to be taken into account. Although MacDougal was more interested in systemic, organismal-level mutants that reproduced themselves over time than in the occasional bud mutant, he recognized that "partial vegetative saltation" (or what de Vries had called "sectorial variation" and others called "bud sports," but MacDougal preferred to call "sectorial bud-mutations") had "much significance as to the localization of mutations, and as to the nature of the stimuli which set the mutatory processes in action."[116]

For all his experimental successes, MacDougal remained radically uncertain just how the direct influence of the environment brought about its mutatory effect. In this he was like the physicists described by Coen who chose to remain agnostic about a precise mechanism for radioactive change: "None of the attempts hitherto made to perfect a theoretical conception which would be useful in interpreting the mechanism of environic responses have had anything more than the most limited usefulness."[117] And as he observed elsewhere, "The real problem is the nature of the alterations induced by the action of the compounds to which the test plants are subjected."[118]

The path of justification was clear: the mutation theory promised a solution to some of the problems facing Darwinian theory. The quest to find and to study examples of mutation led to attempts like MacDougal's to induce mutation artificially, using a naturally transmuting element. The ability to produce such mutations—and to see an ongoing heritable pattern of mutated forms in succeeding generations—was sufficient to prove the theory right.

Calculating the Frequency of Mutation

Although just what counted as a "mutation" was contested, there is no doubt that MacDougal was applying rigorous analysis to his own results. It was clear to him and to his contemporaries that he had produced true mutant forms with the application of radium, just as he had done earlier with other chemicals and mineral salts. Testing the mutation theory raised various other thorny questions, however, including how to identify whether and when an organism was entering a "mutating period" and how long such a period might last. It was these concerns, among others, that led MacDougal to shift from de Vries's concept of a mutating "period" to the novel concept of a mutation "frequency"—or, as he also referred to it, a "coefficient of mutability." In effect, this was a way of establishing a measurement for mutation akin to the measurement of the half-life of radium. As early as 1905 MacDougal had noted that

> in Lamarck's evening-primrose five in every hundred plants are mutants, one in every two hundred of *biennis*, and it is conceivable that the atypic form might not occur more than once in a thousand, or once in ten thousand, or once in a million. These large numbers of plants are not all in existence at any one time, and it might take years, or even decades to bring one mutant within the range of the possible number, in which case a false conception of the mutative period might be gained. It is sug-

> gested therefore that the conception of *frequency of mutation* is the primary idea, although the action might become intensified in certain periods, of more or less definite limits.[119]

By 1907 MacDougal had broadened his concept of the frequency of mutation from his initial focus on the primrose to a more abstract population of any annual species, applying statistical reasoning to ever larger populations. If a population had only a limited size and distribution, it might take a long time for a mutant to emerge, MacDougal noted. But when one finally did, especially if this emergence followed large-scale cultivation of the plant, it would be a mistake to consider this the start of a "mutating period" when the laws of statistics were simply at work. "It appears, therefore," he concluded, "that the real state of affairs is better represented by the phrase 'frequency of mutation,' by which is expressed the number of individuals which must be grown to furnish one mutant, and which is nearly identical with 'the coefficient' of mutability."[120] Studies of radioactive decay and of mutation thus shared a concern with developing novel probabilistic techniques.

Actively concerned to establish more of significance than the mere description of the appearance of new mutants, MacDougal was arguably the first to develop the concept of mutation frequency. Emerging out of a linguistic—if not mathematical—inversion of de Vries's idea of a mutation "period" (frequency is the inverse of period), MacDougal's new concept transformed what had previously been an uncontrollable and unpredictable phenomenon into a theoretical tool that could aid genetic analysis. It also strikingly paralleled the manner in which physicists encountering the uncontrollable and unpredictable behavior of radium had devised the concept of the "half-life." While the half-life had been defined by Rutherford and Soddy as the inverse of λ ("the proportionate fraction changing per second"), MacDougal's mutation frequency was none other than a biological version of the inverse of λ—the proportional fraction of mutants per generation. MacDougal's calculation of mutation frequency was thus the first reference to a biological half-life for radium-induced mutants. (H. J. Muller would later pick up on this particular biological usage of "half-life" in his own work; see chap. 5.)

MacDougal was also concerned at this time with investigating fluctuations in the variability of the newly produced mutants, as he tried to ascertain whether these fluctuations followed any pattern or law of their own accord. He wondered, for instance, whether there was a correlation to be established between how old the mutants were and how much they fluctuated: "We may confidently expect that the species which show the greatest variation, or are eversporting, are the youngest."[121] Or, as he remarked in 1907:

> To the metaphysician who is striving to picture to himself the laws of vital motion, the results of these statistical studies will give an altogether new view of the relation between fluctuation and mutation. . . .
>
> It is now seen that although the individual species may decrease in variability as it grows older, this decrease is compensated for every time a new species springs into existence. Instead of mutations being the cumulative results of ever-increasing fluctuation, they appear now as an initial process of which fluctuation is in part an after effect.[122]

Physicists, likewise, at this time had begun to move away from a lawlike understanding of the nature of the half-life of radium and toward what Coen has called a "probabilistic conception of radioactive decay," one based on understanding the nature of the fluctuations of variation rather than its apparent regularities:

> From radium's discovery at the end of 1898 until 1907, physicists hailed radioactive elements as a supremely constant source of energy. . . . In 1908 when new instruments allowed experimenters to zoom in on the emission of individual particles, the relative fluctuations in the number of particles decaying per second were suddenly too large to ignore.[123]

This led most physicists, Coen notes, "to accept the probabilistic description of the fluctuations," and shortly thereafter, to begin to treat the "predicted average deviation from a constant rate of decay as 'the law of the fluctuations.'"[124] Probabilistic reasoning thus emerged roughly simultaneously in theories of mutation and of transmutation.

: : :

Building on his early work, MacDougal—along with George Harrison Shull at the Station for Experimental Evolution at Cold Spring Harbor—had by 1907 published the results of a collaboration analyzing the relationships of *Oenothera* mutations, claiming that "hereditary characters may be altered by external forces acting directly upon the reproductive mechanism."[125] By 1911 MacDougal had laid out the paths that all later investigators of the mutagenic effects of radium were to follow: first, the "demonstration of induced hereditary alterations and the study of their behavior in pedigreed strains, in hybridizations, and under various environic conditions," and second, considerations "of the mechanism by which an environic agency affects the physical bearers of heredity." Both research agendas, MacDougal thought, "comprise some important interesting possibilities in evolutionary science."[126] MacDougal had already convinced the world that mutations could be induced; what was now desperately needed was an experimentally accessible mechanism of mutation.

MacDougal was not entirely convinced that the action of external agents such as the radioelements on the "reproductive elements" (his term) was *stimulative*. This it may have been, but this was only "one of a number of allowable suppositions." It worked well for W. L. Tower's investigation of potato beetles, MacDougal argued, but it was an assumption that did not necessarily hold with his own plants.[127] Moreover, the theoretical possibility of a mutation frequency independent of population size was one thing, but MacDougal now needed to have enough mutants to calculate it. For this he needed a partner. MacDougal was soon joined in his radium-based experiments by his friend and colleague at the New York Botanical Garden, Charles Stuart Gager, who set out to examine just this question: the nature of the stimulative action.

Charles Stuart Gager: From Burke to De Vries

A colleague of MacDougal's and an exact contemporary of Burke, born in the same year, Charles Stuart Gager served as director of the laboratories at the New York Botanical Garden from 1906 to 1908, where he first began his investigations into the effects of radium. One of a number of scientists who investigated the effects of radium on life in the first decade of the twentieth century, Gager first began experimenting with the effects of the rays of radium on plants in 1904, just after having first arrived at the botanical garden. He summarized the results in his 1908 *Effects of the Rays of Radium on Plants*, his only book and the only monograph of its kind in the first years of the twentieth century.[128]

One of his first ground-clearing tasks in the book, after discussing the existing literature and before detailing the results of his own experiments, was to counter the emerging popular association of radioactivity and life and the perceived metaphorical profligacy that led to such seemingly bizarre experiments as those of Burke. In an intriguingly titled fifth chapter, "Bio-Radioactivity, Eobes, Radiobes," Gager was just as quick to dissociate radium from life as Soddy, Burke, and the popular craze had been to associate them:

> The general conclusion seems to be warranted that radioactivity is not a property of protoplasm nor of living tissues. A clear understanding of the nature of radioactivity would lead, *a priori*, to the same inference. Radioactivity and vital activity are in two respects very roughly, but only very superficially analogous. Both radioactive bodies and living organisms are undergoing a destructive process; atomic disintegration in the one, molecular transformation in the other; both, with exceptions, maintain themselves constantly at a higher temperature than their surroundings. These analogies have in two or three instances proven dangerously attractive.[129]

Gager dismissed Dubois's 1904 attempts to produce "eobes," or living things, in bouillon gelatine with the aid of radium and barium bromide crystals, and to induce "a kind of spontaneous generation by radium."[130] He also claimed that Leduc's related experiments were "controverted" in 1907. But he reserved his greatest criticism for the "voluminous" 351 pages of Burke's *Origin of Life*, berating Burke for his "most extravagant claims" and his book's "dangerously attractive" but "superficially analogous" relations between "radioactivity and vital activity." Gager insisted that Burke's "observations on the spontaneous action of radioactive bodies on gelatine media . . . have little of the scientific importance they have been held to possess in the popular mind,"[131] and he concluded that Rudge's characterization of Burke's results as mere precipitates "clearly indicat[ed] that there is not the slightest connection between the formation of the radiobes and radioactivity."[132]

Gager's main beef with Burke's experiments with bouillon was Burke's utter lack of historical sensibility. Life was not merely the association of parts; it had an evolutionary history: "We can never understand a plant protoplast by studying merely it; we must know something of its genealogy and its past history. . . . What was the origin of life? What is life? No one can give complete answers to these questions; but the purpose of the study of botany is to help fit us to seek the answers intelligently."[133] Gager's reaction to Burke's less historicist approach to life merged with the growing use of radium as a mutagen in an attempt to understand the nature not so much of "life itself," but rather of evolutionary processes.

Like MacDougal, Gager had been taken with de Vries's mutation theory, remarking in one review that "the influence of the mutation-theory (like Darwin's 'Origin') amounts to little less than a rejuvenescence of all biological science":

> The importance of de Vries's work not only lies in the elaboration of the theory of saltation as an adequate method of the origination of new forms in the organic world, but (and more especially) in removing the entire question forever from the realm of ineffectual debate, and establishing it upon the firm basis of experimentation.[134]

Also like MacDougal, Gager was to become a longtime supporter of de Vries, and by 1910 had even served as the American translator of his earlier work, *Intracellulare Pangenesis* (1889), which had begun to outline how changes in the structure and action of the hereditary material were responsible for the emergence of new mutants. Finally, Gager, like MacDougal, saw in de Vries's mutation theory a course for his own research, and he began experimenting with the effects of radium on *Oenothera* in 1904.

Gager first began his researches "in the autumn of 1904, with the intention of making them a minor problem during a year's residence at the New York Botanical Garden." The idea had been suggested to him by the consulting chemist of the botanical garden. Radium was in short supply, however, and he was limited to performing only one or two experiments at a time, so "the work progressed slowly." Fortune suddenly smiled on him, however, and "when it became certain that the facilities of the garden laboratories would be available for an indefinite period, other work was made secondary to the radium problem, for the relatively large quantities of radium and radium preparations placed at my disposal created an opportunity too valuable to let pass unimproved." Mr. Hugo Lieber, of H. Lieber & Co. of New York City, gave Gager "some $3,000 to $4,000 worth of standard preparations of the purest radium bromide yet obtained," and even "devised apparatus," as Gager put it, "without which many of the experiments could not have been performed."[135] Chief among these was the radium pencil, rod, or needle: "small sealed glass tubes, of capillary bore, containing radium emanation" that could expose photographic plates brought near them (figs. 5 and 6). Gager ended up conducting his experiments at the botanical garden over three years, and he reported his preliminary results in his 1908 monograph as well as in the popular press.[136]

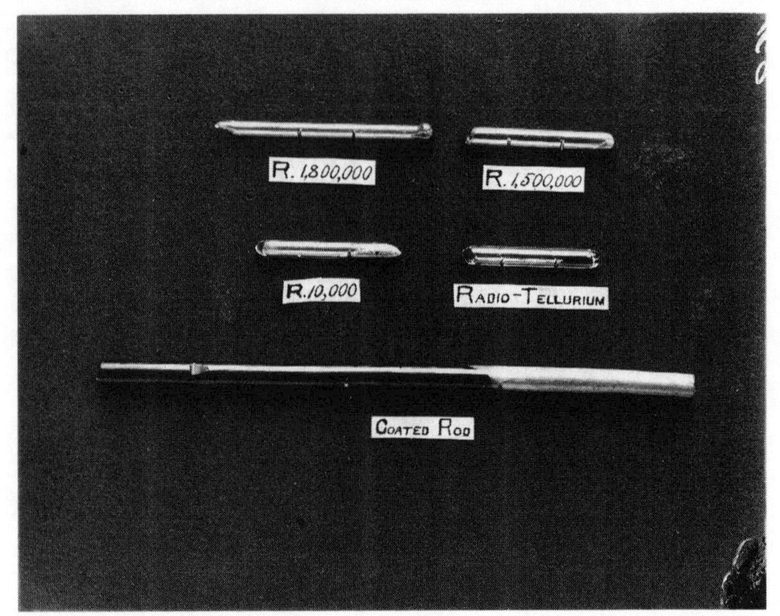

FIGURE 5. "Needles," or sealed glass tubes of radium, used in Gager's 1908 botanical experiments. (Courtesy of the Brooklyn Botanic Garden.)

FIGURE 6. Gager's radioactive writing: "Radium 1,800,000 Negative made with sealed glass tube of $RaBr_2$, wrapped in opaque paper in dark room, and penciled rapidly with the radium tube." (Courtesy of the Brooklyn Botanic Garden.)

With radioactivity such a novel phenomenon that it presumably needed introduction to many of his botanical readers, Gager began his text with a chapter entitled "The Discovery and Nature of Radioactivity." In fact, Gager had to educate himself, and at one point had even initiated correspondence with Rutherford in an attempt to understand more precisely the nature of radioactivity and its infectivity. Rutherford's response bolstered Gager's confidence in describing the results of his experiments as results of the rays of radium.[137] Moreover, Gager maintained a lifelong amateur's interest in the phenomena of atomic physics, and he mounted a one-man crusade against what—in an era of N-rays and other more fantastic claims—he characterized as a persistent "unfortunate misunderstanding" in "physiological circles . . . as to just what radioactivity is."[138]

While critical of Burke's approach, Gager—just like MacDougal—willingly made the link throughout his life between inorganic and organic evolution, viewing them both as part of a larger category of evolution: "When the non-living world is in mind we speak of *Inorganic Evolution*: when referring to the world of living organisms (plants and animals) we speak of *Organic Evolution*."[139] Even well into the 1920s, he described the process of "evolution" in language, astronomy, geology, and chemistry, and he specifically related the idea of the evolution of the elements to the discovery of radioactivity:[140]

> Inorganic Evolution.—The process of evolution is not confined to living things, but, as indicated above, applies to all nature. Even the chemical elements are believed to have been produced by evolutionary changes, and to be even now in process of evolution. This is one of the results of the newly discovered phenomenon of radioactivity, which is essentially the transformation of the atoms of one chemical element into those of another.[141]

The discovery in the earliest years of the century that radioactivity was everywhere present in the environment—from Thomson's early discovery that the air in Cambridge tap water "became decidedly radioactive" to the later findings of various other physicists who, Gager noted, "have taught us that this property belongs to the waters of most deep wells, to mineral waters generally, to freshly fallen rain and snow, to the spray at the foot of waterfalls, to the water of the ocean in certain localities, and quite probably to all spring waters"—convinced Gager that "radio-activity is a factor in the normal environment of plants." One result was that Gager sounded a note of interdisciplinarity similar to what Burke had called for a few years earlier: "Plant physiology and the newer physics join hands. Here, as elsewhere, the boundaries between the different 'sciences' break down."[142] Moreover, Gager suggested, this "general occurrence in nature of free negative electrons . . . not only add to the interest and importance of the study of the physiological rôle of radium rays, but also point out the way for further investigation."[143] In the second chapter of his 1908 monograph, "Radioactivity a Factor of Plant Environment," Gager built on this awareness that the everyday environment contained a certain natural level of radioactivity to suggest that "probably all plants are in a state of *radiotonus*, or adjustment to the radioactive forces of their normal environment."[144] This background assumption was to prove fundamental to Gager's radium work.

Neither Gager nor MacDougal was the first or the only researcher to investigate the effects of the rays of radium on life. The literature of the period is speckled with early experiments and investigations, and Gager provided a near-comprehensive summary of the state of the field in his day, from the 1901 induction of heritable changes in a bacterium by the use of radium (by E. Aschkinass and W. Caspari) to the first recorded observations detailing the effects of radium on plant tissue in 1902, in which it was found that plant radices (roots) were most susceptible.[145] He noted that the effects of radium had been studied on organisms ranging from anthrax, planaria, sea urchins, and tadpoles to caterpillars, mice, and rabbits, and in phenomena ranging from reproduction, regeneration, and egg development to the bactericidal effects of radium rays. Gager even cited an earlier review of the effects of radium on animals that was compiled by Caspari in 1904.[146]

Gager also cited several other botanists who had studied the effects of radium on plants. He concluded that "there seems to be a general agreement among them that the rays exert a retarding or an inhibiting effect, depending upon the activity of the preparation employed and the duration of exposure to the rays," with a sole exception who found "some evidence that acceleration of activity might follow exposure to the rays under suitable conditions."[147] This initial focus on the effects of radium rays and other forms of radiation on plants was intense—a fact that is hardly surprising given plant breeders' desire to produce new varieties with distinct characteristics that could maintain themselves without steady selection. Gager was thus one of many botanists who were interested in exploring the potential evolutionary implications of radium's effect on life at this time.

By 1908 Gager already knew that radium rays had "the power to affect [*sic*] marked change in the chromatin" and that such changes, often identified by anomalous modes of mitosis, could affect the organism as a whole.[148] Only a few years after Burke, strange radium-induced phenomena associated with mitosis were back in the news. (Gager also wrote in the *American Naturalist* that year that "interesting possibilities are here suggested, along the line of experimental mutation."[149])

After detailing the work of previous investigators into the effects of the "Röntgen" and "radium rays" on animals, plants, and plant fibers, and after taking into account earlier reviews of the literature, the remainder of Gager's 1908 monograph is concerned with the manner and results of his own investigations: the nature of the radium preparations and methods of exposure he used, the effects of a radioactive atmosphere and radioactive precipitation on plant growth, and the effects of radiation on seeds, soil, the synthesis of carbohydrates, respiration, alcoholic fermentation, and tropistic responses, and even on the histological, nuclear, and germ-layer effects of radiation. Citing de Vries's 1906 *Species and Varieties*, and inspired by de Vries's idea that mutation was "decided within the seeds," as Gager put it, he explained that "it was from some such point of view as this that I undertook to see what would result on exposure of the germ-cells, male and female, of *Onagra* [*Oenothera*] *biennis* to the rays of radium" (fig. 7). He would ultimately conduct over two hundred such experiments.[150]

In his review of the literature, Gager had noted that "with only one or two exceptions, exposure to radium rays has been found to either retard or completely inhibit all cell-activities. The rays may cause irregularities in mitosis." He similarly noted in his popular article on the same series of experiments "that the rays exert a retarding or an inhibiting effect." And yet he went on to note that exposure to radium "may, under certain suitable conditions of exposure and with certain tissues, be followed by an acceleration. . . . It was found possible to increase the rate of respiration of germinating seeds by means of the rays, and alcoholic fermentation was also accelerated by suitable exposure."[151]

Apart from the humorous reporting of his inability to determine whether some of the plants had a geotropic sensibility ("for histological examination showed the tissues to be so abnormal that it is possible the plants could not have stood erect even if they had been able to detect the stimulus of gravity"),[152] the vast majority of Gager's text reads like an endless parade of sad defects and negative effects on his plants following exposure to radium. Two of the first effects that Gager noted were the "cessation of cell-division and accelerated tissue-differentiation," which, he held, "have both the same significance, that is, early senescence." Curiously, he emphasized that in both cases "the tissue-differentiation has

FIGURE 7. Gager's experimental specimens of soon-to-be irradiated evening primrose (*Oenothera lamarckiana*). (Courtesy of the Brooklyn Botanic Garden.)

not been vigorous, nor normal in any other respect, but the important point to emphasize is that such differentiation has taken place."[153] Other effects of radium included a "marked tendency" toward double nucleoli in exposed cells; the "disintegration" of the cytoplasm; the distortion and general abnormality of "practically all of the mitotic figures [chromosomes] in whatever phase"—which Gager saw as "a morphological expression of physiological disturbance"; functional and morphological asymmetry; and, more generally, "irregularities and complete inhibition" of cell division and nuclear division. Gager concluded that

> whatever picture we may try to form as to just what occurs in the protoplast when it is exposed to the rays of radium, the foregoing histological effects seem clearly to indicate that one of the ultimate results is an acceleration of the period of senescence. If this acceleration proceeds gradually enough, the cells and cell-complexes may assume during growth, the various morphological configurations characteristic of the successive ages; but if the acceleration is too rapid, physiological senescence is reached quickly, without the usually accompanying structural changes, while a sufficiently intense over-stimulus by the rays may be quickly followed by complete loss of vitality and death.[154]

Gager's conclusion that the rays of radium could "excite" or "depress" the processes of assimilation or disassimilation in the living plant in some ways echoed a broader understanding of disease and health presented a decade earlier in Max Nordau's infamous fin de siècle tome *Degeneration*:

> The difference between disease and health is not one of kind, but of quantity. There is only one kind of vital activity of the cell and of the cell-systems or organs. It is the same in disease and in health. It is sometimes accelerated, and sometimes retarded; and when this deviation from the rule is detrimental to the ends of the whole organism, we call it disease.[155]

In other words, despite having delineated and described multiple times the various inhibitory consequences, strange growth patterns, and general bad-news effects resulting from the exposure of living plant matter to radium, Gager concluded *not* that radium was detrimental to life, but rather that the onset of early senescence after treatment with radium rays—that is, "a retardation or even a complete cessation of certain processes"— "may really be an expression of what is fundamentally a stimulation. The facts here reviewed substantiate the conclusion, drawn from other results, that *radium rays act as a stimulus to living protoplasm*." He continued:

> The broadest and at the same time the most definite generalization warranted by the work so far done is that the rays of radium act as a stimulus to metabolism. If this stimulus ranges between minimum and optimum points, all metabolic activities, whether constructive or destructive, are accelerated; but if the stimulus increases from the optimum toward the maximum point it becomes an over-stimulus, and all metabolic activities are depressed and finally completely inhibited. Beyond a certain point of over-stimulus recovery is impossible, and death results.[156]

In other words, in the course of its exposure to radium, *a given plant could be exposed to too much of a good thing.*

Stimulating Life

Gager was not alone in suggesting that radium rays could have a stimulating effect with the right dose and a retarding effect with an overdose. The idea that both progress and decline could be the result of the acceleration of the life process was widely shared, and it appeared most prominently in recapitulationist evolutionary ideas at this time. The *Lancet* had reported on the treatment of cancer with radium in 1904 in similar terms: "This actinic stimulus stirs up the sluggish system to the work of rebuilding healthy and normal tissues, but, like other kinds of stimulus, it ceases to be beneficial when given in too great quantities." Even more closely related to Gager's work was a report in the London *Standard*, which appeared just before the publication of his monograph:

> Another method of affecting plant growth is by watering with water containing a minute quantity of a radio-active substance. Acceleration or retardation of growth is produced according to the particular species of plant and the amount of radio-active material used. Experiments on the exposure of flower pollen and ovules to radio-active influence have resulted in the production of highly abnormal varieties of plants, and it is now being tested whether those variations are permanent and transmis-

> sible, or merely individual, like human deformities. A wide field of research is here opened out.[157]

Debates over whether accelerated vitality could result from the stimulation of radiation continued for some time. With a headline trumpeting "Sleeping Plants Wakened by Radium," the *Literary Digest* reported in 1914 that "many keen minds have been busy in late years on the fascinating study of the effect of radium on cell-growth, whether animal or vegetable."[158] Referring to Gager's earlier "1909" work, the *Digest* also focused on the work of the Viennese investigator Hans Molisch, whose attempts to influence the "period of rest so as to accelerate the sprouting of plants" were further discussed in *Scientific American* in 1912. Reflecting on years of attempts to "abridge the period of rest of vegetables by different means such as refrigeration, etherizing, hot baths, lesions, and injections," the magazine echoed the confused state of the research, noting that "while growth was generally found to be delayed, some cases are also on record of the germination of grains being accelerated by radium."[159] The abuses suffered by plants in these years, in the course of efforts to induce mutants, mirrored those suffered by the fruit fly *Drosophila*, as H. J. Muller was soon to point out.

How could the overwhelming evidence of general dysfunctionality, asymmetry, and early senescence be seen as the result of a positive *stimulant*? Drawing on other authorities, Gager had defined a "stimulus" as any "change in the external agencies that act upon an organism"; any change in radiation from a base level or "tonus" that a given plant experienced in normal life was a stimulus according to this definition. This use of the term "stimulus" obviously bobbed in the wake of the early realization that radioactivity was essentially everywhere and in everything at background levels. With this definition, however, one should expect progressively increasing *or decreasing* amounts of radioactivity acting on a plant—as identical stimuli—to have the same effects. For Gager to be true to his own definition of "stimulus" as meaning simply a change in ambient conditions—rather than connoting a positive, life-affirming, revitalizing influence—he would have had to experiment not only with *increases* in the level of ambient radioactivity, but also with *decreases* in it by shielding plants from ambient radiation in some experiments.

Gager never conducted such experiments. Nor did he even indicate, as he well could have done, the difficulties of performing such a test given the absence of utterly nonradioactive shielding material. This omission, combined with the absence of any significant evidence in Gager's experiments for *positive* effects of radium, implies that Gager could not have interpreted his results as the effects of *positive* overstimulation by radium—thereby implying that there was an optimal level of such positive stimulation—*without already having had a conception of radium as life-giving*. Such a conception, of course, did not come out of thin air.

The study of just this kind of "*radioexpérimentation negative*" was laid out as a theoretical possibility a couple of years later by Hyacinthe Guilleminot in his *Rayons X et Radiations Diverse: Actions sur l'Organisme* (1910).[160] (Guilleminot noted that since an environment with no radiation did not exist, only positive radioexperimentation was possible.) While a theoretical possibility, negative radioexperimentation was not practicable in the first decade of the century. Experiments of this nature had to wait until Blaauw and Heyningen's investigation of the "deradiation response" in 1925[161] and, more famously, Muller's "essential experiment" in 1930, which was explicitly designed to block out natural radiation, among other later investigations.[162]

While MacDougal had found in his experiments that retardation correlated directly with the length of exposure to radiation, the relationship between length of exposure and *acceleration* was "too irregular" to establish any such correlation. He focused instead on establishing the mutation frequency. Gager, drawing on MacDougal's work and his claims—"that many stimuli which retard or stop growth if of high intensity will accelerate if they be weak enough," and that "the maximum acceleration is not as great as the maximum retardation because, in the nature of things, it must be more limited"—had gone a step further to assess the nature of this stimulation. And with this, Gager, now heralded as "the most important botanist" to have studied the stimulating biological effects of radioactivity, had brought what *Science* termed the "pioneer stage" of investigations into the biological effects of radiation to a close.[163]

: : :

At least up to 1915, Gager held his monograph to be "the most extensive treatise on the subject," though he acknowledged the interesting work of other investigators in 1913 and 1914. He was apparently one of only a few investigators able to successfully use radium to accelerate particular life processes. As he wrote to a colleague in 1918, "So far as I know, I am the only one who has experimented with radium who has succeeded in accelerating germination. I think it is in the range of possibility that a mutation might be accomplished by exposure to radium rays, and that such a mutation might take the form of a color change in the flower."[164] Gager was far from the last investigator to research the effects of the rays of radium on plants, however: he later listed a thousand-page treatise by Julius Stoklasa and Josef Pěnkava published in 1932 as "doubtless the most exhaustive publication" on the subject.[165]

Gager held that radium rays were thus capable of causing processes of "assimilation or dissimilation" to "excite or depress." Where did this notion of a stimulus come from? Surprisingly, from the same source that Burke had used in developing his ideas: Verworn's biogen hypothesis. As *Science* noted:

> Gager has adapted Verworn's biogen hypothesis to explain the manner in which radium rays act as a stimulus to organisms, and to provide the mechanism by which the stimulation may be supposed to operate. A stimulus is any change in the external agencies that act upon an organism. Metabolism according to Verworn "depends upon the continual destruction and continual reconstruction of a very labile chemical compound," biogen, which "develops at an intermediate point in metabolism, and by its construction and destruction comprehends the sum total of metabolism." It is not a protein nor living, for a molecule can not be alive.[166]

For all his criticism and dismissal of Burke's radiobes, and for all his attempts to separate the phenomena of "radioactivity" and "life" as only suggestively and not substantially related, Gager's own account was written with the same metaphorical ink. Indeed, drawing on the same roots in Verworn's theory of intramolecular atomic vibration and the biogen, Gager even argued that radium rays "may act, not upon the more immediate physical basis of life, but upon some non-vital constituent other than the biogen, or upon some purely chemical process, thus producing their effects indirectly."[167]

Radium for Gager may have been just as life-affirming and effervescent as it was for Burke, or as it was ever held to be in the popular imagination of the time—a stimulant in much the same way that the radium tonics and radioactive snake oils of his day were. Gager's place in the context of popular understandings of vitalizing radium suggested to him what to look for—and helped to constitute what he found in the absence of negative radioexperimentation: results that seemed to show the stimulating effects of radium. The powerful associations between radium and life in the early twentieth century not only led to the *possibility* of Gager's experiments, but even helped to characterize the nature of the very knowledge he obtained from them. Gager's work thus serves as another example of the ways in which the metaphors and metaphysics of the strong associations of radium with life provided the conditions of possibility for a prominent experimental program of research, drove that research, and conditioned the very ways in which that research was interpreted, helping to shape and constitute its success by providing some of the vital means of understanding and interpreting the program's experimental *results*. For all of Gager's attempts to disarticulate Burke's facile association of radium with the origin of life by amassing harder data on the biological effects of radium, he nevertheless drew on the same framework that Burke had to understand his experimental results.

Gager later summarized his achievement thus:

> It would appear that in this memoir there was reported for the first time experimental evidence that the rays of radium may, under suitable conditions of exposure, induce an acceleration of vital functions. . . . Only retardation or cessation of germination and growth, or the killing of the seeds, had hitherto been reported. . . . The experimental results led to the broad generalization that radium rays act not, for example, like a stroke of lightning or immersion in boiling water, but as a true *stimulus* to metabolism.[168]

Gager's use of the words "stimulating" and "stimulant" were also part and parcel of a widespread pattern of discourse in the early years of the century in which biological application of the term paralleled its social use as a term of praise. Any number of discoveries, interpersonal relationships, promulgations, and other entities were described in this period with the terms "stimulus" or "stimulating." The idea was everywhere at the dawn of the century. Gager himself used it in dedicating one of his works to a former zoology professor of his at Syracuse University, his alma mater, and the professor wrote back, quoting the word in his response.[169] W. E. Castle wrote to Charles Davenport, director of the Station for Experimental Evolution, commenting on a colleague's state of health: "He is entirely normal and you may safely emit a 'normal reaction' to further stimulation. If you decide to apply the stimuli in person, don't fail to come and see us."[170] Even Darwin's legacy was described at this time as "a radiant influence so penetrating and so stimulating that it has been felt in every field of thought."[171] The word even appears twice in a letter Davenport wrote to de Vries, asking him to speak at the dedication of the Station for Experimental Evolution:

> It would be very pleasant for us if the new Station could be formerly [*sic*] opened by you as the one person who has done most to stimulate the line of work which we are to pursue. We should like it if you could give a short address upon that occasion, concerning the aims of work in Experimental Evolution and with special reference to your own researches. I am sure that your participation in the opening of this Station would be a great stimulus to the workers there.[172]

Davenport wrote again a month later: "We are looking forward to something which will give our friends, who may be present, an adequate idea of the importance and dignity of the work and which shall serve as a stimulus to the resident staff whose regular work will begin upon that opening day."[173] Gager later even referred to an article on an impressive radium growth-response curve as "a stimulating paper."[174]

Much as James Secord has shown the term "sensational" to inhabit a peculiar ecology of 1840s print culture in England, a similar story can be told for the early twentieth-century use of the terms "stimulus" and "stimulating."[175] Stimulation was so popular that it even appeared as a typographical error replacing "simulation." It is deeply ironic that one review of Burke's work, calling him to task for having misunderstood (and misspelled) "mytosis," described the radiobes as passing "after their formation through certain stages which stimulate those undergone by the living cell."[176] The stimulating and simulating powers of radium in the investigation of life and the lives of investigators were never far apart.

In fact, Gager's work had more resonances with the claims of Burke than he may have wanted to admit. In addition to drawing on the same generalized radioactive background—physical, metaphysical, and metaphorical—as Burke, Gager later made a structurally identical claim: to have produced the *ancestor* of a living thing, but not necessarily a new form of present-day life itself. As we will see, Gager was ultimately forced to claim that he produced only potential ancestors of newly mutant lines (as Burke had produced "the microbe's ancestor") in order to ensure that he was counting only novel mutants that were truly radium-induced. Just as life had slipped through Burke's fingers (as through his petri dish) despite his best efforts, life remained elusive for Gager. Radium was the key to unveiling the secret of life for Burke, as it was for Gager, but for all its experimental productivity it had yet to deliver up that secret entirely. Life remained, for the moment at least, the sublime object of biology.[177]

MacDougal and Gager

MacDougal wrote to Gager on December 17, 1908, acknowledging receipt of Gager's monograph and saying that he had only "cut the leaves with just a dip into it here and there." MacDougal wondered, however, whether "some of your facts might have been drawn upon to much more extended conclusions," but suggested nevertheless that "this will leave you a fair start off for the next piece of work." Five days later, MacDougal wrote again, reporting, "I've read the book + it is splendid. I believe your method will be a splendid one for finding out something about bud sports and producing them at will. My Baltimore paper considers heritable changes chiefly."[178]

Gager had written to MacDougal late in 1908, letting him know that he was interested in undertaking further radium experiments. Radium was scarce, and Gager needed a source. MacDougal replied, saying this time, "I am delighted to know that there seems to be an open space in your time. I would be glad to take up the matter of getting a tube of radium bromide and putting it in your hands if it is within reach. It is precisely this kind of co-operation that yields the best general results."[179] MacDougal seemed keenly aware of issues of priority in this new realm of research, writing to Gager:

> I wish you would let me know more exactly the lines of work you will carry out with it, and I am wondering whether or not it would conflict at all with my use of it to some extent. Thus for instance, I have secured eight new forms from a Penstemon by chemical treatment and it would be important for me to see if some of the same things might be incited by the use of radium, especially since three of them have been called out by the use of more than one substance.[180]

MacDougal's interest was thus clearly in the use of radium as a mutagen like any of the other chemical mutagens he had thus far employed. Gager's research was directed in a different direction—viewing radium not as an alternative mutagen, a tool in the toolkit for controlling the evolutionary process, but instead as an element of the natural environment of plants whose stimulating power had an effect on the evolutionary history of life on the earth. Gager was fundamentally interested in the *historicity* of life.

Gager eventually found a generous donor in Hugo Lieber once again, but MacDougal wrote back and clarified that it was a matter of sharing the radium in a short time, not priority in its use, that was his primary concern. "However," he noted, "since writing you last, I have pretty definitely made up my mind to add the feature of static or brush discharges from a frictional machine to my list of stimuli and will keep out of radium."[181]

It is unclear exactly why MacDougal decided to leave his radium work. It was partly due to the fact that Gager's work was already building on and extending MacDougal's own findings, a fact MacDougal had readily acknowledged by 1909: Gager's "extensive study" had demonstrated that treatment by radium "afforded a method by which chromatin elements might be eliminated from reproductive cells" during mitosis, leading to the "suppression or substitution of characters."[182] Decades later, in 1941, MacDougal still held that Gager had "used radium emanations more exactly administered with more definite results."[183]

MacDougal handed on the radioactive baton, his researches having catapulted him into even wider scientific recognition. Having been named director of the Carnegie Institution's new Department of Botanical Research Desert Laboratory at Tucson, Arizona, in December 1905, MacDougal had plenty of other research irons in the fire. He reported to Gager in 1908 that he had then "coming into completion, the finest, most diversified lot of results that I have ever seen and this year will be a pretty full one for me."[184] In 1911 MacDougal reported elsewhere that he had undertaken no new treatments, as he was paying "continued attention to a number of slowly maturing species." At any rate, "facilities for cultivation have not yet permitted the sowing of all the seeds from treated ovaries obtained early in 1911."[185] And in later years, MacDougal moved on to a variety of other projects, which included the construction of model artificial cells.

Like MacDougal, Gager succeeded in producing any number of odd forms of plants. While the differences he was noting were obvious phenotypic differences, he was careful to distinguish between those caused by exposure to radium and those that might have occurred spontaneously: "The appearance in the radium-cultures of elementary forms already recognized in normal pedigree cultures was rather to be expected, and the occurrence of such a form is to be attributed to the influence of the rays only with a great caution."[186] To avoid this peril, Gager compared his radium-induced mutants with "standard" mutants in *Oenothera*.[187] While de Vries was free to argue that the primrose mutations he found would have been declared by any systematist to be new species—since *Oenothera* was already in a "mutating period"—Gager had to be more circumspect and allow only truly novel mutants to count as possibly being radium-induced.

What's more, writing at the zenith of the presence-absence theory, and building on de Vries's tripartite distinction of types of mutation—progressive, degressive, and retrogressive—Gager considered that the use of radium did not bring about "a progressive mutation per se," but acted instead to cause "damage or loss," bringing about a shift from "dominant to recessive" or the loss of a character. "It is hardly probable," he concluded, "reasoning from other known facts, that the acquisition of a new factor, could be accomplished by exposure to the radium rays."[188]

Gager set the bar high. The odd results of nuclear division in a fertilized egg cell exposed to radium "might have been the first step in the production" of an anomalous plant of *Oenothera biennis*, but from subsequent developmental anomalies and the apparent inability of radium exposure to *bring about* a new hereditary factor, "it must be concluded, then, that most of the variants were not true mutations, and that further evidence is needed before we may say with entire confidence that mutations may be induced by the stimulus of radium rays."[189] Radium held the secret of life, but it was not giving it up without a fight.

When Is a Mutant Not a Mutant? The Historicity of Life

Could radium truly induce new mutants? MacDougal was convinced that it could. For Gager, by contrast, the matter was more complicated:

> The species question . . . is here regarded as secondary to that of variation. By whatever method or combination of methods species are produced in nature, our immediate and fundamental concern should be with the causes and behavior of variations. . . . I do not believe that I have experimentally produced a new species. Nor indeed do I believe it probable that we shall ever do so in the laboratory, at least with the higher green plants . . . it seems almost self-evident that such a natural group of higher plants may not be artificially produced in the laboratory, nor, indeed, within the narrow confines of an experimental garden.
>
> But whether we may artificially produce a parent or ancestor of a species is quite another question.[190]

While MacDougal had undoubtedly achieved a first—visible success in the experimental induction of mutation, and even in calculating the *frequency* of such mutation—Gager was reluctant to claim that his use of radium had produced new species (much as Burke was reluctant to claim that he had produced a natural form of life in his petri dish). Why? MacDougal gave some indication in 1902, warning that there might be difficulties in having the mutation theory accepted by systematic botanists:

> Few botanists are prepared to assign specific rank to any individual or group of individuals which have been observed to descend from a group of forms constituting a separate species. A somewhat more considerable number accept self-perpetuating hybrids as species, although it is doubtful whether this attitude may become universal. To this greater majority of systematists then the entire matter of origin of species by sports, single variations, or by mutation is entirely out of court.[191]

Gager, however, was not a traditional systematic botanist, nor was he as enthusiastic about the production of new species as MacDougal. Why the reticence?

First of all, Gager seems to have been more concerned with radium's effects on growth—the ways in which radium rays acted as a stimulus to the various physiological processes. But Gager also held that the historicity of the life process was central to any proper understanding of mutation: protoplasm, he said, "is a historical substance." A plant cell has an "ancestry *as a cell*," he noted, and "its protoplasm has what we may call a physiological memory of the past. It is what it is, not merely because of its present condition, but because its ancestral cells have had certain experiences. We can never understand a plant protoplast by studying merely it; we must know something of its genealogy and its past history."[192] (Recall that this emphasis on historicity was one of the main reasons for Gager's strong antipathy to Burke's claims: radiobes had no such history, and without it they could not properly be considered living.)

While for de Vries the mutation theory was an outgrowth of intracellular pangenesis—and so species were defined by the collocation of pangenes present—Gager was less committed to this theory than to the mutation theory.[193] While according to de Vries's theory, *any* small change—he was never entirely clear on the difference between fluctuations and mutations—necessarily implied that a new species had come into existence and therefore ought to be considered a mutation, Gager was more circumspect. He thus actively reinterpreted de Vries's mutation theory, moving from the idea that an actual *plant* grown in the next generation from radium-exposed seeds was a mutant to the idea that such exposure could establish a particular *filial line* of plants.

Any disturbance in the radiotonus of the plant that led to disease or conditions of ill health could at most, Gager held, in some cases, and over time, lead to a potential new mutation and new species. In other words, Gager acknowledged, the modification of one individual "would be quite sufficient material out of which to form a new specific group." Indeed, he noted, "it was from some such point of view as this that I undertook to see what would result on exposure of the germ-cells, male and female, of *Onagra* [*Oenothera*] *biennis* to the rays of radium."[194] Or, as a reviewer noted, even if Gager had not succeeded in inducing mutations, then at least "some intimations were obtained that it may be possible to do so."[195]

Gager's initial interest in the physiological effects of radium rays thus had been rapidly parlayed into a greater concern with the *evolutionary* effects of the rays on plants, even as species were still, for him, group phenomena that shared a fundamentally *historical* aspect. At the most, one could hope to produce a parent or an ancestor of a new species, but not new species themselves. This uneasy reworking of de Vries's theory—both inspired by it and distinctly modifying it—would appear to call for experimental physiological and evolutionary investigation such as Gager's, even while preventing any premature claim to have induced mutation in the space of one generation.[196] MacDougal had picked up on just this implication and, when commenting on Gager's production of "chromosomic irregularities by the exposure of ovaries of *Oenothera* to radium emanations" and the appearance of "aberrant" offspring, pointed to the one obvious lacuna in Gager's research: Gager had not taken the obvious next step and tested the "transmissibility of the new characters."[197]

While it was at least partly for this reason that Gager was never entirely certain whether he had actually produced novel mutants—he remained indecisive on this front well into the 1920s—others at the end of the first decade of the twentieth century clearly thought that he had succeeded. As one reviewer noted, "Nothing can yet be stated as proved, but it appears that when the pollen or the ovary of a plant is exposed to the influence of radium the resulting seed will produce a plant profoundly different in the first generation from the parent."[198] And as William Spillman had noted early on, Gager had "not only produced mutants apparently of the character of those found by de Vries, but has shown that in mitosis in treated specimens there is irregularity in the distribution of chromosomes."[199]

Gager's sophisticated questioning of the results of his own experiments, at least partly the result of his particular views on the historicity of life, led to a decade-long status of indeterminacy for his claims. Had he truly produced mutants or hadn't he? Even as late as 1920, he entertained doubts:

> *Can the inheritance of a strain be artificially altered?* . . . The experiment has been made of exposing the ovaries of flowers to the rays of radium, and of injecting them with various chemical substances, with an idea of altering the physical or chemical nature of the egg-cells, and thus altering the inheritance. The results of such experiments, so far tried, need to be further confirmed before we can say with certainty that the result sought has been accomplished.[200]

For Gager by this time, the associations between evolution in the inorganic and in the organic realms—a metaphysics he had held since at least 1902—were still clear, but their implications had shifted. The question for organic life was not so much about *transmutation* (as it was for the radioelements) as it was about *filial descent*. Even as late as 1926, Gager's colleagues seemed uncertain about claiming success with his method: "Very little, if anything, is known of the mechanics of mutations, that is, of what causes them, or of the nature of the hereditary materials (genes) which are responsible for them."[201]

Oenothera *Reconsidered*

As it turned out, Gager was wise not to immediately attribute transmutational power to radium. Shortly after Gager's experiments, during the second decade of the twentieth century, de Vries's theory began to come into disrepute as it was discovered that the plant he (and Gager) had been studying, *Oenothera lamarckiana*, displayed odd chromosomal dynamics all its own—even without radium treatment. The so-called mutants that de Vries had discovered, which did indeed breed true generation after generation and which many had held to be proof of his theory, were shown by cytologists, such as Reginald Ruggles Gates in the 1910s and later in the 1920s in a masterful synthetic account by Ralph Cleland, to be the result of chromosomes sticking together in rings rather than the homologous pairs separating during meiosis, producing an unequal distribution of chromosomes into daughter cells. The result of this unusual chromosomal behavior was a generation of organisms that bred true—and so appeared to be mutants by the standard test—but that had undergone no actual genic change.[202]

Whether or not these chromosomal abnormalities were "mutations" was debatable. By the traditional observable botanical and morphological criteria, the mutants were (and most botanists would have classified them as belonging to) new species; by the standards of later geneticists who did backcrosses of subsequent generations to map genes on chromosomes, they were not. Levels of analysis of "mutation" began to come into conflict, making radium a less dominant source of metaphysical or metaphorical explanations of biological transmutation. The once powerful associations between radium and life were already beginning to decay.

Gager's work was thus among the earliest studies to explore the chromosomal dynamics of *Oenothera*. Even without his claiming to have produced "mutants" in *Oenothera*, and even without his testing for the transmissibility of new characters, Gager's discovery of irregularities in the distribution of chromosomes following radium treatment dovetailed with nascent microscopic and histological investigations of the nature of chromosome dynamics in *Oenothera*. What's more, his constant qualifying of his own results and his idiosyncratic reworking of de Vries's theories thereby contributed to a broader opening up of the question of heredity and mutation. Mutation was not just what de Vries had said it was—it was a category of biological thought worth further examination. The discovery of chromosomal irregularities related to the production of mutants meant that genic mutation could not be the whole story of genetic change.

Gager's experiments, in short, were among the first intimations that the study of *Oenothera*—that apparently ordinary weed that showed such extraordinary evolutionary behavior—might be a key to unraveling the secrets of heredity in a way that even de Vries had not envisioned, and in ways that the Morgan school of drosophilists was still far from recognizing. To Gager, as to many of his contemporaries, mutation seemed to be primarily a matter of the chromosomes. Indeed, Gates, reflecting on the state of affairs by 1920, observed that "two decades of intensive experimental work with plants and animals has led to a greater diversity of opinion concerning evolutionary factors than ever before." And part of this greater diversity of opinion concerning evolutionary factors was due to a greater diversity of potential experimental mutagens than were ever before considered—from chemicals and starvation to heat and ultraviolet light to radium and X-rays. There was also as yet no "generally and universally applicable method of evolution."[203]

By 1908 Gager knew that radium rays had "the power to effect marked change in the chromatin" and that such changes, often identified by anomalous modes of mitosis, could affect the organism as a whole. Although Gager's discovery that radium could induce the fragmentation of chromosomes confirmed earlier work by Bohn (in 1903), Perthes (in 1904), and Koernicke (in 1905),[204] it was only during Gager's collaborations with Blakeslee in the 1920s that he finally became convinced that radiation could in fact produce mutations. By fine-tuning the method of identifying mutations, these collaborations established a solid link for the first time between the rays of radium and the induction not only of chromosomal mutation but of a separate level of *genic* mutation. Indeed, subsequent uses of radium in genetics, although distinctly less sensational, proved vastly more significant in the long run as radium's "stimulating" power became reinterpreted as its power to induce mutation.

A new phase had begun. Radium was not only part of the evolution of the cosmos, a putative component of the environment of the early earth, and a daily fact of existence for plants whose effects could conceivably be felt for generations yet to come. With the clarification of the nature of chromosomal mutations in the 1920s, Gager helped to finally establish de Vries's dream: that radium could also be used to gain control over evolution by inducing mutation.

4 Radium Genetics

Most geneticists are not radiologists but merely use radiation to get results.
—Albert F. Blakeslee, letter to John T. Buchholz, March 6, 1935

Within a decade after its discovery, radium had "come to life" in several different senses and experimental contexts. The powerful association between radium and life was evident not only in the simple use of radium itself—in radium as a tool for experimentation—but also in the key role it played in interpretation of the results of experiments in which it had served as a mutagen. As a fundamental homology between physical and biological transmutation was perceived, the probabilistic and statistical approaches exemplified in the concept of the half-life came to be common to both the half-living element and the study of the mutants it might induce. Even as MacDougal sought to better assess mutation frequency in his radium work, he drew on methods similar to those developed by physicists of radioactivity to better characterize radioactive decay. And even as Gager carried on MacDougal's work and sought to move beyond the "extravagant," "superficially analogous," and putatively "dangerously attractive" analogies of Burke's radiobes, Gager's central concept of a "radiotonus" emerged at the intersection of biological concerns with the world of the radioactive and drew on the same sorts of powerful associations between radium and life. And yet Gager associated radium not with life in general, but with the specific physiological and mitotic effects radium might induce, and especially its effects on heredity and the chromosomes. Such work opened up a new realm of investigation into the nature of "mutation" and the various modes and mechanisms by which it might operate.

Indeed, from de Vries's inaugural address at Cold Spring Harbor in 1904 through MacDougal's and Gager's pathbreaking experiments, radium was central to many of the earliest efforts to induce mutations artificially, and it played a crucial role in the emerging plurality of meanings of induced mutation. The increasing success of radium-based techniques and radium-suffused interpretations in eliding any distinction between the natural and the artificial—in which Burke had failed and de Vries succeeded, at least for a time—meant that half-living radium not only came to be associated with the experimental induction of mutations, but resonated with the very qualities attributed to those atoms of life, the elements of heredity, themselves. This chapter, then, tells the story of how radium came to life in a fourth way: having entered the rich realm of research into the nature of biological transmutation in the early twentieth century, radium had everything to do not only with experimental attempts to get at the question of the origin of species, but also with the manipulable stuff of heredity itself: first chromosomes and later genes.

From Oenothera *to* Drosophila*: T. H. Morgan and Jacques Loeb*

Although much of the early work on mutation was done in plants, no sooner had Gager's work received prominent attention than other investigators sought to extend his findings to the animal world. In the year following the publication of Gager's monograph, J. Arthur Thomson was already deploring the lack of attention paid to the study of mutation in animals:

> While De Vries has given much convincing evidence in regard to plants, we have as yet very slight evidence of the origin of species of animals by mutation. . . .
>
> It is quite conceivable that a mode of evolution common among plants may be rare among animals. It is difficult at present to apply the mutation concept with security to the animal kingdom.[1]

And yet by 1911, an editorial in the *American Breeders Magazine* (soon to become the *Journal of Heredity*) could remark, "There are signs of a scientific awakening in animal breeding such as occurred in plant breeding a decade ago."[2] Gager was thus far from the only one interested in inducing mutation experimentally, in plants or in animals.

De Vries's idea that "the rays of Roentgen and Curie" could be put to use in the study and control of evolution—itself a particular scientific manifestation of the broader radium craze—proved compelling to many, including the famed geneticist Thomas Hunt Morgan. De Vries and Morgan had been friends for some time, and Morgan had been actively searching for "*Oenothera*-like mutations" in animals throughout 1906 and 1907. In particular, he was searching for cases of mutation in *Drosophila melanogaster* that seemed similar to those de Vries had encountered in *Oenothera*.[3] The fruit fly was the "animal *Oenothera*."[4]

Like MacDougal and many others at the beginning of the twentieth century who were interested in investigating and ultimately controlling the evolutionary process, Morgan was keen to find an experimental mutagen. In contrast to MacDougal's widely reported success in the plant world, however, Morgan regularly failed for a couple of years to produce a single mutation in *Drosophila*. He employed a variety of techniques, including the injection of various substances "into pupae in the regions of the reproductive cells," and he tested his way through "wide ranges of temperature, salts, sugars, acids, alkalis," and other chemicals, and even different kinds of food. He even investigated the effects of changes in temperature. Nothing seemed to work—it was all "without any resulting mutation," he reported.[5]

"Morgan apparently did nothing more with insects until the fall of 1907," according to Robert Kohler, "when he persuaded Fernandus Payne to try inducing mutations experimentally in *Drosophila*." As Payne recalled later in life, "I tried the effects of heat and cold, variations in the food and even X rays. All results were negative with the possible exception of X rays. One variation [with wing modifications] occurred but the strain was weak and after 5 or 6 generations the strain was lost. Of course the variation might not have been a mutation. The Physics department gave me no cooperation, I am sorry to say."[6]

It was during Payne's second year at Columbia, 1908–9, that Morgan began his own attempts with *Drosophila*. "Cells seemed to succeed in setting up barriers against too violent an intrusion of chemical agents," the radiation geneticist Curt Stern later recounted of Morgan's efforts. But another possibility still held hope: "It occurred to T. H. Morgan at the very beginning of the Drosophila work that no cell can refuse the entry of γ-rays from radium."[7] Unsuccessful in his attempts thus far and willing to try almost anything, Morgan began exposing *Drosophila* larvae to the rays of the new element.[8]

Morgan's experiments with radium were a success and proved central to the early history of classical genetics. He discovered his first mutation in May 1910—a mutant with a distinctive wing mutation he labeled "beaded" because of the pattern of the wing margin. It had originated in a culture of flies he had treated with radium.[9] He reported the similar appearance of his first white-eyed mutant in a paper read at the Society for Experimental Biology and Medicine on May 18, 1910, and to Charles Davenport at the Cold Spring Harbor Station for Experimental Evolution on June 11, 1910.[10] Morgan clearly attributed the appearance of this white-eyed mutant to his use of radium, and he published the story of his successes in *Science* and in the *American Naturalist* in 1911.[11] Many of his contemporaries viewed these early experiments as foundational.

Gager's monograph *Effects of the Rays of Radium on Plants* had been published in 1908, the very year Morgan first began working with his flies. By 1910 Morgan was referring to his own work using the same words (as the study of "the effect of the rays of radium" on *Drosophila*). For his part, MacDougal saw Morgan's experiments on the occurrence and transmissibility of mutation as an explicit confirmation and extension of Gager's earlier work: "By the use of similar excitations Morgan has recently induced the appearance of white eyes and short wings in the fly, *Drosophila*, which characters seem to be fixed and fully transmissible."[12]

Sharing a strong interest in the eventual hope of controlling the means and modes of evolution, but ignorant of Morgan's radium research, Jacques Loeb—already famous for his discoveries in artificial parthenogenesis—and his collaborator F. W. Bancroft built on their own interests in MacDougal's mutagenic studies and began to conduct their own experiments with radium. (Loeb and Morgan even had a bit of a priority dispute over whose idea it had first been to use radium to induce mutations; Loeb, for one, seems to have received his supply of radium directly from de Vries, who in turn seems to have brought it from Rutherford's laboratory.[13])

Citing other work by Tower on *Leptinotarsa* and Gager on *Oenothera*, Loeb and Bancroft reported that in their own work they undertook a "very large number of experiments with radium . . . because it happened that the first culture which we treated with radium chanced to give us mutants."[14] They experienced some difficulty repeating their experiments, however, and according to one review, "the short-winged mutants have appeared thus far only in cultures treated with radium, but in only two out of several hundred such cultures . . . the authors appear to doubt whether or not the treatment was responsible for the mutations."[15] Nevertheless, Loeb and Bancroft noted (in a jab at Morgan's reported results) that "as long as the full account of his results is not available, it is not easy to judge to what extent it is possible to produce mutations at desire with his method."[16]

In his personal correspondence from these early years, however, Morgan was clear that his mutations resulted from the application of radium. He had written to Jacques Loeb in the spring of 1911, "As I told you last summer all my wing mutations go back to my flies treated with radium, as do also *at least* two of the eye mutations."[17] Immersed in the radioactively tinged terminology of his day, Morgan even once told Loeb that he had submitted a paper for an upcoming conference entitled "The Disintegration of a Species and Its Reconstruction by Artificial Combinations."[18] And Morgan had reported to the *American Naturalist* in 1914 that "one of the first mutants that I observed in [*Drosophila*] *ampelophila* appeared in the offspring of flies that had been treated with radium."[19]

Morgan also seems to have further encouraged some of his students—including Alfred Sturtevant—to study the effects of radium on *Drosophila*.[20] The importance of this fundamental association of radium with life in Morgan's experimental practice has thus far been generally overlooked by historians. Morgan's promising mutational results led in time to the development of his large-scale program for studying mutation in *Drosophila*—the heart of research in early classical genetics, as Robert Kohler has shown. Morgan's lab was thus dependent not only on a particularly fruitful moral economy surrounding the fruit fly—it was also dependent, initially, on radium.[21] Inspired by de Vries's dedicatory speech, Morgan produced his first mutants by the application of radium, and he characterized these experiments with a phrasing identical to that Gager employed—even describing radium as bringing about the "disintegration" of a species of fruit fly. Morgan thus found the interconnections of radium and life as provocative and as productive as Burke, MacDougal, and Gager had.

Morgan experienced some difficulties in getting his mutant flies to breed, however, and—in a context where radium was readily associated with a stimulating effect—didn't seem to fathom at first that the very radium causing the apparent mutation might itself be a factor in the flies' sterility.[22] Moreover, he soon became increasingly evasive about whether the radium he had used in his earlier experiments had actually been responsible for the mutant flies he observed. He began to claim instead that there was merely a correspondence between the application of the radium and its purported effects. By 1914 he had qualified and "clarified" his findings to such an extent that the exact nature of his earlier claim—and his very accomplishment—were cast into doubt:

> Although there was no proof that the radium has had a specific effect I felt obliged to state the actual case, refraining carefully from any statement of causal connection. Nevertheless, I have been quoted as having produced the first mutants by the use of radium. I may add that repetition of the experiment on a large scale both with the emanations of an X-ray machine and from radium salts has failed to produce any mutations, although the flies were made sterile for a time.[23]

Had he discovered a new and effective experimental mutagen or hadn't he? By 1923, reflecting on early work with the mutant *beaded*, Morgan (with Calvin Bridges) noted that "subsequent work with radium gave no indication that the appearance of Beaded was due to the use of radium."[24] As Morgan would later explain:

> I did quite a lot of work by treating the flies with radium, and as a matter of fact some of the descendants of these flies produced mutants of the type we are now familiar with. But since I did not get them in the immediate offspring of the treated flies I thought the results not worth publishing, and made only a brief statement with regard to the facts in the case. *What I was looking for was to find a specific effect of radium, or some other external agent, that could be repeated.* It is clear now, of course, that I expected too much, because even after x-raying the number of mutants is not very large, and there are many kinds of mutants. In other words: there is no specific effect. I have never put in any claims on this score.[25]

Despite his initial enthusiasm, Morgan thus rapidly backpedaled on his claims for the mutagenic power of radium, even going so far as to emphasize Loeb and Bancroft's own wariness in claiming that they had induced mutants with radium in their experiments: "They found a black mutant type after treatment with radium but since the same type appeared in the control they do not believe that its appearance had any connection with the radium." More significantly, Morgan suggested that the stocks Loeb and Bancroft were working with were "contaminated," even while claiming further priority in the production of the noted mutations:

> Although "two hundred different cultures" were subsequently treated with radium and no short-winged (miniature) flies appeared, I get the impression that Bancroft and Loeb must have had stock that was already contaminated by some recessive mutant factors. All of these mutants had been obtained and described by us, and the stock used by Bancroft and Loeb was obtained in part at least from my friend Dr. Frank E. Lutz, who had at that time in his possession, as a letter I have from him states, certainly two of these mutants, black and miniature, that he had received from me. It seems to me not improbable that the collector, who got the stock from Dr. Lutz for Professor Loeb, included by mistake some flies heterozygous for these two characters; for in our very extensive experience with wild stock from Cold Spring Harbor (the origin of most of Dr. Lutz's stock) and elsewhere these mutants have never arisen again.[26]

Morgan recalled similarly later in life:

> In regard to the experiments of Loeb and Bancroft, I am quite sure that what they got were mutants already present in the stock supplied to them by Lutz—in fact, I have somewhere a letter from Lutz in which he states that the material he supplied them might have been contaminated with the kind of mutants they thought originated from their treatment. It has always seemed to me, therefore, that the least said about those results the better.[27]

In all, neither Loeb and Bancroft nor Morgan believed that the other party had succeeded in truly inducing mutations by means of radium. Curiously, then, radium, which had at first held such promise, rapidly came, after a slew of experimental disappointments, to be questioned as a legitimate mutagen. As mutants became increasingly readily available for genetic study, radium's fall from grace seemed certain. Morgan even cited other investigators (such as Émile Guyénot) who had tried similarly to induce mutations in *Drosophila* with high temperatures, radium, and X-rays and claimed that they had all been "without result" (the only exception being that UV rays gave rise to black eggs). Even Morgan's student Payne came to later dispute the classic story of radium-induced mutation: "In 1910, as you know, a fly with white eyes appeared in one of his cultures," Payne said. "It was not an induced mutation."[28]

Morgan's reluctance to attribute mutagenicity to radium had more to do with questions of experimental setup and the proper understanding of mutation, however, than with any doubt about radium's disequilibrating power. Just as MacDougal had come up with the concept of "mutation frequency" to replace the potentially inaccurate concept of "mutation period," Morgan argued for taking greater care in assessing the causal role of potential mutagens:

> Our own experience with *Drosophila* shows that mutations appear under conditions where all the other flies in the same culture are normal and we have become unduly skeptical perhaps towards evidence which refers a particular mutant to some unusual treatment to which the flies have been subjected. Until we can get definite information as to how mutants arise, whether through external influences, through accidents of mitosis, through hybridizing, or through changes in the chromosomes with its consequent dislocations of the machinery of crossing over, or in some other way, it seems futile to discuss the question.[29]

As the historian of genetics Elof Axel Carlson has claimed, however, at least two of Morgan's famed mutations—*truncate* and *beaded*—in fact *did* come from lines Morgan had exposed to radium: "The rest did not." But, as he concluded, "With so many more spontaneous mutations than allegedly induced ones during this 'mutating period,' Morgan played it safe and dismissed the role of radium as an agent inducing mutations."[30] Radium began to lose its central role as a mutagen of choice in no small part because of such shifts in conceptions of induced mutation frequency.

One of the most revealing things about Morgan's apparent retractions is his conception of "mutation" as an event with any number of possible causes—from the external to the internal, and from hybridization (a theme in the history of speciation at least from Linnaeus to contemporary criticisms of de Vries) to errors in crossing-over (a novel mechanism suggested only in the wake of the Morgan school's own discoveries). Morgan thus shied away from claiming a causal mutagenic role for radium in part because the *mechanism* of mutation was not yet known. Nor was there yet a clearly articulated concept of what would later be understood to be a "point" or "gene" mutation. As Morgan concluded, "We know now that the white-eyed mutant is one of the commonest mutant types; it has recurred again and again, as have also its allelomorphs. Finding it was not so important as the use to which it was put."[31]

The history of the use of radium in the experimental induction of mutations has been largely forgotten not simply because its main practitioners—Loeb and Morgan—doubted each other's findings, but for the rather more significant reason that *doubting radium's mutagenicity seemed the most promising way to gain clarity on the real nature of mutation.* Accordingly, as Morgan gained greater insights into the relationship between hereditary factors, "genes," and chromosomes, understanding in ever greater detail what a mutation might be, he reinterpreted his earlier experiments with radium as having produced mutants by chance rather than by direct action.[32] And although Morgan was soon able to acquire a relatively large number of mutants simply by culturing flies in a system later metaphorically described a "breeder reactor," he concluded in 1914 that "our experience with *Drosophila* has given us the impression that mutations are rare events."[33]

Not all *that* rare, however—Charles Davenport commented in 1922 on the relative abundance of mutations, comparing it directly with the phenomena of radioactivity: "There is certainly much in the phenomena of gene mutation with its prevailing recessive tendency, its measurable rate of occurrence, and its predictability, that shows at least many points of similarity to the gradual changes, by loss, of the salts of the uranium-radium-lead series."[34] And a few years later, Davenport reiterated the connection between mutation and transmutation.[35]

Mutations were rare enough, though, that it wasn't until the work of the Morgan fly group that evidence for mutations in animals went from being "scanty"—in the words of geneticist Reginald Ruggles Gates in 1920—to being established. Gates even went so far as to claim that "the *Drosophila* work has therefore given us a look into the constitution of the germ plasm such as no annual-breeding plant or animal could furnish in a lifetime."[36] Gates was just a few years away from being proved wrong by a simple weed, the study of which also grew in part out of radioactive roots.

From Oenothera *to* Datura*: Albert F. Blakeslee*

Morgan may have had doubts about which experimental mutagens were responsible for which mutants in *Drosophila*, but the situation was far more complicated in plants—and especially in the case of de Vries's own favorite, the much studied and hotly debated *Oenothera*. By the second decade of the twentieth century, it was becoming increasingly difficult to determine just which newly identified mutant forms in *Oenothera* truly deserved to be considered "new species"—with all that that might mean—and which should be seen as merely the result of the plant's newly discovered messy chromosomal dynamics. Various cytological investigators struggled to come to grips with *Oenothera*'s "normal" karyokinetic idiosyncrasies, and with the sheer complexity of the phenomena it presented even in the *absence* of radium treatment.[37]

In the years following the publication of his monograph, Gager did his best to keep on top of advancements in radiation genetics, and he reviewed some of the more significant literature in a 1916 piece.[38] George Harrison Shull, meanwhile, had taken up the challenge of studying *Oenothera* at Cold Spring Harbor, but with his departure in 1916, the future course of research on *Oenothera* at the Station for Experimental Evolution was unclear. Charles Davenport, director of the station, wrote to de Vries directly with his concerns:

> Now that Shull has left us to go to Princeton I fear the oenothera work will suffer. Still I think that Dr. Blakeslee will keep up with some of it and I hope the opportunity will arise for us to have a man who shall devote a good share of his time to the oenothera. I have not forgotten that in your opening address at this Station in 1904 you recommended this plant especially to our care.[39]

De Vries had also recommended the application of "the rays of Röntgen and of Curie" in the study of mutations and in the attempt to accelerate evolution, and although the new man, Blakeslee, would rapidly turn from *Oenothera* toward other species, he collaborated with Gager for a decade in further investigations of the effects of radium on plants.[40]

Having begun his botanical career at Harvard under the mycologist Roland Thaxter in 1904, Albert Francis Blakeslee ("Bert") first encountered de Vries's mutation theory while teaching at the Connecticut Agricultural College in Storrs. As he recalled in an autobiographical account, it was in 1909 that he first had "the thrill" of reading de Vries's theory, "and thought that if I scoured the country I too might be able to find a species in the process of mutation."[41] The mutation theory was at the core of Blakeslee's interest in genetics, and both its promise and its unanswered questions sparked his imagination on more than one occasion. "I have always felt that the Mutation Theory was a strong factor in turning my interests and research toward genetics," Blakeslee later remembered. His interest in de Vries's theory remained strong for the rest of his life.[42] Even as late as 1949, Blakeslee continued to say that de Vries was "perhaps the greatest biologist of all time" and that "the mutation theory is one of the corner stones of genetic research."[43]

Like many others of the period, Blakeslee was keenly interested in the experimental control of evolution, and he gave several lectures over the years with this theme prominently highlighted. "Methods of controlling genetic processes have always been of interest to us," he remarked.[44] At first Blakeslee thought he had found a suitable choice of model organism in the yellow daisy known as the black-eyed Susan (*Rudbeckia hirta*). He was soon forced to move on to another model organism, however, when the daisy proved to be self-sterile and too "reduced in vigor" after two or three crosses to withstand inbreeding—not to mention the generations of inbreeding required for proper detailed research. Blakeslee began to search for other possible organisms.[45] It was at this time, while at Storrs around 1909, that Blakeslee received from the United States Department of Agriculture "a batch of seeds of *Datura stramonium* as an example of an economic weed." The seed "happened to give both purple- and white-flowered seedlings," Blakeslee recalled, "and for several years this species was used to demonstrate Mendel's laws of inheritance" in his teaching.[46]

On leave from the college during the 1912–13 year, Blakeslee worked at Cold Spring Harbor, finally joining the staff as a resident investigator in genetics to replace the departing Shull in 1915. Having devoted considerable attention to genetics in his botany work—Blakeslee had offered what was "probably the first organized course in genetics in the United States in 1914–1915"—it was only natural that he chose to bring his work on the "coarse, weedy plant with its beautiful flowers" with him when he moved to Cold Spring Harbor permanently.[47] Once there, and at last giving up the multifaceted teaching load he carried at Storrs, where his courses included (among other things) freehand drawing, Blakeslee was free to begin work as a full-time geneticist with access to superb greenhouse and garden facilities. He was on the hunt, as he put it, for "the best possible '*Versuchstier*'" and for the best possible means to do research with it.[48] Over the next twenty-seven years, Blakeslee would make full use of six greenhouses and various agricultural test fields, running experiments on a grand scale. These were resources that Gager, having difficulties even finding eight to ten feet of bench space, simply could not match.[49]

Blakeslee noted that *Datura* (jimson weed) "was the best organism I could find in the botanical line."[50] He had been drawn to *Datura* for a variety of reasons, including its hardiness, the ease with which it could be grown, and the fact that four generations could be grown per year in greenhouse environments, making the results of his evolutionary experiments (radium-based and otherwise) that much quicker to uncover.[51] "At first," Blakeslee recalled, echoing newly emerging concerns about *Oenothera*, his own choice, *Datura*, "seemed to have too many chromosomes, but we kept at it as a side problem since it was so easy to work with."[52]

The decision paid off. Blakeslee's assistant, B. T. Avery, found the first novel type in *Datura*—the so-called Globe mutant—in the summer of 1915.[53] As Blakeslee later reported, "The Globe mutant differs from normals apparently in all parts of the plant. It forms a complex of characters readily recognized whether the plants in question have purple or white flowers, many or few nodes, or spiny or smooth capsules."[54] This was no ordinary mutation like those found in *Drosophila*. Much more than one factor had been affected: the entire plant was different from its ancestor, in a whole suite of traits.

Blakeslee became convinced that he had found a new species, and he labeled the original new plant specimen as such ("N.S."), including a photograph of the plant in the 1919 paper reporting the discovery (fig. 8). Although the plant proved sterile with "normal" plants, it could be self-pollinated successfully and produced progeny that bred true, resulting in further generations with "depressed globose capsules." Blakeslee concluded that it "seems to have established itself as a distinct new race."[55] He continued:

> This physiological incompatibility between a mutation and the parent species from which it arose suggests that we have actually been witnessing in our controlled pedigrees the birth of a new species which may be capable of maintaining itself in a mixed population uncontaminated by crossing with its ancestral line. The race is relatively vigorous.[56]

In the caption to the photograph included in the paper, Blakeslee put the point more plainly: "Tests have shown that this mutant differs from all others investigated in that it breeds true as a distinct new race. Here we appear to be witnessing the birth of a new species."[57]

As Blakeslee, Avery, and his other assistants bred the "Globe" mutants, they rapidly discovered that still "other types appeared as mutants

FIGURE 8. One of Blakeslee's "new species" of jimsonweed (*Datura stramonium*). (From Blakeslee and Avery, "Mutations in the Jimsonweed," *Journal of Heredity* 10 [March 3, 1919]: 119.)

in our cultures, and *Datura* soon became practically our sole object of investigation."[58] As one observer at the station recalled:

> One new form after another began to appear in his cultures. Some were gene mutations but many were evidently different. These produced some offspring like themselves but threw many normal plants. For an outsider to recognize these forms was difficult, since most of their differences were subtle ones. It was the despair of his colleagues to see Blakeslee go down a row of plants and pick out these mutants unerringly. This he could do partly because of his acute powers of observation and partly because he was personally familiar with his material and did not leave the observing and recording to his assistants alone. . . . The size of the *Datura* cultures increased and in the summer as many as 70,000 plants were grown. Work was actively carried on in the winter, as well, in the six greenhouses and laboratories.[59]

Blakeslee was even able to identify types that while "indistinguishable in gross appearance from each other," were nevertheless "in respect to a whole series of characteristics strikingly different from the normal Jimson Weed from which they have been made up to order, as it were, with definite plan and purpose." Blakeslee eventually found three types in particular that he thought "perhaps merit the term of synthesized new 'species,' since they satisfy the criterion of breeding true and are more different from the normal type than some of the species which already have been described in the genus *Datura*."[60] He took these newly encountered mutants to be indicative that his team had encountered a situation in *Datura* similar to that which de Vries had encountered in *Oenothera* (this was particularly important because de Vries's theory faced increased criticism and skepticism by the time Blakeslee reported his discovery in 1919):

> During the past few years . . . we have discovered in our cultures a number of mutative variants of greater or less distinctiveness which, so far as studied, seem to be inherited in a manner different from that shown by simple Mendelian characters. . . . These mutations are of sudden, though rather rare occurrence and transmit their characters—chiefly through the female sex—to only a part of their offspring. . . . The mutations are distinguished from the normal plants from which they arise, not merely by single visible differences, but by a complex of characters which seem to be inherited as a whole when transmitted to their offspring. Leaf and capsule characters are perhaps the most conspicuously affected, although the growth habits and flowers are also involved in the mutations. . . . The mutations in *Datura* are distinguished by the same kind of differences apparently that characterize mutations in the classical genus *Oenothera*.[61]

A year after his initial discoveries, Blakeslee made further explicit reference to the "increasing rôle in experimental evolution" of the de Vriesian "theory of mutations" that had first been laid out two decades earlier. Understanding the exact nature of mutation in plants—which for Blakeslee meant understanding much more than simply gene mutations—piqued his interest and became his central goal.

Chromosomes Regnant

Unlike the drosophilists, who fairly readily shared their stocks and data across the fly room and with other centers of fly research, Blakeslee kept full control of his *Datura* data. But Blakeslee was nothing if not collaborative: having collected the seeds of ten different species of *Datura* from around the world, he engaged in a series of ongoing collaborative ventures over the years, working with the geneticist Edmund W. Sinnott, an expert in the internal anatomy of the *Datura*s who could recognize most mutants from tissue samples alone (and who also happened to come from Blakeslee's old stamping ground in Storrs), and John T. Buchholz, an expert on "the growth of pollen-tubes and the abortion of ovules as problems in developmental selection," among others.[62]

One of Blakeslee's earliest ongoing collaborations was with the cytologist John Belling, who had joined Blakeslee's group in 1920 and helped him in his "study of the nuclear condition of our mutants."[63] Blakeslee, Belling, and greenhouse manager M. E. Farnham published a "preliminary report" of their findings in *Science* in 1920.[64] And it was Belling's cytological work—on the appearance and behavior of chromosomes—that was later held to have given "the greatest possible assistance in the interpretation of the originally baffling phenomenon of mutation in *Datura*."[65] Indeed, it was largely as a result of this "fruitful association" with Belling—as well as the invention of the acetocarmine staining method that permitted chromosomes to be readily enumerated in "smear preparations"—that Blakeslee was rapidly able to establish that "each mutant was the result not of a gene difference but of a third chromosome added to a particular pair of the twelve in this plant." Such mutants were termed "trisomics" or "$2n + 1$" types. More generally, this discovery enabled Blakeslee at last to interpret his results: he had found mutant plants that differed by a whole "complex of characters" that were "transmitted collectively" and which segregated "in a very unusual fashion"—on a *chromosomal*, rather than a *genic*, basis that would presumably otherwise have required the simultaneous mutation of a number of different genes.[66]

While acknowledging that it was "sudden germinal changes, large or small in amount" that were the basis of "perhaps the most fundamental work in modern genetics," Blakeslee noted that "mutations could not be confined to cells associated with sexual reproduction." In an apparent reference to the remarkably productive and groundbreaking work of the drosophilists and other more gene-oriented investigators, Blakeslee's remarks emphasized that botany had already applied the mutation concept in ways that extended far beyond the genes that many animal geneticists were most concerned with. *Somatic mutations*, for instance, were those mutations that took place in cells in which sexual processes were not involved. While fairly "less common phenomena in animals," such somatic mutations—or "bud sports," as they were also frequently called—were common in plants, and many were even quite well known (recall MacDougal's interests). After examining some cases in the "nonsexually propagated races of *Mucor genevensis*," Blakeslee had concluded that mutation also took place "in lowly organized plants and animals in which nonsexual reproduction is the rule or in which sexual reproduction is not known to occur."[67] In other words, mutation need not be restricted to the gene or the soma; it could also take place in the context of reproduction, even if that reproduction was *not* itself sexual. Such instances of mutation were real, and yet they were clearly beyond the ken and the techniques of the drosophilists—no matter how powerful and innovative these investigators were in identifying and mapping mutant genes. Blakeslee argued that all these categories, including those whose "inheritance could not be established by breeding experiments," had been and should continue to be called "mutations."[68]

Blakeslee held that the effects of chromosomal duplication or other alterations in producing phenotypic change were also valid additional instances of mutation:

> To us, one of the most interesting features of the *Datura* work is the possibility afforded of analyzing the influence of individual chromosomes upon both the morphology and physiology of the plant *without waiting for gene mutations*. . . . Our work so far we believe adds evidence to the conclusion that the mature organism—plant or animal—is not a structure like a child's house of blocks, made up of separate unit characters, nor is it determined by separate and unrelated unit factors. It is rather the resultant of a whole series of interacting and more or less conflicting forces contained in the individual chromosomes.[69]

Blakeslee fully acknowledged that classical Mendelian research up to this time had "dealt almost exclusively with disomic inheritance."[70] But he noted that "distinct variations, provisionally termed mutations . . . [have] regularly recurred whenever a sufficiently large number of plants have been subjected to observation," and that these, "so far as investigated . . . have been found to be connected with a duplication of one or more of the normal chromosomes."[71] Blakeslee's mutant plants thus revealed that *phenotypically distinct* mutations could result from *genically identical types*, simply with different arrangements or numbers of chromosomes.[72] Mutation could thus take place at a level that was neither organismal *nor* genic, but chromosomal.

Charles Davenport, director of the Cold Spring Harbor Station for

Experimental Evolution, was completely convinced. With Blakeslee, he held that it was through the study of the nature and structure of chromosomes and their alterations, and not just genes, that the phenomena of heredity would be properly understood. In 1921 Davenport had already prominently noted the work of his researchers in "demonstrating the close relationship between variations in chromosome number and specific variations in the form and other qualities of the body," and there is little doubt that it was Blakeslee's work in particular that drove this new understanding: "The work on *Datura stramonium* is offering remarkable explanations of the complexities of de Vriesian mutation, a form of mutation of possibly not less general significance than Mendelian mutation."[73] By 1922, Davenport noted:

> As modern genetics has been bringing to light the dependence of somatic form and structure on the architecture and number of chromosomes, the urgency of the problem of the experimental control of the structure and number of chromosomes has become more pressing. *Indeed, not until such control is secured may the era of experimental evolution strictly be said to have been entered upon.*[74]

In the hands of Blakeslee, Davenport, and many others, de Vriesian mutation—like atomic physics before it—was "inward bound," from the organismal level to the chromosomal level, while still remaining distinct from the Mendelian, "factorial," or genic mutations that were of such interest to the drosophilists. Here at the station, one of the very centers of early experimental genetics, a distinction was being drawn not only between organismal mutants and chromosomal mutations—a distinction unknown to de Vries's original mutation theory—but also between what Davenport called more generally "extrachromosomal changes," or "changes in numbers of chromosomes" (such as the phenomena that Blakeslee had discovered), and "intrachromosomal changes," or "changes in the genes."[75] Davenport even explicitly referred to Blakeslee's mutative variants as cases of "interchromosomal mutation." A mutation did not need to be *genic* in order to be *genetic*.

In short order, Blakeslee and his collaborators, colleagues, and competitors identified many other varieties of "chromosomal mutants," including reciprocal translocation among trisomics, the existence of haploids in higher plants (theretofore unknown), and even mutants with chromosomes arranged in sets and rings (precisely that phenomenon determined to be responsible for the seemingly endless bedeviling of an earlier generation of investigators of *Oenothera*). While the drosophilists acknowledged the phenomenon of nondisjunction at the microscopic level, it was Blakeslee who connected the dots to its effects at the phenotypic level and brought nondisjunction and other related chromosomal phenomena into the realm of "mutation" proper. Davenport agreed: "It has remained for *Datura* to reveal in the hands of Blakeslee and his associates, Belling, Farnham, and others, an extensive system of inter-chromosomal mutation and corresponding somatic change the like of which had been entirely unknown."[76]

With *Datura*, Blakeslee had found a "genetically simpler mutating plant material" entirely relevant to unraveling the more complicated mysteries of the chromosomal dynamics of *Oenothera*. As Davenport had once reported to de Vries, "Here we are, as you know, submerged in *Datura* and feel, as you feel yourself, that it throws light also upon *Oenothera*."[77] And as he reported elsewhere:

> The outstanding feature of the species is that it, like *Oenothera*, is undergoing a variation in its chromosome-complex; and with every variation in its chromosome-complex goes a special somatic form. This department is now fully launched on a program of work with this valuable form, and we trust that with appropriate support the analysis of De Vriesian mutations can be carried beyond anything hitherto accomplished.[78]

Although undoubtedly invested in the success of the research program conducted at his own station, Davenport could still hardly praise the significance of the *Datura* work enough: "The *Datura* work is of such great theoretical importance that it deserves all the cooperation that can be secured for it." Time and again he trumpeted the significance of the "variations of the chromosomal complexes and their corresponding somatic mutation."[79] The genic mutations of the drosophilists were important, Davenport argued, but were properly understood as complementary to the work coming out of Cold Spring Harbor and its focus on the chromosome as a primary agent in evolution:

> The studies of Morgan, Sturtevant, Bridges, and Muller of gene mutation and of Blakeslee and his associates on holochromosomal mutation, as well as those of Metz on chromosomal homologies and chromosomal fragmentation, elevate *the chromosome* to the position of the principal mechanism of heredity and evolution. It illustrates the slowness with which new discoveries

> filter into popular knowledge that the very name of the chromosome—so fateful for mankind and civilization—should still be almost unknown outside of genetic circles and sometimes insufficiently regarded and recognized even by active biologists. To the geneticist, however, the chromosome with its genes affords another precious link between the complex phenomena of the development of the individual on the one hand and the constitution of matter on the other.[80]

The chromosome was central. *Oenothera* may have once upon a time struck investigators as peculiarly problematic, but with a better understanding of chromosomal dynamics and the effects this had on the production of actual mutant organisms, *Datura* was saved from a similar fate. As Blakeslee noted sotto voce: "We do not believe . . . that the jimson weed is peculiar among plants in giving rise to chromosomal mutants."[81]

Elemental Heredity: Atomizing the Chromosome

Even as the precise mechanisms of heredity remained unknown and the understanding of mutation continued to evolve in these early years, important areas of overlap and resonance between the phenomena of mutation and the phenomena of radioactivity continued. Sir George Darwin's presidential address to the BAAS was one prominent example, but there are countless others, even decades after the initial radium craze. To H. G. Wells and Julian Huxley in 1931, for example, "it seemed that mutations were like the transformation of radio-active elements—something truly spontaneous, in the sense of being determined from within, not to be influenced in their rate of occurrence by any treatment which could be devised."[82] J. Arthur Thomson proclaimed in the same year that "in the domain of things the processes that come nearest those of organic evolution are to be found in radio-active changes," and he described the transmutation of uranium as "in some ways like the transformation of species; but, nowadays, the *known* chemical-physical clocks are all running down, whereas the vital clocks are able to wind themselves up."[83] And Morgan, even years after his preliminary experimentation with radium, continued to stake out a position regarding radium's relationship to life, on one occasion even taking Henry Fairfield Osborn to task for quasi-vitalist claims relating radium to life processes. (Morgan characterized Osborn's claims as to "the atomic constitution of the chromatin" and its possible constitution by as yet "undetected chemical elements" as nothing more than "a sort of poetic outburst."[84])

Poetic it may have been, but such "outbursts" specifically relating atoms to the phenomena of the cell were commonplace. "There is something about the . . . declarations of Professor Edmund B. Wilson regarding the structure of the cell that reminds one of Sir Ernest Rutherford's description and Bohr's graphs of the atoms," wrote F. M. Getzendaner in the *American Naturalist* in 1924. He suggested that the journal's readership explore what he called the "periodic differences in species," even going so far as to identify eight "super phyla" of animal life and explicitly noting that this was the same number of groups as in the periodic table.[85] Indeed, the relationship of the physicochemical to the realm of heritable variation had always been provocative. According to one early historian of genetics, Bateson had earlier "derived the discontinuity of substantive variations . . . from chemical differences which were determined by a chemical stability," as opposed to meristic variations, which were determined by a more "purely mechanical" stability.[86] Inverting the anxiety of influence, meanwhile, MacDougal claimed that it was de Vries's "speculative insights" and his theory of heredity that had led to the "present conception of the ions of the physical chemist"![87]

Either way, mutations were strongly associated with changes in chemical elements, and the implications of a biology analogous to chemistry or physics were widely commented on. As R. C. Punnett noted as early as 1909:

> The position of the biologist to-day is much the same as that of the chemist a century ago, when Dalton enunciated the law of constant proportions. In either case the keynote has been Discontinuity—discontinuity of the atom, and the discontinuity of the variations in living forms. With a clear perception of this principle, and after a long and laborious period of analysis, the imposing superstructure of modern chemistry has been raised upon the foundation of the atom. Not otherwise may it be with biology; though here, perforce, the analytical process must be lengthier, both from the more complex nature of the material, and from the greater time involved in experiments on living forms.[88]

With only the most nascent of cytogenetics to depend on, it was not beyond the pale in 1913 to consider mutation as having to do primarily with change in the number of chromosomes in an organism, given that chromosomes were understood to be the primary vehicles of heredity. In fact, de Vries's *Oenothera* mutants were being explained in just this way. Chromosomes were, for a time, the undisputed atoms of heredity in many quarters. As such, one commentator in 1913 noted:

> If this new departure [mutation] depends on a modification of the number of chromosomes in the nuclei of the reproductive cells, we have discovered, if not the cause, at least an early effect of the still hidden cause; and we cannot fail to be struck by the analogy with the theory which finds in the differing numbers of corpuscles the cause of the differences between the atoms of the chemical elements.[89]

Gates also explicitly linked the phenomena of heredity with those of the atom:

> It must not be inferred from the preceding remark that the whole mystery of heredity is believed to have been solved . . . each discovery represents a further step in analysis, whether it be in the processes of inheritance or in the structure of an atom. The evidence for the independent identity of chromosomes is at the very least equal to that for the existence of electrons, emanations and other particles constituting the atom. Fortunately, physicists are not worried by the argument that until the exact nature of electrons and corpuscles is known it is unsafe to recognize their existence in formulating a hypothesis of atomic structure. But this is the type of argument with which the cytologist is frequently confronted, coming from biologists whose knowledge often does not extend to the chromosomes.[90]

Gates even called this view of heredity the "elementalist" or "particulate" view.[91]

Such thoughts conjured visions of a new kind of periodic table of the chromosomes—presaging the kinds of charts of "chromosomal types" that Blakeslee was to produce in the 1920s. Even Loeb, referring to Morgan's mapping work, noted that "biology has thus reached in the chromosome theory of heredity an atomistic conception, according to which independent material determiners for hereditary characters exist in a linear arrangement in the chromosomes."[92] Morgan was mapping genes, but it was references to *chromosomes* being the units of hereditary mutation—just as atoms were the units of radioactive transmutation—that were legion. Lancelot Hogben made the connection clear in the late 1930s:

> Like the individuality of the modern atom the individuality of the chromosome must be conceived in statistical terms. For the discussion of the more familiar chemical reactions the statical atom of traditional chemistry is adequate. For the interpretation of hybridization experiments the diagrammatic chromosome of the text-book suffices. In the field of radioactivity the statical atom makes way for a dynamical model. So also in the domain of cell physiology we conceive the chromosome as an ever-changing entity. The logical situation is analogous in the two cases.[93]

The Meaning of Mutation

The wider community of geneticists and other students of heredity were already well aware that it appeared possible to make a distinction between genic and chromosomal mutation. Another important and complicated shift in the meaning of mutation was taking place at the same time, however: a distinction was also emerging between the *process* and the *object* of mutation (arguably, "mutant" vs. "mutation"). This latter distinction led to a carefully reasoned exchange between Shull, then the editor of the *Journal of Genetics*, and Blakeslee following Shull's editing of the title of one of Blakeslee's submissions to read "mutation" in place of "mutant."[94] Blakeslee wrote to Shull to complain about the change, and the two engaged in an exchange that captures an important moment of transition in the terminology of the period. Blakeslee first complained to Shull on April 15, 1921:

> I feel that the title in the MS is better than the one you have given the paper. To discover how others would react, I read the two titles to Drs. Davenport, Banta, Metz, Little, Mr Belling, and Mr Farnham and, without telling them the reason for my request, asked them to say which they preferred. All preferred my title except Mr Belling who said he liked the word mutation tho he thought my title more logical. We all feel a difference in meaning between the words mutant and mutation.[95]

Shull replied:

> You seem to think that the change in the title was based upon an assumed synonymy between "mutant" and "mutation" which synonymy does not exist, but I believe that best usage of these terms is in agreement with the feeling of yourself and your colleagues that the two words are *not* synonymous. I *intended* to *change* the meaning of your title because I thought the new title a more adequate and more telling indicator of the contents of your article. *You go so far in the solution of the change which brings about the occurrence of the Globe mutants that it seemed to me you were justified in applying the more fundamental term "mutation" as a title of your contribution* [emphasis added]. The mutation is the *change*, the *process*, the mutant is the *changed*, the *product*. Your article deals with both the process and the product and therefore might with propriety have *either title*. I think your unfavorable reaction to the title I proposed,—and probably that also of most of your colleagues—has been due to your supposition that the two titles were intended to have the same meaning. I must confess that I am not so much surprised at the reaction of those to whom you read the two titles, for I remember that we used to use the word "mutant" very often at the Station when we should have used "mutations" and perhaps the distinction between the two words is even yet not as precisely recognized there as it might be. I have often wished that some friendly Editor had done for one of my papers what I have done for yours though I should probably have made a much louder noise than you have over the unjustifiable interference of the Editor, with my title;—I refer to my paper on "Reversible sex mutants in Lychnis dioica." A mutation is reversible, but hardly a mutant.[96]

Not everyone was in agreement with Shull, nor would they necessarily be in the years to come. James Neel remarked later on a further distinction: "Mutant vs. mutation. I have polled the geneticists here, and they seem to agree that it is unfortunate but true that the term mutation covers both the changed condition in the genome and the process of change." A note attempting to standardize genetic nomenclature that had been published in the 1921 *American Naturalist*, Neel remarked, "does not help much" as it did not touch on the issue.[97]

Shull was also fully aware of the first axis in the meaning of mutation, and whether chromosomal variations were "mutations" became a matter of debate in the field. Although Shull initially seemed to agree with the designation of chromosomal aberrations as mutations—as his initial reply to Blakeslee's complaint shows—a week later he had coined a new word for such chromosomal mutations and tried to get Blakeslee to use it. The word was "anomozeuxis":

> I note that you are seeking for a term to express peculiarities brought about by chromosomal irregularities. I have formulated the fundamental categories of heredity in a paper which I am about to submit for publication in "Science." I am a little fearful that the pouring of much cold water would give me "cold feet" on this question, so hesitate sending you the term in question. It takes considerable courage to invent new terms and my chief motive in making the invention to which I refer is to call attention to the fundamental categories to which I am applying names. The term which I have selected for the chromosome-exceptional type of heredity is "anomozeuxis," with the corresponding adjective "anomozygous." I feel fairly certain that your first reaction to these words will be unfavorable, but they are words which grow easier to say and pleasanter to look at as you become more familiar with them. The other words in the series to which these belong are "monozeuxis" and "monozygous," "pleiozeuxis" and "pleiozygous," "exozeuxis" and "exozygous." The meaning of the words will doubtless be sufficiently obvious to need no special explanation here. I shall be very much interested to learn of your reaction to this suggested terminology.[98]

Shull's two responses to Blakeslee's work are illustrative of how the differing levels of analysis employed by competing groups of biologists ensured competing definitions of "mutation." For example, by traditional observable botanical and morphological criteria, and by the simple fact that they bred true, Blakeslee's plants were clearly mutants, and any botanist (as de Vries himself had often remarked) would have classified such new organisms discovered in the field as mutants belonging to a new species. By the standards of the drosophilists who used genetic mapping techniques and some other geneticists, however, these were clearly not new mutants (or mutations), but merely chromosomal aberrants.

Within a decade this unresolved issue—were mutations genic or chromosomal?—would start to resolve itself in ways that had a longstanding negative effect on the assessment of the significance, scope, and legacy of Blakeslee's work. As "mutation" increasingly became genic, as Morgan's recantation of the mutagenic power of radium settled in, and as Blakeslee's work increasingly came to be overlooked or devalued as irrelevant to a proper understanding of mutation, the important ways in which radium had been instrumental and even central in the early study of heredity would come to be forgotten. Even the important and pathbreaking work with radium that Blakeslee was to undertake in the 1920s in collaboration with Gager would, as a consequence, be almost entirely forgotten by the mid-1930s in the wake of exciting new findings about inducible genic mutation.

: : :

In the early 1920s Blakeslee was fully aware of the polyvalent meaning of "mutation" and of the declining influence of de Vries's theory among biologists of all stripes. In an article entitled "Types of Mutations and Their Possible Significance in Evolution," he compared the influence of Mendel with that of De Vries:

> While the garden pea stands intimately associated with a conception of inheritance of wider application than was at first imagined, the evening primrose and the theory of mutation connected with it are by many considered to furnish an example of a valuable theory founded upon incorrect interpretations. The belief is growing that most of the new forms which have appeared in cultures of the *Oenotheras* are not mutations at all and that the evening primroses, as an abnormal group of plants, are not to be seriously considered as representative of the processes of evolution in normal forms.[99]

Having laid out the relevant details—from the drosophilist H. J. Muller's work on balanced lethals in the 1910s to the importance of the study of the behavior, association, and mechanism of chromosomes and chromosomal duplication and polyploidy—Blakeslee asked:

> What then is a mutation? I do not feel we need to be bound by its application to the evening primrose for reasons of priority, since Waagen . . . had previously used the term in paleontology in an entirely different sense. I believe, with the idea that mutations must involve a qualitative change, that we shall ultimately confine the term to mutations of genes, although such muta-

> tions may later be shown to be as different from our present conceptions of them as are mutations in the Oenotheras from the conceptions in de Vries's classical publication, "The Mutation Theory." It may still be desirable to employ the word *mutation* as a collective term to designate the sudden appearance of any apparent genetic novelty—whatever its real cause—until we know better.[100]

In all, Blakeslee's approach represented a distinct modification and reworking of de Vries's theory.[101] Although Blakeslee acknowledged that "strictly speaking I should not call chromosomal aberrations mutations when the changes are purely quantitative," the table accompanying his 1921 article labeled those forms in precisely that way.[102] The situation was further complicated by the fact that not only could phenotypically different species be genically identical while differing at the chromosomal level (as Blakeslee had shown), but phenotypically *and* genically identical species could still be *chromosomally* distinct through simple and well-known processes such as translocation.

Blakeslee was fully aware of the drosophilists' genocentric focus, and he gave their understanding of mutation a certain priority in his 1921 article in the *American Naturalist*: "We have seen that chromosomal duplications and related phenomena may simulate gene mutations in their effects upon the individual." And yet his focus always remained on understanding the nature of chromosomal mutations: "What is their possible significance in evolution?" he asked, since this is where the fundamental question of speciation ultimately resided. (He also noted that "sudden genetic changes are not necessarily associated with sexual processes," meaning genic changes, while chromosomal changes often were.)[103] Blakeslee sidestepped any firm answer on the nature of mutation in 1921:

> There is not time at my disposal to discuss mutations of genes. . . . It has not been possible in this brief presentation to give an extended classification of mutations, nor to discuss in detail their possible significance in evolution. It will be sufficient if I have made clear the distinction which must be kept in mind, in any discussion of the subject, between mutations in individual genes and those brought about by chromosomal aberrations.[104]

Chromosomal mutations, or "chromosomations," thus served as a halfway point between the classic de Vriesian organismal mutants and the drosophilists' clear identification of gene mutations. (By 1933 Hurst had proposed an alternative coinage with radioactive roots: changes "due to chromosome transmutations and not to gene mutations . . . may be distinguished as transmutants."[105]) Only when the effects of a mutagenic treatment that produced phenotypic candidates for new species could be shown *not* to be the result of chromosomal mutation, and to have resulted in the formation of a new mendelizing character, was Blakeslee willing to attribute the visible aberrant effects to gene mutations. In a new collaboration with Gager, Blakeslee was soon to use radium to explore the natural history of the chromosome in greater depth. The secret of life, like the secret of radioactivity before it, was inward bound.

Making a Go of It

Although Blakeslee initially planned to stay at the Station for Experimental Evolution only two years before returning to full-time teaching at Storrs, fate intervened, and he became assistant director in 1923. He took over the reins as acting director after Davenport's retirement in 1934 and finally became director in 1935. By the time he retired in 1941, Blakeslee had spent twenty-seven years at Cold Spring Harbor, during which time he uncovered so many new and important data from his work on *Datura* that other geneticists began to refer to "the *Datura* Klondike"—a gold mine that revolved in no small measure around his experiments with radium.[106]

Blakeslee laid out the problem: If plant mutants were due to alterations in chromosomes and not just in genes, then "it should be possible by breeding tests to connect up mutants with as many chromosome sets as there are known Mendelian factors, or factor groups." This, however, was not always the case, as there were unusual situations (such as various forms of chromosome duplication) in which varied effects also needed to be taken into account. The discovery of what were termed "balanced" and "unbalanced" types—that is, mutant types with all paired chromosomes and types in which an additional chromosome was left unpaired—provided a new means of exploring the influence of mutation (see figs. 9, 10, and 11 for later visualizations of this phenomenon). In effect, Blakeslee argued, it meant that there was now a means to avoid depending on the random appearance of mutations in a population:

> The unbalanced condition gives us an opportunity, never before realized, of analyzing the influence of individual chromosomes without waiting for the appearance of gene mutations. Hereto-

> fore, the number of factors determined in the chromosomes has been dependent upon the number of mutated genes available for crossing with the normal type. In the jimsons, however, we may study the sum total of all the factors in individual chromosomes by the unbalancing effect upon the structure and physiology of the plant when a single specific chromosomal set has 1 or 2 extra chromosomes.[107]

"Knowing the mechanism to be affected," Blakeslee concluded—that is, the behavior, mechanism, and association of the chromosomes—"we may be able ultimately to induce chromosomal mutations by the application of appropriate stimuli."[108] Radium was one of the first of those stimuli to which Blakeslee turned.

Although Blakeslee was unaware of de Vries's inaugural remarks at the station, his use of radium echoed de Vries's hope that it might be used to induce artificial mutation in the chromatin. As noted in chapter 3, Gager had been the last major figure to investigate the effects of radium rays on plants. After finishing his first round of experiments at the New York Botanical Garden, Gager had gone back to his benefactor, Hugo Lieber, in 1909 to request a further 10 mg of radium for a further series of experiments (doubling the 5 mg he had previously used). As he explained to Lieber, "I hope to give a little more finish to some of my first work, and then to specialize more along the lines of the next to the last chapter in the Memoir, i.e., experimental heredity by means of radium rays."[109]

While other researchers had investigated the effects of radium in inducing mutations after Gager's initial work (such as Emmy Stein, working with the snapdragon *Antirrhinum* in August 1921, "exposing the vegetative tip of shoots . . . and by exposing seeds to radium rays"), little of significance was found that had not already been reported in Gager's work. "The net results of the experiments reported . . . leave it wholly an open question as to whether mutation can be caused by exposure to radium rays," Gager and Blakeslee would later report.[110]

In fact, it wasn't until the winter of 1921 that "the three essentials" needed to properly investigate and finally establish the effects of radium on plants came together once again: "a supply of radium preparations, carefully pedigreed plant materials, and sufficient time and cooperation to make the exposures and to follow the behavior of the plants developed from seeds produced from ovules that had been exposed to radium rays either during gametogenesis, fertilization, or the development of the fertilised egg."[111] Although it seems unlikely that Gager got the

FIGURE 9. Varieties of mutant capsules of jimson weed (*Datura stramonium*) similar to those induced by radium treatment. (From Albert F. Blakeslee, "New Jimson Weeds from Old Chromosomes," *Journal of Heredity* 25 [March 1934]: 89.)

additional radium from Lieber this time around, his interests in experimental heredity drew him to Cold Spring Harbor, where he brought his experience in irradiation to a full-scale collaboration with Blakeslee and his *Datura*. Gager was soon hard at work at the station, "investigating the possible effect of radium emanations upon gene and chromosomal mutations."[112]

Theirs was a close professional friendship: Gager and Blakeslee often shared train cabins and hotel rooms at botanical conferences, and they visited each other's homes with their wives. Having begun their collaboration in 1921, Gager and Blakeslee were already well aware of reports that *Oenothera*'s "mutating period" might be due to complicated chromosomal dynamics. And yet they were committed to seeing with a sort of double vision, not only recognizing phenotypic mutants but also retaining a conceptual space for "mutations" that were neither (now disparaged) cases of ring chromosome nondisjunction, as in

Oenothera, nor strictly genocentric, as the drosophilists were wont to hold. As Blakeslee himself reported to the Botanical Society of America on December 28, 1921, the year before he and Gager presented the first fruits of their collaborative work, "The fact has recently been emphasized that two distinct types of mutation may occur in plants—those which are due to the change of a single factor or gene and those which are due to the addition of one or more entire chromosomes."[113] Blakeslee claimed to have thus far discovered three "factor" mutations and twelve "chromosome" mutations in the jimson weed, all of which were "identified by various external characters." It was in this presentation that Blakeslee first publicly outlined the goals of his collaboration with Gager:

> To study and compare the *structure* of these mutant forms, both as to gross external morphology and as to internal anatomy; and thus to determine the structural effects produced by a single

TABLE II

TYPES OF CHROMOSOMAL DUPLICATION, GAMETIC AND SOMATIC FORMULAE FOR PLANTS HETEROZYGOUS FOR FACTOR PAIR A AND a AND RATIOS OBTAINED WHEN SUCH PLANTS ARE SELFED, TOGETHER WITH DIAGRAMS ILLUSTRATING THE CHROMOSOMAL CONDITION IN SOMATIC CELLS

No. of Extra Chromosomes in Set	No. of Sets Affected	Gametic Formula	Selfed Ratios	Somatic Formula	Somatic Diagram
2	12	AA + Aa AA + 4Aa + aa Aa + aa (12 + 12)	1A : 0a 35A : 1a 3A : 1a	AAAa AAaa Aaaa (12 + 12) + (12 + 12)	
1	1	2A + a + AA + 2Aa A + 2a + 2Aa + aa 12, (12 + 1)	NOR. 8A : 1a MUT. 9A : 0a NOR. 5A : 4a MUT. 7A : 2a	} AAa } Aaa (12 + 12) + 1	
1	12			(12 + 12) + 12	

No. of chromosomes 12 +	0	1	2	3	4	5	6	7	8	9	10	11	12
Frequencies	1	12	66	220	495	792	924	792	495	220	66	12	1

FIGURE 10. Gametic and somatic formulas, with diagrammatic karyotypes of different "chromosomal types" in *Datura*. (From Blakeslee, "Types of Mutations and Their Possible Significance in Evolution," *American Naturalist* 55 [1921]: 257.)

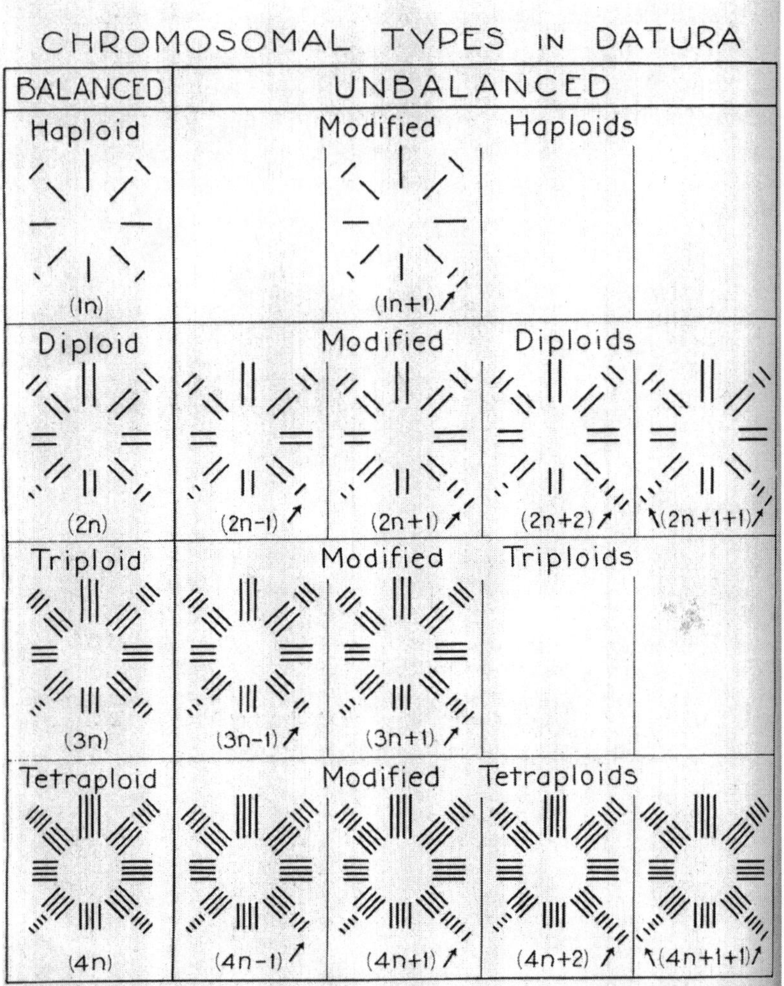

CHROMOSOMAL TYPES IN DATURA

BALANCED	UNBALANCED			
Haploid	Modified Haploids			
(1n)		(1n+1)		
Diploid	Modified Diploids			
(2n)	(2n-1)	(2n+1)	(2n+2)	(2n+1+1)
Triploid	Modified Triploids			
(3n)	(3n-1)	(3n+1)		
Tetraploid	Modified Tetraploids			
(4n)	(4n-1)	(4n+1)	(4n+2)	(4n+1+1)

DIAGRAMS OF SOME CHROMOSOMAL TYPES FOUND IN THE JIMSON WEED

Figure 7

These diagrams show only some of the kinds of chromosomal mutations which have been found in the course of these studies. For a more complete list see Table III.

FIGURE 11. Diagram of chromosomal arrangements in the jimson weed. (From Albert F. Blakeslee, "New Jimson Weeds from Old Chromosomes," *Journal of Heredity* 25 [March 1934]: 88.)

> factor and those produced by a single entire chromosome. In this way it may be possible to begin an analysis of the factorial constitution of each of the chromosomes.[114]

Blakeslee's approach to mutation studies was thus intended to complement other studies in the field and to better highlight the different ways mutations could be produced—by both chromosomal and genic factors.

With most radium in the hands of physicists and hospitals at this time, Blakeslee found a benefactor in a certain Halsey J. Bagg, of the Memorial Hospital of New York City. (As Blakeslee explained, "Anything that you can do for us in getting hold of the rays, will be greatly appreciated. The Jimsons are becoming more interesting and ought I feel to be attacked from every standpoint possible."[115]) Blakeslee then wrote a note to Gager on a copy of the letter: "Hope we can make a go of the radium work. I can bring in the plants any time you are ready for them."

Blakeslee expected that the application of radium to the plants would have one of two major categories of effects. As he once asked Gager, "Just what do you anticipate the results will most likely be—induction of the mutations or effect upon somatic growth?"[116] These two possible outcomes were not entirely distinguished in Gager's earlier 1908 work—in many cases, effects on somatic growth *were* mutations in the early days of the century.[117]

In mid-April Blakeslee brought potted *Datura* plants from Cold Spring Harbor to the Brooklyn Botanic Garden, where Gager was now director. Gager then either took the plants to the hospital for irradiation at some point in the spring of 1921 or, more likely, borrowed radium needles to expose them himself.[118] (As the glass of the radium needles would absorb most of the α-and β-rays emitted by the element, it was primarily the γ-rays that could be held responsible for the effects observed.) The plants were then transferred back to Cold Spring Harbor for the summer growing season, where both Gager and Blakeslee tended them.

The experimentation was far from easy, and Gager himself was not particularly sanguine that the results would be all that revealing: "My forecast is that we shall probably not get any results that can be attributed to the radium unless possibly a dwarfing."[119] Gager seemed continually beset by obstacles, and he found by the end of June 1921 that his schedule had been so overtaken with other work that he was unable to carry out further experiments. He had hoped to be able to continue with the experiments the next spring, but was again unable to do so, and he found in subsequent summers that his busy schedule, hay fever, and other matters interfered with further experiments.[120] Gager called the disappointing results of that first summer useful only for studying the "the best method of procedure," and he considered that the work done thus far served only "generally to indicate errors in method to be avoided."[121]

Blakeslee was considerably more enthusiastic about the prospects of their work, writing in 1922 to Gager, "This spring the offspring of the mutants which came from your rayed capsules showed albino seedlings in the proportion of one albino to three normally green plants." If only he could find capsules heterozygous for that albino character, Blakeslee concluded, "we will have proven the presence of a new mendelian character for the Jimson Weed and can feel the probability that the cause of its appearance was the radium treatment."[122] Taking Gager's own initial investigations and carrying them further, Blakeslee continued to come up with other experimental possibilities to try out, drawing on his increasing knowledge of chromosomal (and not just genic) mutations. In one case he proposed assessing the "differential mortality or stimulation of the various mutants caused by a single extra chromosome" when exposed to radium in order to better study the "stimulation and retardation" of the rays of radium "upon the various physiological processes of mutants."[123] Gager, for his part, was more concerned to know whether any of the mutations were totally new or whether they were simply more of the same kinds that had already been identified (MacDougal had made a point of measuring the frequency only of *new* mutations).[124]

Gager's earlier 1908 work had found effects that "seemed confined to purely somatic characters of the offspring and did not appear to affect their genetic constitution."[125] He had concluded, therefore, "that most of the variants were not true mutations, and that further evidence is needed before we may say . . . that mutation may be induced by exposure to radium rays."[126] His collaboration with Blakeslee extended this earlier work in a new direction, and with provocative new results. Together they hoped "that it may be possible, soon, to take up again the study of the effects of radium treatment upon the genetic constitution of the offspring and to determine more precisely at what stage or stages the stimulus is effective."[127] In other words, even in the wake of Gager's earlier claims to have failed to produce "true mutations," Blakeslee's awareness that mutational changes could be somatic or chromosomal while still not being genic led to a new understanding crediting radium with mutagenic power.

Blakeslee only occasionally expressed minor impatience with the speed of their progress. As he wrote to Gager late in 1923, "Wish we had some more radium work to report. You must find a cure for the Ragweed and come down and play with them again next summer."[128] Nevertheless, by this time, the two had already presented their preliminary results at the 1922 meeting of the Botanical Society of America in Boston. And already by 1921 they had encountered a peculiar mutant, "Nubbin," that had clearly arisen from a "radium-treated parent" and was probably the result of ray-induced "breaking up and the reattachment of parts of non-homologous chromosomes."[129] (As Blakeslee later reported in the *CIW Year Book*, some of the "three chromosomes were fragments, and the fragments of one were attached each to a fragment of the other two."[130]) Blakeslee thought that "Nubbin," with its interchanged chromosomes, was thus "probably the first induced chromosomal mutation."[131] He held that an albino character might also have been due to radium treatment.[132] In short, Blakeslee believed that the radium treatment increased the proportion of mutants, but he remained open-minded as to whether it could cause new gene mutations—such as the albino mutant—and waited for evidence that such traits acted as Mendelian characters.[133]

By the following year, the two had begun to draft a paper, eventually to be published in the *Proceedings of the National Academy of Sciences*. Gager took the time in this paper to explain in detail the "exact nature of the stimulus," by which he meant "the different kinds of rays given off by radium," expecting his audience to be botanists not well acquainted with the properties of the various radioactive elements. (Gager wondered aloud to Blakeslee as he sent the draft "whether I have said more than is desirable" on this front.) Gager understood well, however, that his previously published monograph

> has had an exceedingly limited circulation, so that it is probably wholly unknown that any such work has been done to the majority of workers and institutions, or even if not unknown, the Memoir has not been seen. That is why I thought it might perhaps be a good plan to make a little fuller statement in the new paper than would otherwise be desirable.[134]

In fact, neither Blakeslee nor Cold Spring Harbor had copies of Gager's work on the shelf.[135]

Production of their paper became bogged down for years, both because of the inherent difficulties of the project and because Gager's other commitments kept him away from the radium work. By the dawn of 1927, Gager wrote to Blakeslee, "I have just glanced the paper through. Apparently, it will need very considerable revision, if not rewriting. Among other things, it might be desirable to mention the results of Mavor on the production of nondisjunction and crossing over . . . by X-rays, though reference to those papers should, I think, be very brief."[136] James Mavor's results, published in *Science* in 1922 under the title "The Production of Non-Disjunction by X-Rays," had indicated that the phenomena of nondisjunction, first identified in *Drosophila* by Calvin Bridges as the cause of various heritable traits, could be induced artificially.[137]

Fully aware that some of MacDougal's earlier successes in inducing mutations had come into question, Blakeslee and Gager were concerned that their own work not fall prey to the same criticisms. Though certain that they had discovered two radium-induced mutations, Blakeslee nonetheless advocated caution: "It seems to me that in view of the trouble which McDougall [*sic*] got into with his induction of mtations [*sic*] it behooves us to be extremely cautious, perhaps unnecessarily so, in claiming much for our preliminary experiment."[138]

The idea that new mutations (Mavor's cases of nondisjunction) could be induced with X-rays had found little favor at this time. As Blakeslee noted:

> You may have noted the critical attitude of [C. C.] Little and of Schull [*sic*] and when Mavor read his paper the critical attitude of Bridges toward the induction work with Xrays. Personally I believe that in one experimentthe [*sic*] treatment caused an increase in the proportion of mutant forms but for the albino I am open minded until we can get albinos *which act as Mendelian characters* from more treated plants.[139]

Blakeslee was not alone in seeing difficulties in attributing the results in fruit flies to the effects of irradiation. In one of his early unpublished manuscripts, A. H. Sturtevant recorded his own inconclusive experiments on the effects of radiation on *Drosophila funebris*, in which he exposed some 902 flies to radium and compared them with 2,348 control flies. As Sturtevant's notes reveal:

> Nothing like a mutation was obtained in the control. In the radium lot occurred a larger percentage of imperfectly developed wings, and two distinct types of wings which did not look as though due to any accident. One of these may have been inherited, but only in a very small proportion of the descendants.

> The other, if caused at all by radium, must have been the result of action upon the somatic cells alone. Therefore it was not to be expected that it would be transmitted, and there is evidence that it was not.[140]

Encouraging a bit of devil's advocacy, Blakeslee therefore recommended to Gager that

> we conjure up all the opposition which can be brought to bare [*sic*] against our belief that in expt 1 the high percentage of mutant forms was actually due to the radium treatment. Schull [*sic*] says the number of mutants was due possibly to the small number of seeds in the capsule.

Blakeslee was concerned about a number of issues, from the lack of an ideal control (which he took to be their own fault) to the effects of cold temperature in producing mutants (as was "known from other experiments"). And he raised another potential objection: "we shouldhave [*sic*] been able to control the production of mutants and we get 2 cases where the radium had no effect to only one inwhich [*sic*] it seemed to have an effect." All in all, he concluded, "I am wondering if we ought not to do a little more work with the radium and get more than an isolated capsule effected [*sic*] before we get out a formal paper."[141] A little over a month later, he wrote to Gager again:

> I think we want to be a little cautious about speaking of these mutants as resulting from the radium treatment. I hope that you will be able to get at this again soon and that we will be able to publish these two along with a considerable number of others in the *Journal of Heredity* and feel some confidence that the radium treatment would have or would not have an influence in their production.[142]

Radium-Induced Chromosome and Gene Mutations

Blakeslee and Gager rapidly "made a go" of their radium work. Their collaboration also ultimately led, among other things, to the discovery of various abnormalities besides visible changes in the chromosomes, such as "definite proportions of aborted pollen . . . abnormalities in pollen-tube growth . . . including non-germination of half the pollen grains, bursting of half the pollen tubes and bimodal curves of pollen-tube growth."[143] More significantly, after years of delay, their joint paper "Chromosome and Gene Mutations in *Datura* Following Exposure to Radium Rays" finally appeared in the *Proceedings of the National Academy of Sciences* in February 1927.[144] While they acknowledged that when they first presented their results in 1922 they had not yet "a sufficient body of data in regard to the mutability of untreated parents to permit us properly to evaluate the significance of the results," they now claimed to have accumulated "considerable" data regarding *both* "gene and chromosomal mutations in closely comparable normal material which can be handled as control to the treated material."[145] Finding great surprise in their success, they reported that they had discovered a variety of what they called "chromosomal mutants," mostly of the $2n + 1$ form—having a complete diploid set of chromosomes with an additional chromosome.

Although these types of chromosomal mutants had first been mentioned in the *Anatomical Record* as early as 1923, what was significant in Blakeslee and Gager's new publication was the sheer rate of production of these mutants.[146] While overall they had discovered some 73 "$2n + 1$" forms among 15,417 progeny in the controls (a rate of 0.47 percent), in one case they found "[a] percentage of 17.7 chromosomal mutants in over 100 offspring from a single capsule" of one of the treated plants—a rate they described as "enormously greater than we have ever obtained before or since." They concluded, "In view of the above figures, we believe the radium treatment was responsible for the increased proportion of chromosomal mutations, as also for the appearance of the compound chromosomal type Nubbin."[147]

While Mavor had noted in 1925 that it was still unclear just how X-rays and other "modifying agents" affected the germ cell, saying that "it is quite possible that . . . there may have occurred only a loss or an abnormal distribution of chromosomes,"[148] by 1927 Blakeslee and Gager had given such cases of chromosomal mutation clear and proper standing, and they cited Mavor's own production of nondisjunction as an explanation of their own 17.7 percent rate of chromosomal mutants attained from a single capsule. They indicated that the number of these chromosomal mutants, which were chiefly nondisjunctional forms, represented "a much higher percentage than ever obtained from untreated capsules."[149] Radium, they made clear, could induce chromosomal mutations and, as such, was an important first tool in the experimental control of heredity. After all, as Blakeslee had noted time and again, gaining control of chromosomal mutation was one key way not to have to wait for unpredictable mutations in genes. (Control was a key desideratum of the station's efforts at "experimental evolution"—what in a few years Blakeslee would begin to characterize as the work of a "genetics engineer" after he turned to using colchicine as a mutagen.)[150]

Even years before the publication of his 1927 *PNAS* paper with Gager, Blakeslee was convinced that he had found both chromosomal and gene mutants as a result of the radium experiments, as when he had discovered the *swollen* mutant (which came from an earlier generation of Gager's irradiated capsules).[151] As Blakeslee had written to Gager in 1923, "You may be interested to know that a new mutant which we had called *swollen* seems to be a gene mutant rather than a chromosomal mutant as we had at first believed it to be." The evidence at hand "strongly indicates" that this was the case, Blakeslee remarked, which meant that "we will have had two gene mutations following radium treatment and these are the only gene mutants we have ever identified in all our cultures."[152]

By 1927, in a new round of radium treatment, Blakeslee and Gager had discovered two more induced gene mutations among the offspring of eighteen irradiated individuals.[153] While the discovery of two new genes might be a small number in absolute terms, they argued, it "is very large if the proportion of gene mutations can be considered significant with so few individuals tested." (As they did not save seeds of the normal offspring from the treated capsule to test for heterozygosity in other new genes, however—they were looking only for chromosomal mutations—they were unable to work up these results on genic mutation further.)[154]

The end result of their collaboration was clear. There was no longer any doubt that radium could transmute species, and that it did so in at least two different ways: "It is our belief that for most, if not for all, of these three types of results"—the compound chromosomal type Nubbin, the chromosomal mutants, and the gene mutants—"the radium treatment may be held largely responsible," they concluded.[155] Blakeslee summarized the significant findings of their collaboration thus:

> In regard to our radiation work, I might say that we probably have done as extensive work as anyone so far as the chromosomal analysis is concerned. We have not put a great deal of it in long publications. Many of the results are summarized, however, in the *Anatomical Record* and in *Science* and also in the series of annual reports from our Department of Genetics, starting with the *Year Book No. 27* for the year 1927–28, which was issued December 13th, 1928.[156]

Blakeslee identified three main "chromosomal types" that they encountered in their radiation research on *Datura*: "prime types," which were the result of segmental interchange and other related forms of translocation; "compensating types" such as "Nubbin," with interchanged chromosomes; and the third and perhaps most interesting, "synthesized pure breeding types, which correspond to synthesized new 'species'" resulting from radiation treatment. Blakeslee was firmly convinced that these synthesized pure breeding types—the result of chromosomal and not merely genic mutation—were indeed new species in an evolutionary sense: they bred true, generation after generation, and they were recognized as new types by botanists (the traditional criteria for demarcating new species).[157] Blakeslee was also aware of other effects that were clearly the result of gene mutations, though their direct relevance to evolutionary processes—the emergence and maintenance of new species—was not as readily apparent. Though they included altered pollen tube growth, the non-germination of pollen, and the early or late abortion of pollen grains, "visible gene effects of radium treatment" were, at any rate "not yet . . . common in *Datura*."[158] Nevertheless, Blakeslee's research program proved tremendously successful: over the years, he found 541 gene mutations, 81 of which he was able to map to specific chromosomes.[159] Intriguingly, outside the world of drosophilists, it was not at all clear that gene mutations were in any way more fundamental to the nature of evolution and the origin of species than the chromosomal mutations Gager and Blakeslee were uncovering.

Blakeslee's emphasis on the significance of chromosomal mutation was long-standing. He had written to MacDougal as early as 1923, "I feel very strongly that a study of the chromosomal distribution is likely to explain irregularities in behavior in other plants than the *Datura* and that chromosomal changes in number have been responsible for evolution."[160] Blakeslee was also aware, however, and most especially at the Boston meeting in 1922, "that I have been obliged to caution people with whom I have talked about the *Datura* work from being over-enthusiastic and thinking the chromosome irregularities would explain phenomena which appeared to be explainable on ordinary factorial basis."[161]

Blakeslee's work was warmly and widely received, and many of his contemporaries were impressed with the scope and significance of his many discoveries. Lewis J. Stadler, who was working on maize, and who would later share some portion of credit for the subsequent discovery of X-ray-induced mutation, reviewed the "several investigations of the genetic effects of penetrating radiation in plants in progress" that had already attained "some positive results" and found the work of Gager and Blakeslee to be "especially noteworthy." He described their findings at length, noting in particular that a "single treated capsule of *Datura* included variants resulting from three diverse types of germinal variation, namely, change in the distribution of chromosomes, internal reorganization of chromosomes, and changes in individual genes."[162] As he wrote to a colleague in 1931, "there are undoubtedly diverse types of mutation."[163]

After Blakeslee's death, Edmund Sinnott, who had coauthored a paper with Blakeslee in 1922 entitled "Structural Changes Associated with Factor Mutations and with Chromosome Mutations in Datura," wrote that "botany has lost one of its notable leaders." Milislav Demerec likewise eulogized Blakeslee's decades of work that had "brought forth spectacular results," especially in "our understanding of polyploidy, polysomic types, segmental interchange . . . and chromosomal differences between geographically separated strains or different species of Datura." Another remarked that Blakeslee's "investigations on extra-chromosomal types and the role played by each chromosome in inheritance are genetical classics."[164]

Both chromosomal and gene mutations were important to Gager and Blakeslee, as they were to Stadler and many others, especially those interested in botanical cytogenetics.[165] Gager and Blakeslee's efforts successfully demonstrated the multifaceted nature of radiation-induced hereditary changes (mutations in several different levels and senses). But they were well aware of MacDougal's earlier reception, and they double-checked their results and qualified most of their claims in the few articles they published, including their *PNAS* paper, which was finally published in February 1927.

They paid for their caution with their future fame. By July 22, 1927, *Science* had published other results on the induction of mutations in *Drosophila* under the provocative title "Artificial Transmutation of the Gene."[166] The author was none other than Hermann J. Muller, who was one of the century's most remarkable and brilliant geneticists and would soon become one of its most famous. On July 31, Blakeslee wrote to Gager and mentioned only in passing what was later to be taken as the most momentous news in the history of mutation research:

> You may have seen the paper by Muller on the induction of an enormous number of gene mutations by the use of X rays that has come out in the meantime in *Science*. . . . You see the desirability of going ahead this summer in view of Mullers [*sic*]

> work. The results he has found make it more probable that what we have found may have been a true case of induction and it is fortunate that we published when we did.[167]

After all of Blakeslee's and Gager's agonizing over whether what they had were examples of radium-induced mutation, thinking of all possible counterarguments, acknowledging that they lacked proper quality controls in some cases, and worrying that the argument was not as strong as it could be, the publication of Muller's work suddenly cast their work in a new light—at least initially. Blakeslee, who had been reluctant to publish before additional research could be done—research that Gager could never quite seem to get around to doing—was now glad that they had published when they did. Blakeslee saw Muller's new research as a call to the renewal of his own—if not with Gager, then with Buchholz.[168]

Muller, like Blakeslee, saw powerful associations between radium and life in his studies of the fruit fly *Drosophila melanogaster*. But with Muller, the story of radium's association with life was bound further inward, toward the gene (rather than the chromosome) as containing within itself the secret of life, the mutation as the quantum of evolution, and ultimately the X-ray as the ideal means to explore the relationship between the two. While Blakeslee recalled that the gene had once been "considered an imaginary concept like the equator,"[169] Muller and his Nobel Prize–winning experiments on gene mutations would prove to be a further profound transmutation of the powerful associations between radium and life.

5

The Gene Irradiated

Mutation and Transmutation—the two key ~~words processes~~ stones of our rainbow bridges to power!

—Hermann J. Muller

Hermann J. Muller was the consummate experimentalist, spending his college years in Morgan's Columbia University fly lab with banana peels on the floor and flies buzzing around his head. He is perhaps best known as one of the major participants in Morgan's fly group from 1910 to 1916 and 1918 to 1920, for his analyses of crossing-over and of genetic "interference" and "coincidence," for his development of markers to trace the inheritance of chromosomes, and for his compelling quantitative inclinations—such as his estimates of the size and numbers of genes, and the frequency of mutation—in some of the greatest discoveries of early classical genetics.[1] Born in 1890 and growing up in the midst of the radium craze, Muller was also, however, nothing if not steeped in the widespread conceptual, discursive, and experimental associations of radium with life. Such sensibilities developed well before Muller joined Morgan's team; his earliest writings, from a very young age, are full of such references.[2] Indeed, important aspects of his later work on the gene and his lectures on the nature of life generally can be traced back to these associations.

Muller credited his 1908 reading of Robert H. Lock's seminal *Variation, Heredity, and Evolution* (1906)—along with the influence of Loeb and the cytologist E. B. Wilson—as a decisive influence in his undergraduate years, helping to convince him that genes were "the primary steps of evolution."[3] (At Wilson's suggestion, Muller read Lock's book over the course of a summer, finding time "while working as a bellhop, by ducking under a flight of stairs during his break time."[4]) Widely read by many preeminent and up-and-coming researchers and "judged at the time to be the most up-to-date work relating evolution and heredity," as Garland Allen has noted, Lock's text was also one of the earliest to characterize the similarities in the purported revolutions in thought taking place in biology and in physics, and at nearly the same time.[5] The book is as notable for its treatment of radium as it is for its up-to-date coverage of induced mutations (albeit mutations not yet the result of treatment by radium).

Intriguingly, an entry for "radium" in Lock's index referred the reader to the first page of the second chapter, "Evolution," where reference to radium is nowhere to be found. On subsequent pages of the chapter, however, Lock discussed the newly discovered phenomenon of radioactivity and, like many contemporaries, characterized its discovery as making it now possible "to establish a theory of the evolution of the chemical elements themselves."[6] Lock went even further to unambiguously place the atomic and the living side by side: "The change in our ideas regarding the method of hereditary transmission of characters, which has resulted from these experiments, has been aptly compared with the change brought about in men's understanding of the science of chemistry by Dalton's conception of the atom." Similarly, he noted elsewhere, "we may compare the difference which exists between deviations and stable forms, arising by fluctuating and by definite variation respectively, with the behaviour of the atoms of chemistry, as expressed in the account of their structure recently given by Professor Sir J. J. Thomson."[7] After describing MacDougal's experiments in the artificial induction of mutation, Lock even concluded that "in the course of another decade we may reasonably hope to find out something more about the natural and artificial production of mutations."[8]

Muller took his reading of Lock to heart. It established a clear path from MacDougal's earlier work with radium-induced mutations to Muller's own emerging program of research: "From then on," Muller wrote in a biographical note, "I concentrated more and more upon the as yet almost unworked problems of the gene and its mutation, and of their relation to evolution."[9] He did so by paying special attention to physics. Although Muller's interest in the new physics developed early—he sat in on a course on radioactivity during what he called his "formative years"[10]—he remained keen for the rest of his life to find ways that physical scientists and biologists could work together to uncover the secrets of life.

Indeed, in his subsequent teaching, Muller regularly emphasized the importance of the physical sciences, often devoting as much as several days of introductory lectures to basic physical concepts relevant to biology. When later in life he made an explicit comparison between the organic and the inorganic to sum up his decades of work—"The real, material existence of the gene is as thoroughly established a fact as that of the atom"[11]—he was drawing on a trope that was already well established among his contemporaries, and one he had first encountered during the height of the radium craze, in his reading of Lock.

After having been headhunted by Julian Huxley, Muller left the Columbia fly room in 1916 for a position at the Rice Institute in Texas. While at Rice, Muller gave a public lecture—one of several that he delivered to share the implications of the new genetics with the public—contending that evolution took place by the accumulation of mutations that, Muller's notes record, "happen at random and uncontrollably, like Ra[dium]."[12] In an early draft of an essay written in the same year, he made this association between radium and mutation even more explicit:

> I may digress here to draw attention to the curious similarity which exists between two of the main problems of physics and of biology. The central problem of biological evolution is the nature of *mutation*, but hitherto the occurrence of this has been wholly refractory & impossible to influence by artificial means, tho ~~such control~~ a control of it ~~would~~ might obviously place the process of evolution in our hands. Likewise, in physics, one of the most important problems is that of the *transmutation* of the elements, as illustrated especially by radium, but as yet this transmutation goes on quite ~~uncontrollably~~ unalterably and of its own accord, tho if a means were found of influencing it we might have inanimate matter practically at our disposal and would hold the key to unthinkable stores of ~~energy by which~~ concentrated energy that would render possible *any* achievement with inanimate [~~illegible~~] things. Mutation and Transmutation—the two key ~~words processes~~ stones of our rainbow bridges to power![13]

Muller was clearly supporting a kind of quasi-alchemical transmutational bridge between two powerful concepts in the realms of physics and of biology—an association clearly shared by many of his contemporaries. From early on, then, Muller was eager to associate the elements of heredity with radium and to transform the study of *transmission* genetics into the study of *transmutation* genetics. In 1918 he overtly compared the advances in genetics toward a "general theory of heredity" with those in radioactivity:

> ~~Just as the discovery of radium brought to prominence~~ Recent popular books on the subject have, it is true, laid emphasis on "Mendel's laws," but we have now gone beyond and beneath this law, and into an analysis of the constitution and behavior of the germ hereditary material, in much the same way as the physicists, not long after the discovery of radium, went beyond a mere description of the startling properties of this substance to make the most remarkable progress in their electron theory of matter.[14]

Indeed, from near the very beginning of his career in Morgan's laboratory, Muller time and again viewed genetics in ways that associated it with radium (and with radiation more generally). This association proved central in the larger association of the earlier "living atoms" discourse surrounding radium with the newly transformed "atoms of life" tradition surrounding the elements of heredity.

As early as 1916, Muller found it "evident" that the "problems of life," and, indeed, even the very "'secrets' of life itself will be within each single cell." Focusing at first on the constitution of the hereditary material at the chromosomal level, Muller drew on the storehouses of metaphor filled by his own exposure to the popular radium craze in his youth in giving his own accounts of the wonders of heredity. Patently drawing on Soddy's tropes of potent forces trapped in small elements and "unthinkable stores of concentrated energy," Muller declared in a lecture delivered at Cold Spring Harbor in 1921:

> Here, then, is a wonderful stuff—the most wonderful stuff in the world, barring none. This much nitroglycerine . . . could cause a sizeable explosion . . . this much radium . . . would have stores of subtle energy, could we but release it,—enough to drive many fleets across the Atlantic and back, or to blow up a city,—but just this much of our germ cell material can unfold into a whole

> generation of men, build countless cities, level forests, transform the earth, and speculate upon its own destiny, and keep on spreading and reaching out into the universe indefinitely, at an accumulating pace. . . . This substance, then, is perhaps worth studying if any substance is.[15]

Muller spoke of the hereditary material as—like radium—incalculably rare and valuable.[16] Equally striking are Muller's references—shared by others of his day—to the "elements" of the chromatin and their description in half-chemical, vaguely radioactive terms:

> Ordinarily it [the thread of destiny] lies scattered carelessly in the central portion of the nucleus—of the cell—but its activity does not depend upon the way its loops are coiled, but upon the nature of the chemical influences which emanate from each tiny particle of its length.[17]

The "emanations" from these threads of life were responsible for all the phenomena of life, just as radium's emanations were thought to accounts for its many effects.

Although means of affecting heredity by artificial methods remained beyond reach, Muller exclaimed:

> Could we but find a way to influence its changes we would have our grip right on the heart of the machine—for this is the stuff at the bottom of it all—it is by far *the most wonderful and potent stuff we know about*—which in even infinitesimal quantities has the machinery to make a man—stuff that can grow, stuff that can multiply, stuff that can make a brain which thinks—the stuff that evolution is made of! This thread is the thread of destiny and it's woven thru every fibre of your frame.[18]

In later years, although Muller continued to relate the power of radium to the power of the stuff of life, he increasingly shifted his focus from the chromosomes to the genes. In a lecture delivered in 1927–28, he asked:

> Must we always remain aloof from the challenge thrown to us by our own genes? If we could collect together all the human genes now in the world . . . they would form a mess of material about equal in bulk to a drop of water. *What powerful stuff.*

> *How incomparably more potent, and more important, than that much platinum, diamond, radium, nitroglycerine, or anything else you can think of.* This is a drop of matter that we cannot neglect.[19]

Muller moved seamlessly from describing the nuclear hereditary threads of the chromosomes in radioactive terms to describing genes as the quasi-radioactive atoms of heredity themselves.[20]

The tendency of some atoms to spontaneously transmute, while their neighbors did not, had led to the development of the concept of the half-life in radioactivity. Since at least 1923, Muller had seen a parallel in the tendency of some *genes* to spontaneously mutate while their neighbors did not.[21] Accordingly, he proposed that particular "mutable genes" had a new and peculiar sort of *genic* half-life. Genes were no longer merely "half-alive" in the sense of existing at the edge between life and nonlife (as he had repeatedly emphasized in his characterization of "the gene as the basis of life"). Rather, they were *also* able to be understood as having a half-life in the accepted *radiological* sense, in having a collective property that could be *measured*. As Muller noted after reporting the "mean life" for various mutable genes, "we are here using the physicists' index of stability, which seems most appropriate."[22] The realization that genes could, like atoms of radium, differ in their stability thus paralleled the realization that one could study and measure not just mutation rates, but the very mutability of genes in populations themselves.[23] As Muller turned to study the effects, first of temperature and then of other potential mutagens, on the mutation rate, he made this analogy between the gene and the self-splitting atom increasingly more overt. As he noted in 1927:

> As the changes in atoms, quantitatively analyzed, have given us pictures of their structure, so here this work appears to be resulting in a dissection of the genes in question into smaller elements—"gene-elements"—and the rearrangements of these elements may, it is hoped, be subjected to study even as before we studied the rearrangements of the genes as a whole within the chromosome.[24]

As Nathaniel Comfort has noted, new challenges to "the theory of the gene," involving suggestions that "the gene, like the atom before it, might be divisible," had begun to emerge in the 1920s. William Henry Eyster's "genomere hypothesis," holding that the gene was made of smaller elements and first proposed in 1924, was one such suggestion and one with which Muller flirted for a time.[25] This inward drive of the associations of radium and life, from the chromosomal to the genic and ultimately to the atomic level, aided further speculation.

Some genes might even, as one commentator theorized in the *American Naturalist*, "have affinity or 'valence' for certain other kinds of genes or their components . . . combining with them in definite proportions and varying grouping combinations having varying degrees of stability, just as we find among the electrons and protons in the atoms."[26] In fact, by 1922, Muller had already related the mutability of genes to the stability of radium atoms. In noting that some "specially mutable genes" had been recently discovered, he wrote that in some cases "the rate of change is found to be so rapid that at the end of a few decades half of the genes descended from those originally present would have become changed." The "ordinary genes" of *Drosophila*, by contrast, "would usually require at least a thousand years—probably very much more—before half of them became changed. This puts their stability about on a par with, if not much higher than, that of atoms of radium—to use a fairly familiar analogy."[27] Even H. G. Wells and Huxley, in their best-selling introductory biology textbook *Science of Life*, sought to make the link between the atom and the gene explicit.[28] Muller was thus an important part of a broader shift from an earlier pre-radium discourse of living centers and radiating forces to a new and powerful association of the smallest atoms of heredity uncovered to date with the atoms of radioactivity.[29]

Muller was convinced that the secret of life lay within the genes. In his landmark address of 1926, "The Gene as the Basis of Life," he called on his fellow investigators to explore these submicroscopic levels further to discover

> the *secret* of this immutable (but mutation-permitting) autocatalytic arrangement of gene parts. . . . We cannot leave forever inviolate in their recondite recesses those invisible small yet fundamental particles, the genes, for from these genes, strung as they are in myriad succession upon their tiny chains, there *radiate* continually those forces, far-reaching, orderly, but elusive, that make and unmake our living worlds.[30]

A year later he continued the refrain: "What gives the gene this peculiar ability [to mutate] we do not know, but we biologists would very much like to be able to look inside the gene and find out the cause of this, for we suspect strongly that in this ability of the gene to reproduce mutations lies the most essential secret of life itself, and of living matter as compared to lifeless."[31] By 1936 this trope of the genes as the source of mutation, and containing within them the secret of life, was firmly ensconced.[32]

Genes were no longer simply the fundamental particles of inheritance. Like radium before them, and in many more powerful ways than with chromosomes, genes had become for Muller a propitious site for the reworking and confluence of a host of other elements. Like radium, genes transmuted; like radium, genes sometimes differentially reproduced, giving rise to transmuted daughter genes that were similar to, but not quite the same as, themselves; like radium, genes existed near the border between the living and the nonliving. They both emanated, they both radiated, they both contained secrets within themselves. Genes reflected the elements of radium in all these ways, and all at once. Genes were the new radium.[33]

The idea of living atoms—now transmuted into atoms of life—continued to reverberate even as critiques of the idea of a single particulate basis of life began to emerge in the 1910s and 1920s. Joseph Needham, for example, cautioned in 1936 that "the comparison between the atom and the organism is in essence an analogy, and analogies are notoriously liable to snap under the weight placed upon them by uncautious thinkers."[34] But by the time of Julian Huxley's *Evolution: The Modern Synthesis* (1942), the link between genes and atomic theory was fully entrenched: "Genes are in many ways as unitary as atoms, although we cannot isolate single genes. They do not grade into each other: but they vary in their action in accordance with their mutual relations. In this they are again like atoms." When Huxley concluded that "the building-blocks of evolution, in the shape of mutations, are, to be sure, discrete quanta of change," this was an idea that came directly from Muller, and Muller's early exposure to radium.[35]

The Quantum of Evolution

Even in the midst of all this radioactively tinged discourse surrounding the atoms of heredity, things had remained relatively mundane in the experimental study of the fruit fly *Drosophila*. But, like Burke and MacDougal before him, Muller envisioned a new form of interdisciplinary endeavor "for those biological, especially cytological, experimenters who are equipped with a good knowledge of physics and chemistry."[36] What Muller had in mind was no mere parallel between the physicists' discovery of radium and the geneticists' analysis of the gene, but an active project of collaboration. If science were ever to successfully answer the question "What is life?," Muller held, it would be necessary to

> eventually arrive at the elementary principles which move the elementary particles. And what are these first principles and particles but the laws and molecules of physics and chemistry? The biologist tunnelling down, and the physical scientist tunnelling up, must finally meet. The rock between is very hard and irregularly grained, but it does show the marks of our instruments, and science is but in its infancy.[37]

By 1919 Muller had found his point of contact between biology and physics, between organic and inorganic evolution, between mutation and transmutation: the *quantum.* Used most notably by the physicist Max Planck in 1901 to describe the smallest possible shift in energy states of an orbiting electron, the quantum was adapted by Muller for his own genetic purposes. As he noted in 1919 in one of his data notebooks on temperature-induced mutation, "*The quantum of evol. is the single mutation.*"[38] In another data notebook (labeled "Chem") from 1921–22, Muller noted:

> It is not physics alone which has its quantum theory. ~~The quantum of biological evolution is the individual mutation.~~ . . . Biological evolution too has its quanta—these are the individual mutations. For biological evolution, like energy and like matter, is now found to be not indefinitely subdivisible, into an ever increasing number of vanishing infinitesimals, but to be the resultant of a vast, finite number of ~~irreducible~~ definite ~~steps~~ units which are the mutations.[39]

In this draft we can see Lock's use of "steps" turning into a thoroughly Mullerian understanding of mutational units. The de Vriesian mutation theory Lock had encapsulated for his readers was being rewritten, and a transmutational radioactive branch was now being grafted onto the study of mutation.

Muller quickly moved from provocative hints to outright declarations. He more fully explained the key role of the quantum and summarized his views in a lecture entitled "The Present Status of the Mutation Theory," delivered in 1935:

> The sudden and discrete character of gene changes, and the long interphases of stability, suggested practically from the first that they probably were chemical changes, so that indeed the genetic quantum theory might really be regarded as a not very remote expression in genetics of the quantum theory in physics itself.
>
> What the quantum theory is to modern physics, the mutation theory is to modern genetics; for mutations provide the fundamental units of change lying at the basis of all genetic differences, including even the grand differences between distant evolutionary divisions, even as quanta lie at the bottom of all greater differences of energy content. Moreover, as in the case of quantum changes, so too in the case of mutations, the changes are sudden and discrete, and are punctuated by interphases of stability, often of a very high order. We owe it to De Vries to have definitely set us on the path of this quantum theory of biology.[40]

He noted again, in 1942, that it was "amusing to play with the idea that the mutation, the quantum of evolution, should be indeterminate in occurrence because it depended upon the activation of an atom, involving the indeterminate physical process of the exchange of a physical quantum."[41] And as late as November 1945, Muller returned to this theme in his Pilgrim Trust Lecture at the Royal Society, adding a further radioactive twist for good measure:

> Thus the quanta of physics become the quanta of evolution . . . and the ultramicroscopic events, with all the possibilities born of their statistical randomness and even of their ulterior physical indeterminacy, become translated into macroscopic ones with a magnification vastly surpassing that of such an instrument as the Geiger counter.[42]

Muller was working at the edge of the metaphysics of metaphor: it was "chiefly these and related considerations" regarding the power of radium to bring about "individual quantum changes of atoms and molecules," he noted—its *transmutational* power—that led him "to the testing out of the possibility that ionizing radiation produces *mutation*" in genes.[43]

Radium was undoubtedly at the heart of Muller's turn to mutation: indeed, he had wondered whether "mutation is unique among biological processes in being itself outside the reach of modification or control,—that it occupies a position similar to that till recently characteristic of atomic transmutation in physical science, in being purely spontaneous, 'from within,' and not subject to influences commonly dealt with? Must it be beyond the range of scientific tools?"[44] Allying the discourse of radium to the "secret of life" that he felt lay in the genes, Muller made the same sort of productive associations between radium and life that had underlain the emergence of earlier work on the effects of radium on heredity. Those metaphorical and metaphysical resonances that conditioned the emerging possibilities of research for Burke, MacDougal, Gager, Blakeslee, and others also conditioned Muller's own. But while Burke had held that "molecular physics will doubtless yet become a branch of biology," Muller was attempting to make biology a branch of molecular physics.

: : :

Mutations were complicated things, and one of the main problems facing Muller was none other than *Oenothera*. De Vries's favorite organism was a plant "in which the processes of genetics are maximally intricate," Muller noted, "and which therefore provided an unsuitable basis for the elucidation of the underlying principles that apply to the primary processes of mutation." Muller wanted to understand the fundamentals of mutation—not what he took to be the idiosyncratic processes of evolution in one particular organism. As he later explicitly put it, he wanted a *logical* rather than a *chronological* understanding of mutation (an almost complete inversion of Gager's earlier concerns with the historicity of life, and closer to Burke's physicalist thinking).[45]

And yet Muller had clearly been conversant with a tradition of older, de Vriesian and phenotype-level descriptions of mutation. He clearly knew that by the 1910s, a good number of investigators had already succeeded in artificially producing mutations by means of chemical agents (like MacDougal) and by means of radiation (like Gager, Morgan, and Blakeslee). Not only was Muller clearly aware of these two decades of research, but before beginning a series of new mutagenic experiments with Edgar Altenburg in the late 1910s and early 1920s—experiments that led up his remarkable success in 1926—he had conducted an extensive literature review and routinely cited these earlier works.[46]

Muller forthrightly acknowledged that "various investigations were undertaken between 1905 and 1925, to find out whether hereditary abnormalities could be induced by irradiation."[47] He credited the "first sound evidence for some sort of effect of irradiation upon the hereditary material to C. R. Bardeen," and mentioned other early attempts, but thought all the same that none of these had "include[d] observation on the transmission of the abnormalities to later generations."[48] As early as 1915, Muller had compiled a list of nearly thirty different "classes of environmental conditions which might affect the gene plasm," which included both substances (items 1–16) and conditions (items 17–29) such as magnetic fields, X-rays and radium rays (as one category), and ultraviolet light. And by the mid-1920s, Muller was already fully aware of reports that radium and X-rays had produced heritable effects on chromosomes: his own notes show direct reference to Blakeslee's "acolytes" (the name given to the varieties of chromosomal mutants of *Datura*); he made a big check mark in his notes next to MacDougal's *Popular Science Monthly* article on mutation; and he was aware of MacDougal's chemical injections into plant ovaries. It is even apparent that he took fairly significant notes on Gager's 1908 monograph (the reference for which is written in relatively large script) and recorded the different forms Gager had found.[49] Muller's surviving notes show at least three references to Gager's monograph and at least two to Gager's article in *Science*.[50] He was also aware of Loeb and Bancroft's attempts, and he took notes at least twice on Gager and Blakeslee's "Induction of Gene and Chromosome Mutations in *Datura* by Exposure to Radium Rays," noting "inbred 12 generation" and "→great increase in no. of mutants, 33⅓% for ovary cell." (Muller summarized much of his literature review later.[51])

In short, Muller took notes on the results of just about every exposure of organisms to radiation that he could find, both physiological *and* hereditary.[52] He later summarized previous attempts by other researchers to cause what he termed "visible mutations" by techniques that included, as he put it, "all sorts of maltreatment":

> Animals and plants have been drugged, poisoned, intoxicated, etherized, illuminated, kept in darkness, half-smothered, painted inside and out, whirled round and round, shaken violently, vaccinated, mutilated, educated and treated with everything except affection, from generation to generation.[53]

Muller even acknowledged the priority of Gager and Blakeslee in having successfully induced mutants by radium treatment in *Datura*.[54] And, in fact, Muller had *himself* produced similar hereditary changes even long before the 1927 results that would make him world famous. On pages dated April 27, 1923, he had written, "Therefore we have a very clear case of an external agent modifying the mechanism of inheritance in such a way that a permanent effect is produced on the germ cells."[55] Muller even proposed experiments on plants, such as *Mirabilis jalapa*, that sounded not terribly dissimilar to those undertaken by Gager and Blakeslee, asking if mutability varied in "different regions of the plant" or "at different times in the history of the plant," whether "the amount of mutation in [the] branches could be affected by external influence," and whether mutations could "be influenced by subjecting to treatment young stages such as seedlings, seeds and pollen grains?"[56] (Muller also later acknowledged that "it has been known for many years that chromosomes exposed to X-rays or radium exhibit a tendency to fragment, and also to clump or adhere together."[57])

Despite being clearly aware of and conversant with the literature of early radiation-heredity research, and despite having consistently acknowledged and summarized such experiments and in some cases even conducted them himself, Muller from early on rejected these efforts as necessarily inconclusive.[58] He even proclaimed it odd that so little was known about the induction of mutation. As he wrote with his longtime friend and colleague Edgar Altenburg in 1920:

> Strange as it may seem, practically no work has yet been done on this fundamental problem [of mutation], in spite of six decades of lively argument concerning evolution, and many volumes of imposing literature concerning "variation." Despite the material existence of these weighty tomes, our knowledge of the rate and conditions of change in the factors of heredity—the changes that really make evolution—is almost a blank.[59]

Muller's radioactive musings on the nature of the gene and the secret of life had brought him to the question of the quantum of evolution. Disappointed to find that others had not approached the question with the level of precision he had in mind, Muller embarked on a new set of experiments using radium (and later, X-rays) that would bring the secret of life to light.

"A Beastly Tedious Routine"

Given that the Morgan school had examined some 20 million fruit flies over the course of several years, but had identified only a few hundred visible mutations, Muller concluded that the techniques thus far employed were insufficient to demonstrate convincingly that any hereditary abnormalities encountered had in fact been *induced* by the radium treatment. Though Blakeslee and Gager may have been successful in producing mutations of a sort, Muller thought, and perhaps even gene mutations, it was almost impossible to establish probabilistically whether they were in fact *produced* by the radium treatment without first having an understanding of the natural *rate* of mutation (by which Muller generally meant the rate of mutation of *genes*).[60] It was therefore clear to him that special methods and techniques would need to be developed in order to properly assess the rate of genic mutation. Only then would he be able to study how frequency might be measurably influenced by changes in environmental conditions and mutagenic agents. As Muller complained to Huxley in 1919:

> There's absolutely no work on genetic variation as influenced by environmental conditions which has been done in such a way as to be interpretable under the factorial theory. Almost the same can be said with regard to the study of the normal rate and incidence of variation, as genetic have seldom been distinguished from "phenotypic" changes. Where they have, observations *e.g.* in selection experiments have usually been limited to so small a number of characters and *parents* (each individual must be a parent, to be studied genetically for variation) that only an inappreciable number of genetic variations (mutations) have come to light. We simply know that mutation occurs, & occurs "rarely," whatever that means, tho on its rate & mode of incidence depend evolution.[61]

Three years later Muller wrote in the *American Naturalist*, "In the past, a mutation was considered a windfall, and the expression 'mutation frequency' would have seemed a contradiction in terms. To attempt to study it would have seemed as absurd as to study the conditions affecting the distribution of dollar bills on the sidewalk. You were simply fortunate if you found one."[62] He clarified this criticism in later years:

> The task of actually counting mutations in ordinary cultures, in order to compare their frequencies of occurrence there with those under other, contrasting conditions would have seemed almost like that of counting needles in haystacks, to compare their frequencies, or like making graphs to show the rates of occurrence of gold pieces on streets of different types. The objective of most genetic counts, therefore, was the determination of the frequencies of crossing over, of chromosome reassortment,

> of non-disjunction, and of other phenomena connected with the transmission rather than with the origination of gene variations. Meanwhile, mutations were of course recorded as they arose, but the numbers in which they were found were insignificant in any one given experiment, and still meant little even when many different experiments were totaled, because of the fact that the conditions for their detection varied so greatly from one experiment to another—owing to personal equation, to the differing characters being considered, to the different methods of breeding used, the varying external conditions, diverse stocks, etc.[63]

In other words, Muller held that the things that Blakeslee and Gager were discovering in their cultures—things like "the frequencies of crossing over, of chromosome reassortment, of non-disjunction, and of other phenomena"—were not properly considered mutations.[64] While it was undoubtedly "easier to see and deal with" certain visible types of mutants, as Gager and Blakeslee had done, Muller held that there were plenty of "physicochemical changes" that were "probably as frequent as changes in visible structures," but which geneticists, with their morphological training, were not yet prepared to identify.[65] Gene mutations, in Muller's view, were simply too exceedingly rare to have been adequately detected and analyzed in previous work on mutation frequency. No matter the mutagen employed, purported mutants found in such circumstances could not be clearly shown to be the result of induction by radiation. One ought not to rely only on the "conspicuous and definite morphologic abnormalities" heretofore held to be characteristic of mutation, as this would be to miss "the great majority both of recessive [and] perhaps even more so of dominant gene mutations." In short, studying visible mutants or chromosomal changes was simply not good enough; the proper study of mutation would mean the study of innermost *genes*.[66] Seemingly dismissive of MacDougal's earlier calculations and his reasoning, Muller—like Blakeslee before him—would no longer be content to wait for mutations to occur spontaneously. He immediately began to construct special stocks of flies and experimental protocols in the hope of finally ascertaining the nature of mutation proper and gaining some sort of preliminary control over evolution.

In 1917 Muller had published an article entitled "An Oenothera-Like Case in Drosophila," in which he laid out a "remarkable genetic situation, wherein both types of homozygotes are prevented from appearing by the action of lethal factors in opposite chromosomes," a condition he termed one of "balanced lethal factors."[67] Muller argued from this case for the successful "simulation" of mutation, claiming that his results "parallel[ed]" reports of *Oenothera*'s behavior. (De Vries was unimpressed, and he remarked to Loeb that he "much regrets that Morgan has had much nonsense published by one of his pupils, as you may have read in Muller's article in *Genetics*."[68] Indeed, not everyone took kindly to Muller's attempts to recast the mutation theory as a theory of gene-level events.) But by the mid-1920s Muller had already found that such lethal mutations were in fact some five to ten times *more* numerous than "visible" mutations—all the more reason, he felt, that they should be recognized and counted in any proper study of the frequency of mutation. Early on, therefore, Muller and Altenburg concluded that it would be easiest to measure the effects of mutagens on mutation frequency by measuring the presence of lethal factors in the X chromosomes. Muller explained his reasoning thus:

> Previous attempts to demonstrate the production of mutations in *Drosophila*—and, in my opinion, in all other organisms—have hitherto failed, but I believe that this is because, by the methods previously used, only a small fraction of the mutations that might have occurred could ever have been detected. Morgan tried radium in 1909, Altenburg tried high pressure and low pressure of oxygen and centrifuging in 1914, Duncan tried alcohol and hybridization, and Morgan, not long after, tried ether, ultraviolet light, and simple mechanical agitation; then Altenburg tried X-rays ["and again radium" added]—all without apparent effect on mutation. But I was convinced on *a priori* grounds, that lethal genes—which kill the flies and so cannot be detected except by means of special genetic tests—are the most frequent types of mutation, and that therefore any effect that had been produced had not been seen.[69]

But hadn't radium time and again—in the hands of MacDougal, Morgan, Gager, and Blakeslee—proved to be a powerful mutagen? MacDougal's findings were later contested; Gager, like Burke, claimed to have produced only the *ancestor* of a new species; and Morgan later retracted his story of radium-induced mutants. Blakeslee may have appeared to have succeeded—and indeed, as with MacDougal and Morgan, many of Blakeslee's contemporaries were quite impressed with the scope and significance of his discoveries. But in Muller's view, although some earlier experimental work may have produced *apparent* mutations, "when they are analyzed genetically, it becomes evident that the relatively few *inheritable* variations satisfactorily demonstrated may have been present in concealed form in the stocks long before treatment, while on the other hand any new gene mutations that might have been produced would probably have escaped detection by the methods used." Since most gene mutations are recessive and "cannot be detected until at least the third generation after they have arisen, and then only if brother has been bred by sister in the second generation," Muller discounted the work of earlier researchers as insufficient for not having carried out a "satisfactory" genetic analysis based on *genes*.[70] Such purported mutations were probably the result of transmission genetics, while the secrets of transmutation genetics remained undiscovered. For Muller, as for Morgan before him, doubting previous accounts of radium's experimental mutagenicity—while nevertheless using radium and the techniques of probabilistic statistics developed in concert with the elucidation of radioactive phenomena—seemed the most promising way to gain clarity on the real nature of mutation.

Just as Morgan discounted Loeb's flies as having been contaminated, Muller discounted MacDougal's, Morgan's, and Blakeslee's findings either as the result of undetected genetic contamination or as merely serendipitous and with no statistically significant connection to the mutagenic techniques employed. While some had discovered irradiation effects that included the "loss or irregular distribution of whole chromosomes," Muller also held that these effects were not yet "permanent changes in [the] internal physical or chemical composition of individual chromosomes or genes." As he concluded, "It still remained to be determined whether irradiation could produce gene mutations or any other inheritable variations of such kinds *as usually distinguish individuals or races in nature from one another*."[71]

Muller was thus interested in a comprehensive accounting of mutations and their frequency—including those that might be microscopically small or recessive in nature, and not just the readily noticed morphological phenotypic mutations that might happen to attract an observer's attention. Muller was offering an implicit critique of what we might term the "mutant gaze"—the ability of skilled researchers to readily identify novel mutants in their pedigree cultures or experiments.[72] While deciding whether a particular wing formation was a proper instance of mutation, as opposed to a mere variation, may have left too much room for debate, the existence of a dead male fruit fly—which careful stock-keeping and genetic analysis could prove had inherited a lethal factor on its only X chromosome—seemed fairly unambiguous. As Muller told Huxley:

> Altenburg and I are now "making a stab" at this problem in Drosophila, by studying the frequency of origin of lethal factors thru specially prepared crosses. This offers more hope of offering a frequency of mutation [*sic*] great enough to be handleable, as lethals we find to occur far more often than all other *detectable* mutations put together. . . . Meanwhile, it entails a beastly tedious routine, which is, however, I hope, worth the candle. It'll take several years to do the first things now projected, but the method is capable of being used to study the effects of all sorts of influences on real genetic change.[73]

Muller first began gathering data on mutation frequency in "untreated material" in 1918 and 1919. His preliminary results, in collaboration with Edgar Altenburg, seemed to indicate an effect of temperature on the mutation rate. After gathering his findings together in a paper called "Mutation," which he presented at the International Congress of Eugenics in September 1921,[74] Muller continued his "quantitative study of mutation" for many more years in repeated collaborations with Altenburg on the effects of temperature and other mutagenic agents.

Throughout the early 1920s Muller was hard at work improving his techniques and developing new stocks (especially of balanced lethals, which he started in New York City and at Woods Hole in 1919, and continued working with after moving to Texas).[75] It was a long process, he reported in 1920: "I shall continue the mutation work I spoke of—but I do not expect really to begin to keep results till next winter. Altenburg & I will work together at Woods Hole."[76] And two years later: "I am getting my mutation methods gradually improved & should soon be in a position to work for some real results. It's hard going, but I'm determined to get there."[77] As he later recalled, "I continued for eight years to carry forward the quantitative study of mutation. It was very laborious and involved a number of time-consuming breakdowns in elaborate experiments which, chain-wise, were lost with the breaking of a single link."[78]

While proponents of the mutation theory and later investigators such as Blakeslee had collected and selectively propagated mutants in order to prove species continuity—an important argument for species types rested on the ability of organisms within a species to function evolutionary as units—Muller rejected this approach as simplistic:

> Accumulating large numbers of abnormal or inferior individuals by selective propagation of one or two of the treated lines—as has been done in some cases—adds nothing to the significance of the results. At best . . . these genetically unrefined methods would be quite insensitive to mutations occurring at anything like ordinary frequency, or to such differences in mutation rate as have already been found in the analytical experiments on mutation frequency.[79]

Previous ways of demarcating mutants were no longer acceptable to Muller. They not only confused phenotypic variation with genotypic variation, but also missed large numbers of changes that could, theoretically at least, be captured by his new genetic methods. In a remarkable move, Muller recast MacDougal's original concept of "mutation frequency" to create *by definition* a class of otherwise unobservable mutation that could by identified only by techniques that Muller had invented. This statistical redefinition of "mutation" would have consequences not only for later appreciation of the value and internal consistency of earlier botanical studies of mutation that used radium, but even for recollection of their very existence.

In wiping the slate clean, Muller wanted to "employ organisms in which the genetic composition can be controlled and analyzed, and to use genetic methods that are sensitive enough to disclose mutations occurring in the control as well as in the treated individuals" in order to calculate what he called "relatively slight variations in mutation frequency, caused by the special treatments."[80] In other words, he wanted to find all the hereditary effects of a radium needle on *all* the genes of a fly, not only the ones with the most visible effects.

In proposing his concept of mutation in 1923, Muller had added "fourteen points" on the nature of the gene. Strikingly, the first of these began by analogizing the stability of the gene to the stability of an atom of radium—describing both as having probabilities of "decay" measured in the span of "a few thousand years."[81] And, like investigators before him, Muller accordingly turned to radium as his mutagen of choice.

How could a mutagen cause a mutation at one locus and not an immediately adjacent one, or at the same locus on the homologous chromosome? How was it possible, Muller asked, for two genes to exist side by side and yet for only one and not the other to be affected by radiation? "Why do not the same general conditions, acting on the same materials, produce everywhere the same results?" he asked. "If events in this sphere are apparently so indeterministic, is it any wonder that we could not in previous trials, by the application of definite conditions, produce definite mutational results?"[82] Muller concluded not that the mutations were "causeless," or as he put it, "expressions of 'the natural cussedness of things,' or of the devil," but that they were due to "the results of individual ultramicroscopic accidents."[83] This needlelike precision of the incidence of mutation signified to Muller that "the mutation is due to an event of such minute proportions, so circumscribed, that it strikes only a single one of two nearby, similar loci in the same nucleus."[84] Stored in what were themselves widely referred to at this time as "needles," minuscule amounts of radium thus dovetailed nicely with Muller's interest in finding a surgical "tool" that could "dissect" the chromosomes.[85]

Over a dozen years after Gager's earliest work, radium needles remained among the foremost ways to "stimulate" the appearance of novel mutations. By the autumn of 1924, Muller had suggested to Altenburg that the β-rays from radium could produce lethal mutations while the stronger—and fly-sterilizing—γ-rays could be screened out. He also suggested experiments "to see . . . how strong a treatment with naked Ra the [*Drosophila*] eggs would stand," and he recommended the book *Radium, X Rays & the Living Cell* (1915) to Altenburg. He also suggested that a carefully bred strain of flies might serve well for testing: "Maybe *ClB* would do for your radium tests! Don't forget to see if you can get the emanation."[86] By October, Muller had even asked Altenburg to "bring along the Radium" during a Thanksgiving visit.[87] The duo were still hard at work in November 1924, when Muller wrote to Altenburg, "It might be well to defer your visit till the week end before Xmas . . . if by that time you have the Ra. & have experimented enough with the dosage to know how to apply it best to our flies here."[88] As Elof Axel Carlson, one of Muller's students, recounts, the radium was

> kept in a glass vial embedded snugly in a lead container. [Muller] shipped this package to Woods Hole, Massachusetts, where he and Altenburg hoped to try out the special stocks which Muller had constructed for the quantitative detection of radiation induced lethal mutations. Somewhere between Austin, with its hot June weather, and their destination in the milder climate of Cape Cod, the thermal expansion of the lead container caused the glass vial to shatter and their precious store of radium disappeared in the lining of the container.[89]

It seems that as a result of this entirely contingent circumstance, Muller turned instead to using X-rays.[90]

From Radium to X-Rays

Discovered only a year apart, radium and X-rays had been linked since the earliest days; in fact, Henri Becquerel had been looking for Wilhelm Roentgen's newly discovered rays when he discovered the radioactivity of uranium. Both "Becquerel rays" and "Roentgen rays" were known to have photographic effects, to ionize gases and make them conductors of electricity, to cause fog in moist air, and to be subject to neither reflection nor refraction. At the dawn of the century, however, the two sets of rays were seen as distinct phenomena, and it remained unclear for some time whether the two forms of radiation could be equated.[91] Part of the reason for the uncertainty was that X-rays were rather difficult to work with in those early years, in medicine as in physics. As John Perceval Lord noted in 1910, "X-rays, valuable as they are, have their strict limitations" in medicine: "The apparatus is of necessity cumbersome, and the bulb is large and requires considerable skill to manipulate effectively."[92] Radium, by contrast, could do for medicine in a few days what it took "long weeks at least" for X-rays to accomplish. Its strength, Lord noted, "was to all appearance greater than that of the X-rays. Here, then, was a convenient method of applying invisible rays, similar to Röntgen rays."[93]

Although second-born, radium thus had a number of apparent advantages in the early years. The *Times Literary Supplement* of London reported in 1903 that small, easy-to-use tubes of radium gave off a "beautiful, constant and uniform supply of rays," giving radium an "obvious advantage over the x rays, which are capricious and often harmful."[94] Compact, powerful, and easily packaged, radium was also thought to be a much more reliable source of ionizing rays.[95] Its rays continued to outperform X-rays in medicine as late as 1912, despite ongoing uncertainty about which of the three types of rays it emitted—α, β, and γ—was responsible for the observed effects. Gamma rays were the most penetrating and the most like X-rays. But Lord had wondered even in 1910: "Was the curative effect due solely to these gamma rays? If so, then the X-rays could be made to cure all cases, which experience showed that they could not do."[96] Soddy held by 1920 that the γ-rays from radium were by far "the most penetrating rays known and are really X-rays, but far more penetrating than any that can be artificially produced" with contemporary X-ray technology.[97]

Radium was not without its disadvantages, however. As noted in chapter 3, no precise measure of the strength of radioactivity had yet been devised. Radium also remained extraordinarily rare and valuable, costing roughly $40,000 per gram according to one estimate, and there were only about three pounds total in existence "in all the hospitals and sanitariums in the world."[98] Physics departments and hospitals proved to be the most common sources, but gaining access required connections, and the papers of Gager and Blakeslee are speckled with requests—written either by themselves or by others to them—attempting to locate the elusive radium for biological experimentation.[99]

While radium forever remained both scarce and expensive, X-rays were cheap and readily available, especially after the invention of the improved "Coolidge tube" by William David Coolidge at General Electric in 1913. But while the International Radium Standard was in place as early as 1910, it was only in the 1920s that financial incentives were sufficient to ensure the standardization of X-ray technology.[100] Physicists would soon come to characterize both types of rays as simply different intensities of the same species of radiation and, eventually, would fully equate the γ-rays of radium with X-rays.[101] Some medical and biological investigators, however, remained unconvinced that the effects of the two types of rays were identical. (One medical investigator even felt that powerful X-ray tubes could never serve as replacements for radium: "No indeed! That's emphatic!"[102])

By the mid-1920s X-ray instrumentation had improved to the point that it became competitive as a source of radiation. Muller, who was well aware of their physical similarity, was treating X-rays and γ-rays from radium as essentially interchangeable for most purposes by that time. In a draft piece entitled "The Effect of X-Rays and Radium on the Germ Plasm," he asked, "In how far does the biologist possess, in X-rays and the rays from radioactive substances, penetrating needles of exquisite fineness, capable of use in the analysis, or even in the control, of the germ plasm?"[103]

The shift from radium to X-rays is visible even in the abbreviations that Muller chose to use to describe his work. In letters from Altenburg to Muller, "radium" is at first spelled out, but soon after in some letters becomes abbreviated as "Ra." and "rad.," which at several points stand indeterminately (without further context) for either radium or radiation. As X-rays and the γ-rays from radium were increasingly identified as interchangeable forms of ionizing radiation, the word "radiated" also came to be replaced in their correspondence with "irradiated." In physics, in biology, in the popular realm, and even in the very words researchers used to describe their techniques and their findings to one another, radium and X-rays were becoming equated.

Muller was thus part of a much larger general shift toward equating radium with X-rays. Indeed, by the early 1920s, a variety of texts had already begun linking the two radiant phenomena in terms of their general effects: "radiotherapy," a concept involving both X-rays and explicit mention of "radium-therapy," increasingly gave way in popular and medical discourse to a more general "radiation therapy." Both X-rays and radium had also been commonly used in studies of physiological and hereditary effects on organisms from early on. Experiments conducted by A. H. Blaauw and W. van Heyningen in 1925 finally demonstrated that the rays of radium that had effects on organisms were almost exclusively γ-rays—in an apparent confirmation of Gager's earlier 1908 claims that "the γ-rays behave as X rays" in their physiological effect, and that "if X-rays have such a property, then we should theoretically expect radium rays to possess it as well."[104] Moreover, many researchers noted that both X-rays and radium rays seemed to have an intimate association with the phenomena of reproduction. Charles Packard, of the Institute for Cancer Research at Columbia University, had noted as early as 1916 that "chromatin, when in its most condensed stage at the time of the metaphase, is particularly susceptible to radium radiations."[105] In an echo of the transmutational power of radium, another researcher claimed in 1926 that after exposure, "cells with injured chromosomes do not divide, but slowly undergo *disintegration*."[106] Even as late as 1927, Morgan had noted in his *Experimental Embryology* that "X-rays and radium emanations appear to be almost specific agents for sperm cells; at least they are more quickly injured than the other cells of the animal."[107] Such findings furthered the equation of X-rays with radium in the study of heredity.

Muller's use of X-rays on his fruit flies thus depended not only on significant improvements in X-ray technology, but also on the changing understanding of the physical and ultimately the biological equivalence of X-rays with the γ-rays of radium. A set of intercalated factors—discursive, technological, financial, and experimental—had thus linked up by the mid-1920s to make radium and X-rays seem increasingly interchangeable to both physicists and biologists as well as to those, like Muller, who would bridge the two realms. X-rays were not just another mutagen in the arsenal; they came to be seen as producing radiation virtually indistinguishable from that of radium.[108] Two years after having lost his radium source, Muller reported that he had, at last,

> turned to the use of X-rays in this quantitative study, since, on physical grounds, they must certainly penetrate to the genes and should there cause alterations powerful enough to affect them. . . . Utilizing X-rays as a tool for obtaining chromosome rearrangements and mutations, the chromosome has, as it were, been dissected.[109]

As Muller also recalled:

> If we had the ability to change individual genes we should have, in effect, a scalpel or an injecting needle of ultramicroscopic nicety, wherewith to conduct the most refined kind of vivisection or biochemical experiments. . . . Hence the question of the production of changes in the hereditary material by means of roentgen or radium rays becomes all the more urgent.[110]

X-rays thus came to replace radium needles as the tool of choice for ultramicroscopic dissection.

Structural, experimental, discursive, proximate, and contingent factors all contributed to Muller's shift from radium to X-rays. Muller himself would rapidly come to equate the two forms of radiation:

> Among the agents of an ultramicroscopically random character that can strike willy nilly through living things causing drastic atomic changes here and passing everything by unaltered there—not a ten thousandth of a millimeter away, there stand preeminently the X- or γ-ray and its accomplice, the speeding electron.[111]

The "trail of havoc" left behind by these quanta of energy was nothing other, Muller proposed, than the origination of mutation, the quantum of evolution. (The *New York Times Magazine* would later describe Muller as having "decided to adopt the methods of the atomic physicists . . . [if] X-rays can tear an electron from an atom and thus convert it into so very excited a bit of matter that it glows, what if they were turned on the genes?"[112])

It was with this equation of X-rays with the γ-rays of radium in mind that Muller began his last round of mutagenic experiments with

X-rays in 1926.[113] Now cheaper, standardized, safer, and easier to use in some circumstances than radium needles, X-rays brought Muller preliminary success in breaking the "double X chromosome," in affecting crossing-over in the middle of the third chromosome, and in causing nondisjunction in the fourth chromosome.[114] As he reported in May 1927:

> My work has gone as never before. If anyone else had told me he had gotten the results that I have, I'd have thought him crazy. Keep it dark, as I don't want to let it out or publish till I have a Big Bertha-full, but I've finally got the trick of *producing* mutations, and it's really very simple—by means of X-rays—and astounding that no one has been able to prove it before. I can make better than *every other* functional germ cell a mutant, and the mutations are of all kinds, including many found before, & also new ones, & chromosome abnormalities too. It is opening up a rich field of "gene physiology."[115]

Having now established a novel genetic technique to detect mutations and study the rate of mutation in what he considered a "quantitatively exact way," Muller began his famous "X-Ray Mutation Experiment" on November 3, 1926, in his Texas laboratory (figs. 12 and 13). With the help of a local radiologist and using dental equipment, he experimented with doses of 12, 24, 36, and 48 minutes and found that he was able to induce over a hundred mutations. He conducted the second run of his experiments in the spring and summer of 1927. Muller concluded that his technique was sensitive enough that he could not only determine whether a chromosome had undergone some kind of rearrangement or breakage, but also detect mutations in particular positions on particular chromosomes, and even "which parent a mutation came from." While chromosomal changes were exceedingly abundant, Muller was more concerned with the production of point mutations—that is, "true" (gene) mutation—which he found occurred 150 times more often in his treated flies than in his controls. During one busy night of experimentation, Muller is said to have literally jumped out of his chair, shouting out the window to the botanist in the downstairs lab, "I got another!" after each new find, and to have proceeded to sleep in the laboratory for the next day and night. "By the time the experiment was over less than two months later," one historian has noted, "his notebook was crammed with entries on over 100 mutations. This was half the number that had been found in the entire 16-year history of

FIGURE 12. H. J. Muller at work at the University of Texas, where he conducted his work on the "artificial transmutation of the gene." (Prints and Photograph Collection, di_03919, The Dolph Briscoe Center for American History, The University of Texas at Austin.)

work on *Drosophila*."[116] As Carlson has noted, "From 1910 to 1926, during the entire history of the *Drosophila* work, some 200 mutations had been found by all *Drosophila* workers combined. In less than two months [Muller] had found half that number by himself, and all from the astounding mutagenic effect of X rays."[117]

Muller was able to track some heritable mutations for 180 generations, many more than Blakeslee could ever have done. This analysis over time meant that rates of mutation in successive generations could be compared with those resulting from the original treatment. Using such large numbers of cultures meant that Muller could quantitatively establish a mutation frequency for his experimental flies and compare it with that of the controls to see "to what extent the usual mutation frequency had been increased or diminished by the treatment." Inspired by Rutherford's induction of artificial transmutation in physical elements, Muller had used X-rays—"essentially the identical tool that Rutherford had used"—to do more effectively in biology what Rutherford had already done in physics.[118] Rutherford had discovered the atomic nucleus in 1911, and he announced the first successful artificial transmutation of the atom eight years later, in 1919; Muller had likewise taken eight years. It was with all deliberate intent that Muller called his discovery "the artificial transmutation of the gene."[119]

The Promised Land of Mutations

After repeated rounds of zapping fruit flies with X-rays, Muller found he could "obtain some hundreds of mutations—as many as have pre-

FIGURE 13. H. J. Muller (*standing*) in his laboratory. (Courtesy of Helen Muller.)

viously been found, without treatment in the whole twenty-year history of fly study, in which dozens of investigators have examined some 25 million flies." Muller called his remarkable result nothing less than "the Promised Land of Mutations," where "all types of mutations, large and small, ugly and beautiful, burst upon the gaze":

> Flies with bulging eyes or with flat or dented eyes; flies with white, purple, yellow or brown eyes or no eyes at all; flies with curly hair, with ruffled hair, with parted hair, with fine and with coarse hair, and bald flies; flies with swollen antennae, or extra antennae, or legs in place of antennae; flies with broad wings, with narrow wings, with upturned wings, with downturned wings, with outstretched wings, with truncated wings, with split wings, with spotted wings, with bloated wings and with virtually no wings at all. Big flies and little ones, dark ones and light ones, active and sluggish ones, fertile and sterile ones, long-lived and short-lived ones. Flies that preferred to stay on the ground, flies that did not care about the light, flies with a mixture of sex characters, flies that were especially sensitive to warm weather. They were a motley throng. What had happened? The roots of life—the genes—had indeed been struck, and had yielded.[120]

If radium was the new Adam, giving sight to the blind and resurrecting the dead, Muller was the new Moses, leading geneticists who had been wandering far too long in the desert of confusion over the rainbow bridge of power to the promised land of (trans)mutations. Muller himself described his results as "startling and unequivocal": not only was the increase in the rate of mutation on the order of 15,000 percent—a 150-fold increase, with forty mutations found on one Sunday afternoon alone—but these new mutations were *real*: they didn't revert, they had their own norms, and they were, as Muller put it, "real, new variants . . . permanent in so far as the word permanent may be applied legitimately to living things."[121] They even followed Mendelian laws of crossing and showed chromosomal inheritance as existing varieties. While some of Muller's mutations were identical with older recognized ones, some were new, and some were found to breed true even beyond the hundredth generation. (Muller here used the same criteria of "breeding true" that breeders and botanists routinely used, even as he disputed the ability of such techniques to adequately determine the true frequency of gene mutations.) These newly discovered gene mutations, however, were in addition to a slew of chromosomal rearrangements and reattachments that cytologists were able to confirm, and they were determined despite the difficulties of a "tenfold difference in control rates."[122]

Within weeks of the publication of Muller's July 1927 paper, word of his triumphal "artificial transmutation of the gene" had spread far and wide, and the scientific and popular media alike leapt all over the discovery. By any measure, Muller's success was remarkable. In 1931 the *Botanical Gazette* announced that Muller's discovery had "opened a new epoch in the history of genetical research. Never has a new technique proved more rapidly prolific of instructive results."[123] While a generation earlier it was *radium* that was held to be responsible for the production of life in the laboratory, with Muller's pathbreaking work on the mutagenic effects of X-rays, journalists now began to report that "X-rays can remake living things," or as one headline put it even more strikingly, "X-Rays Form New Life."[124] Wells and Huxley held that Muller's accomplishment meant that "it may even prove possible to operate directly on the germ-plasm. . . . Man has conquered the hardness of steel; he cuts and twists it and builds with it as he pleases; today he is learning a new art, with living protoplasm as his medium."[125] Huxley also felt that it opened "a very interesting new chapter in evolutionary theory . . . since up to that time mutations had appeared to be a spontaneous process, which, like the transformation of the radio-active elements, was not to be controlled by artificial means."[126] As Muller himself put it, "A new field—the physiology of gene mutation—seems to now open for quantitative study."[127]

Muller's discovery of the effect of X-rays on mutation was indeed astounding, and it was almost something to be expected when he received the Nobel Prize in Physiology or Medicine in 1946 for his pathbreaking work. Within a relatively short time, however, a peculiar thing happened: Muller's discovery began to take on a life all its own, and what he had *actually* done (dramatically increasing the frequency of genic mutation by exposing carefully constructed stocks of fruit flies to radium, and later to X-rays) came to be eclipsed by ever more grandiose claims about the nature and significance of his achievement: for example, that he had been the first ever to induce any kind of mutation with radiation. In a retrospective account written in 1946, *Nature* reported that "in any treatise on modern genetics, H. J. Muller figures as the man who discovered the action of X-rays on chromosomes and genes," a rather significant amplification. *Nature* went on to state that while "a number of workers" had undertaken "similar attempts," these efforts had been "without clear success." It was "not the bare discovery of the metagenic [*sic*] action of X-rays which revolutionized genetics, but the manner in which Muller's previous work had paved the way for the use of it, and the genius with which he exploited it."[128]

These subtleties were lost, however, in other accounts that condensed what was an important advance in a well-worked field into an epochal moment of revolutionary change. Just after Muller's groundbreaking paper, fellow radiation investigators Frank Blair Hanson and Florence M. Heys claimed that "Muller's recent discovery that X-rays produce gene mutations in fruit flies is one of the most notable events in the field of pure biology in this century," calling it "a milestone in biological progress."[129] Almost a year later, reporting on an experiment they conducted at Muller's urging, they intensified their claim, noting that before Muller, almost nothing was known "about agents causing variability. . . . And this situation remained practically unchanged from 1859, the date of the publication of the 'Origin of Species,' until the summer of 1927."[130] Muller's had evidently become the most monumental achievement in biology since Darwin.

Within a few years Otto Mohr had proclaimed that "H. J. Muller's work inaugurates a new epoch," while the *Journal of Heredity*, in the largest issue (64 pages) it had yet produced, claimed that what had been earlier successes in altering chromosomes with radium were *extensions* of Muller's work: "So startling a discovery has resulted in intense activity in this new field of research. This has amply verified Muller's results, and has brought to light new ways to use X-rays in facilitating genetic research, through altering and rearranging chromosomes as well as genes."[131] (Elsewhere in the same issue, other researchers asked, "How can nature's film be speeded up? All attempts failed to accomplish this in anything like an adequate manner until Muller applied X-rays to the fruit fly."[132])

Praise of Muller's success was widespread.[133] "This is the biggest thing since de Vries, maybe bigger than de Vries," wrote one colleague from Johns Hopkins. "I should like to listen in at Woods Hole discussions of your thunderbolt!"[134] In their textbook *Principles of Genetics*, Edmund Sinnott and L. C. Dunn noted that Muller "offered the first convincing proof in 1927 that the frequency of mutation could be increased several hundred per cent above the normal rate by X-ray treatment."[135] Even Tracy Sonneborn, whose work on genetic systems above the level of the gene would stand somewhat apart from Muller's intense focus on the gene, recounted in a radio address, "When I first got acquainted with genetics as a student in the middle 1920's, the first really exciting event was Muller's X-ray induction of mutations in 1927."[136] Alexander Hollaender, director of the biology division at Oak

Ridge National Laboratory and whose name is synonymous with the radiation biology Muller helped to inaugurate, chose, like many others, to *begin* his history of the field with Muller's research on *Drosophila* and Stadler's related research on maize and barley. By 1953 Muller's achievement—recast and glorified—had taken on canonical status: Curt Stern began his review declaring, "Ever since *Muller*, in 1927. . . ."[137]

The sentiment that Muller had opened a new era in the science of genetics and the study of the effects of radiation on it has endured. Even today, Muller's legendary "artificial transmutation of the gene" by the use of X-rays in 1927 continues to be represented in biology textbooks and journals as the origin of the modern study of induced mutation.[138] It also receives central attention in many histories of genetics. The very first line in Carlson's biography of Muller reads, "H. J. Muller is best known as the recipient of a Nobel Prize for being the first to show that mutations can occur in living organisms after exposure to X rays." Carlson later notes that Muller "had startled his colleagues by announcing that *for the first time in the history of life on earth*, the hereditary material of living organisms—in this case, the genes of fruit flies—had been artificially mutated." And he opined that Muller was nothing less than a "genius" who had developed an astonishing variety of genetic tools that "elevated *Drosophila* to an almost exclusive role in the study of the gene concept."[139] Even Jan Sapp, in his 2003 history of biology primer *Genesis: The Evolution of Biology*, concludes:

> The artificial production of mutations gave genetics a new lease on life. It provided genetics with one of its most important analytic devices and one of its most important sources of material for investigation. Geneticists no longer had to wait for mutations to arise spontaneously. The study of mutations was soon extended from X-rays to gamma rays, beta rays, cathode rays, and ultraviolet light.[140]

Truth be told, however, in 1927 it had already been over twenty years since geneticists last believed they needed to wait for a mutation to arise spontaneously.

As the singling out of Muller's experiment time and again as an originary moment in the history of genetics has created a dominant collective mythology, the *actual* details of what Muller accomplished in relation to his contemporaries have been lost in the general afterglow of the Promised Land. As we have seen, the early decades of the twentieth century *before* Muller's research were replete with efforts by MacDougal, Gager, Morgan, Loeb, Bancroft, Blakeslee, and others to produce mutants using various forms of radiation, altered temperature states, and chemical mutagens.[141] "Ever since Muller," and as a result of his redefinition of mutation, this rich history of radiation mutation research (the subject of chaps. 3 and 4) came to be completely obscured, to the point that Muller's work has struck journalists, biographers, historians, and fellow scientists as the work of unprecedented and unparalleled genius. Some have even gone so far as to invert the historical trajectory of this research by placing Muller at the *beginning* of such work. That such statements are so easy to find, and that they recur so frequently over the decades, says much about the rapid rise and persistence of adulation surrounding Muller and his 1927 work even years later (fig. 14). It also points to the difficulty of summarizing in an easy sound bite just what Muller did and how it differed from earlier researches.

This near-excision of decades of earlier work from the historical record—no matter the technical achievements of the discovery underlying it—is one of the more remarkable phenomena in the history of genetics and, indeed, in the history of biology more generally. How did this curious apparent rewriting of history come to pass? One key component

FIGURE 14. A humorous take on Muller's artificial transmutation of the gene, drawn in 1939, and one of the many ways in which his 1927 work was canonized. Various induced mutants of *Drosophila* sit at the "X-Ray Bar," with Muller as the bartender, tapping his famous X-ray tube. Flies say (*l–r*): "This is awkward!! My wings are shrinking." "Help! I'm going black all over. What will the wife say?" "Gee my wings are curling." "Hey, H. J. Anything queer about my eyes or is it only the drink?" Drinks on offer are "Gene & Lime," "Deletion Knock-Out," "Transo-Cocktail," and "Inversion Head over Heels." (Courtesy of Helen Muller.)

in Muller's meteoric rise to fame was undoubtedly his reconsideration of how best to measure the induction of mutation and the concomitant shift in the meaning of "mutation" that this caused—a shift fundamentally rooted in the inward-bound trajectory of the ongoing powerful associations of radium with life.

The Birth of a Myth

"The bandwagon Muller started was soon out of hand," Carlson has noted. In the immediate wake of Muller's achievement, X-rays rapidly became a widespread technique for inducing mutation. In 1928 Muller received word that "X rays are all the rage at Woods Hole."[142] Most genetics laboratories across the world soon "had X-ray machines and were buzzing with dwarfed, twisted, crippled or half-alive fruit flies whose ancestors had been X-rayed."[143] Intense interest in the use of X-rays continued throughout the 1930s. By 1940 James Neel complained to Curt Stern, "Sometimes wonder whether most of the people here [at Cold Spring Harbor] ever thought of a gene as anything except a some-thing that you push around with X-rays. That's an overstatement, but I think you see what I mean."[144] By 1949 one biologist was still expressing his delight: "It's fantastic. This youngster has multiplied the capabilities of my laboratory for me by a hundred and fifty times."[145] X-rays had been known for nearly three decades, and the fruit fly had been a model organism for geneticists for twenty years already. Muller's notes raise the obvious question: "If so easy, why not done before?"[146]

The simple answer is that although many mutations were indeed known, many more flies needed to be studied, and special methods and controls needed to be used, to fully establish the nature and frequency of mutation at the genic level to Muller's satisfaction. But there is an even more complicated answer that is worth exploring as well. Muller had undoubtedly produced a remarkable number of new mutations, but just what had he done that was truly novel? Many of the things that Muller was praised for initiating did not begin with him. Muller wanted to calculate the mutation frequency, but this was a concept and an effort first undertaken by MacDougal decades earlier. Similarly, Muller didn't want to have to "wait for mutations" to occur—but then again, neither did Blakeslee. (Blakeslee's techniques, based on chromosomal dynamics, were even repeatedly described as a form of "evolution to order."[147]) Nor, of course, was Muller the first to induce mutations—chromosomal *or* genic. Blakeslee and Gager had clearly done so, and Muller had initially acknowledged as much (even if he soon came to present their work as confirming his own). And even if his effort had been original, as T. H. Goodspeed noted in a special issue of the *Journal of Heredity* in 1929, it was an obvious development and hardly worthy of breathless claims of "genius": a series of experiments related to Muller's "have been in progress for many years," the critic noted, and "it seems possible that if certain at least of these physiological investigations had been conducted with even a minimum of genetic imagination the discoveries of the last two years with which Muller's name will always be associated, might have been anticipated."[148]

Indeed, Muller himself acknowledged as much in 1932: "At first it seemed a quite understandable result—even one to have been anticipated that high-energy radiation should change the gene. For the atoms of genes cannot be immune from activation either by the X-ray quanta themselves or by the fast-moving electrons released by the latter." With so many other experiments seeming to confirm that the quanta of radiation translated directly into the quanta of evolution—induced mutations—Muller concluded that "it was tempting, especially for the physicist, to believe in this relatively simple explanation of the induced mutations."[149] But as tempting as it was for Muller as well, he needed to *prove* it.

Muller's work was unquestionably brilliant—his contemporaries commented in particular on the close attention he paid to excluding potential confounding causes, the nature of the special stocks he constructed, and the measurements of the mutation rate he undertook in the species he treated. Dunn saw the greatest significance of Muller's work in his contribution to later approaches: "The experimental study of mutation induction" not only "clearly changed the course of the growth of genetics in the direction of physical and chemical methods and ideas," Dunn noted, but "in this sense was a cause of the development which culminated in present views of the basis of the gene in DNA structure."[150] And indeed, Muller's focus on the use of radiation to study the nature and structure of the gene eventually became so all-encompassing that it became a kind of background radiation all its own. The radiation geneticist Curt Stern even had to remind himself later that "not all radiation injuries are genic"![151] Such a reminder would have been inconceivable to Blakeslee and his contemporaries.

In so spectacularly showing that he could induce otherwise undetectable mutations in extraordinary abundance and with tremendous ease, Muller's success therefore contributed to a major but heretofore unanalyzed shift in the meaning and referent of "mutation," itself an unexpected further transmutation in the long half-life of the powerful associations between radium and life. But while his quantitative study of the frequency of lethal gene mutations easily removed any problem of subjectivity in recognizing mutations—what Muller, by analogy with astronomical observation techniques, had referred to as the "personal equation"—his technique discounted a wide variety of *other* sorts of non-genic variations caused by radiation. Chromosomal aberrations, which were admittedly much more frequent than gene mutations, were not counted as "mutations," at least in part because Muller was not able to measure them in the same way as he could X-ray-induced lethals. Other semi-lethal or non-lethal variations were also left uncounted, even though some of these still clearly theoretically fit under Muller's idea of true mutation as genic mutation.[152] (He discounted their numbers as relatively insignificant.)

While Blakeslee himself agreed in 1921 "that we shall ultimately confine the term to mutations of genes," Blakeslee and Gager had their own pioneering methods for determining whether a genic mutation had taken place in their plants and for relating it to the other levels and meanings of "mutation" current at the time. Blakeslee remained willing to preserve the phenotypic and karyotypic attribution of "mutation" as well as the genic. Muller, however, took as proof only evidence from his carefully designed *ClB* stocks of *Drosophila* and carefully thought-out crosses. (Somatic mutations, of interest to Blakeslee, would also not be picked up by Muller's approach, with its carefully constructed sexually reproducing stocks with lethal genes.)[153] Indeed, any approach that kept a supragenic understanding of mutation in addition to a genic understanding, Muller felt, must necessarily use such a small sample size that it was inherently suspect and could give no proper indication of gene mutation frequency. Even a 100 percent or 500 percent effect on the rate of mutation, he held, was "likely to escape detection." One would be lucky to find one mutant in 50,000 flies, he said, and yet experiments that involved only 10,000 were usually considered respectable. And yet Blakeslee and Gager were clearly convinced from the samples they worked with—Blakeslee grew 70,000 plants in one summer—that they had found mutations, and that they had found them in unparalleled abundance. So were many of their contemporaries.[154] Indeed, to miss many of these *other* kinds of mutation would have been, for many investigators at the time, to miss an important category of heritable variation with evolutionary significance. Both Muller and Blakeslee wanted to put mutation on a firm factorial basis, but the gene was not the only evolutionarily significant unit of analysis for Blakeslee and other investigators of induced mutation.

For all its later force, Muller's definition of "mutation" as a chemical change within the gene differed from many earlier investigators' working definitions. His reinterpretation of the term to fit the questions he thought most important and the experimental techniques he had derived was thus a signal moment in recasting the history of mutation research. What "mutation" had *actually* meant to other biologists, even other geneticists, mattered less than the "logical" (rather than merely "chronological" or historical) determination of what a mutation was, which was what could be studied by Muller's carefully developed techniques. Muller's shift in the meaning of mutation thus ultimately contributed to the "forgetting" of decades of early work with radium to investigate and to induce mutation.

That the ascendance of Muller's meaning of mutation is a fundamentally historical phenomenon, tied as much to this discursive shift as to his experimental work, is evident in the fact that Muller's experiments were not initially seen to be quite as singular as they would later become. In the years just following Muller's experiments, many still sang the praises of other investigators who continued producing mutants by means of temperature, radium, UV rays, and X-rays. As late as 1931, Gager continued to report that he and Blakeslee "were the first to induce mutation in plants by exposing the germ-cells of the Jimson Weed (*Datura*) to radium rays," and in the same year, even *Science* described Gager and Blakeslee's work in the 1920s as producing "probably for the first time inheritable changes in living organisms by exposing their living cells to penetrating radiation. It is epoch-making work and is a field worthy of most careful study."[155] Claims like these sound distinctly similar to those that were later made more exclusively for Muller. This complexity in the historical record was obscured as time went by and as the focus of attention shifted increasingly to the gene as, quite literally, the target for mutation by X-rays.

: : :

Muller's technique—what he thought was necessary to prove that a gene mutation had in fact been artificially produced by radiation—was complicated. His definition of mutation, however, was elegant, simple, and infused with the discourse of radium. Blakeslee's understanding of mutation was more broadly rooted in ongoing discussions in the 1910s and 1920s over what "mutation" could mean—its relationship to a species' evolutionary history and to questions of hybridity, as well as its multiple levels of signification.[156] Accordingly, Blakeslee grappled constantly with how to interpret radiation-induced mutation genotypically, phenotypically, and at in-between levels such as the chromosomes. He was also aware of how the choice of model organism affected the definition of mutation preferred. There was not only a phenotypic-karyotypic-genotypic dimension in the shifting meaning of mutation, but also a zoological-botanical dimension. As one commentator noted, polyploidy was "widespread in plants, but rare and unimportant in animals."[157] Working with *Drosophila*, Muller discounted polyploidy and emphasized instead that mutation was fundamentally the same, and genic in nature, in all species of *Drosophila* and "in all other organisms."[158] But as Blakeslee's supporters noted, "No other organism, not even *Drosophila*, has involved so many aspects of biology in its genetic investigation as has *Datura*."[159]

From a genic point of view, Muller was no doubt justified in arguing that a new method was called for to better determine the frequency of gene mutations. But even as he conducted such research, he knew as well as his contemporaries that he was far from being the first to use radiation to induce mutations, and that there had been years of earlier (and sometimes famed) work along just these lines. But as mutation became a gene-level phenomenon, and as induced gene mutation began to be seen as detectable and provable only through Muller's carefully devised system, Muller *emerged* as the first to actually succeed in inducing mutations by means of radiation—the first to transmute the gene artificially. Indeed, the further clarification of the unusual chromosomal behavior of *Oenothera* in the 1920s dovetailed with Muller's legacy to confirm the redefinition of mutation and to "refute" the de Vriesian understanding, rather than to modify it in the ways that Blakeslee sought to do. Muller even co-opted de Vries's own explanations of hereditary behavior in his new account of balanced lethals, calling de Vries's account "identical with" and an "important confirmation" of his own work.[160] With the elucidation of the mysteries of *Oenothera*, and as the mutation theory came under increasing attack, some proponents of a multivalenced view of mutation found it increasingly difficult to speak clearly about mutations on several different levels without seeming fuzzy-headed latecomers to seemingly discarded de Vriesian views.

In Muller's wake, even Blakeslee seemed to find it increasingly difficult to defend his own novel reworkings of de Vriesianism. Although in 1923 Blakeslee had readily referred to "gene mutants" and "chromosomal mutants," after 1927 he rapidly began to refrain from using the word "mutation" for anything but a gene mutation, talking instead about "changes" or "off types" and speaking of changes in chromosomes increasingly as "chromosomal aberrations" rather than "chromosomal mutations," abandoning his own new coinage of "chromosomations." In 1936 Blakeslee said that "the chief credit for this discovery properly belongs to Muller but like most discoveries there is a considerable history back of it."[161] Still later, in 1940, in a recollection published in *Science*, Blakeslee discounted his own early successes in inducing mutation, saying that "others independently had already started radiation experiments for the same purpose."[162] In another recollection from 1949, Blakeslee returned his own name to his account of events, but once again gave Muller top billing: "It was nearly twenty-five years before Muller *and some others of us* began the successful use of radiation in the induction of gene and chromosome mutations without having known of the earlier suggestion of de Vries."[163] And by 1951 he would even go so far as to say that while a plant treated by Gager with radium emanation was "probably the first induced chromosomal mutation," it was Muller's later "brilliant work with Drosophila" that "gave ample evidence that genes may be caused to mutate by radiation treatment." (Notably, Blakeslee doesn't say Muller was the *first* to cause genetic mutation by radiation, but he does give Muller credit for having *amply* demonstrated it.)[164] Blakeslee's role in the emergence of a genic view of mutation was thus a curious one. He, too, seems to have played a role in the constriction of the meaning of mutation from any novel heritable variation at any level to a fundamental change in a gene.

Moreover, Gager and Blakeslee's increasing caution in declaring that they had encountered radium-induced mutations in *Datura* contrasted with Muller's straightforward claims, his own remarkable experimental successes, and his production of an untold number of new mutants in a single batch—with the sheer and striking productivity of his novel approach. The clarity offered by Muller's constricted concept of genic mutation and the ability to calculate its occurrence with such precision were clearly enticing. With the passage of time, and with the fading of a strongly expressed alternative, Muller's approach came to seem not only more and more dominant, but more and more obvious. In a matter of a few short years, then, Blakeslee's and others' work would be entirely forgotten, and the connection between radiation and induced mutation would be attributed almost wholly to Muller and to his use of X-rays.

Gager and Blakeslee's caution had another important consequence: their delays in reporting their findings in the late 1920s meant that some of their work was published well after Muller's—leading many to naturally conclude that their work had been inspired by Muller's discovery.[165] The continued use of radium in experimental heredity was ultimately reported as *following* the prior long-standing use of X-rays. Muller contributed to this confusion by writing the *Datura* work of Blakeslee, Buchholz, and others into his narrative not as the precursors they actually were, or even as contemporary endeavors, but as confirmations among dicotyledonous plants of his own work in flies.[166] Even Stadler, whose work on the X-ray induction of mutations in corn had been independently conducted—a fact that Muller acknowledged—reported that his experiments, "which were independent of and coincident with those of Muller, though by no means so comprehensive and thorough, confirm Muller's discovery of the power of X-rays to induce mutation and show its application to plants. They show also that mutations may be induced similarly by radium treatment."[167] Gager and Blakeslee were thus far from alone in continuing to grapple with the proper meanings of mutation. Stadler, too, struggled with finding adequate and appropriate terminology for ever more complicated hereditary phenomena and for parsing the precise differences between changes in genes and changes in chromosomes.[168]

Several factors thus contributed to the birth of this myth of Muller having been the first to experimentally induce mutation, a story that was widely trumpeted by the press and which in time even Muller himself came to believe. These factors included an unstable sense of just what constituted a mutant or a mutation and how best to find one or induce one; differences between botanists and drosophilists in their methods, techniques, and questions; Gager and Blakeslee's caution in continuing to use the language of "chromosomal mutation"; the putative "explanation" of the seemingly aberrant behavior of *Oenothera*; empirical and discursive parallels between the γ-rays of radium and X-rays among physicists and biologists; improvements in X-ray technology; the remarkable precision and fecundity of Muller's approach; and the ways in which Muller's approach could be adopted and adapted by teams of researchers everywhere, as it soon was. All these factors and more contributed to the elision of an entire realm of mutation discourse, even among botanical investigators who had long held pluralistic concepts of heredity and evolution and who had offered the only counternarratives to the emerging dominance of a genic of view of mutation largely based on work in *Drosophila*.

The discovery that X-rays had been used to induce mutations well before Muller came as a surprise even to those who should have known better. Upon stumbling across de Vries's 1904 inaugural address at Cold Spring Harbor, Blakeslee wrote to Morgan in 1935 with surprise, saying, "I find also there that there was quite a little history back of the

X-ray work."[169] The power of the rewritten history surrounding Muller's discovery was such that de Vries's inspirational address, which had motivated Morgan's and MacDougal's initial work with radium, was all but unknown to a later generation—and even to Blakeslee himself.

: : :

In 1920 Muller had noted that "despite the material existence of . . . weighty tomes, our knowledge of the rate and conditions of change in the factors of heredity—the changes that really make evolution—is almost a blank."[170] So what did Muller think his original contribution was? Or, to ask the question once more in his own words, "If so easy, why not done before?" In some respects, the shifting meaning of mutation meant that the fact that some form of experimentally induced mutation *had* been done before, and successfully, could be forgotten. But perhaps the most intriguing answer to these questions is found in a draft from Muller's papers consisting of a plain sheet that is labeled with item number 5: "Muller found that x-rays"—and the rest of the page is left tantalizingly blank.[171] Just what had Muller discovered, indeed?

This tantalizing blank symbolizes the aporia between what Muller's experiments *actually* did and what he and others increasingly took them to have done, a distinction that is possible only because of a dramatic shift inward in the meaning of mutation (toward the gene), which was itself intricately intertwined with Muller's radium-infused understanding of heredity.

Hardly deliberate, the blankness of this page—in the midst of an article claiming non-mutation mutations of the germ line—is even more suggestive of the historical transmutations of Muller's own claims over the years: his earliest claims to have used X-rays to produce inherited permanent modifications of the germ cells (such as nondisjunction), which he refused to label mutations (such modifications were similar to those induced by countless others); his intermediate claims that X-rays could undo what he, for a time, described as *mutations* in the *chromosome*; and his final, later claims that it was not until his 1927 work that X-rays were successful in producing mutations. The shifting meaning of mutation not only affected the understanding (both scientific and popular) of Muller's accomplishment, but is also visible between the lines of Muller's own struggle to characterize the exact nature of his accomplishment, writing against his own words and the established history of the field.[172] In short, in reviewing the history of earlier experiments with radium, Muller ended up—like Morgan—casting the mutagenicity of radium into doubt in order to better understand the nature of mutation on his own terms.

In sum, then, Muller's landmark 1927 announcement marked the beginning of the end of a multilevel, nuanced understanding of mutation and its replacement with a fundamentally genic theory of mutation based on the spectacularly precise and detailed new methods that ultimately earned him the Nobel Prize. Moreover, these new methods, which highlighted the centrality of the gene in calculating the frequency of mutation, came from Muller's own reworking of the powerful associations between radium and life. The birth of the myth of Muller's 1927 achievement thus brings to light the ways in which the ongoing associations of radium with life in the newly emerging science of genetics continued to transmute as that science developed.

The gene, therefore, was not the only thing transmuted in Muller's experiments and in his shift from radium to X-rays. Indeed, as the powerful associations between radium and life continued to permit novel experimental systems like Muller's, the increasing experimentalization and technologization of these associations—including the expedient replacement of radium with X-rays, the deployment of sophisticated statistical reasoning in experimental design, and the tremendous capacity of a powerful metaphysics of metaphor—contributed to still further transmutations and ultimately to some of the initial signs of decay in the once all-powerful associations themselves.

Radium vs. X-Rays and the "Deradiation Response"

Muller had clearly proved the effectiveness of X-rays as a mutagen, and his production of mutant fruit flies had everything to do with the powerful associations of radium and life he was already well versed in. Radium and X-rays were deployed side by side by many researchers in the field in the late 1920s, and the extra-large issue of *Journal of Heredity* published following Muller's discovery reported on a variety of research that made use of both X-rays and radium.[173]

Muller, for one, had never seemed terribly discriminating as to the source of his radiations. As he reported to the *Scientific Monthly* in 1929, "Radium rays, like X-rays, produce mutations, because they too, being short-wave-length high-frequency electromagnetic waves of great energy content, release high-speed electrons."[174] Muller's former colleague in the fly room at Columbia, Alfred H. Sturtevant, likewise agreed that "radium has the same effect" as X-rays, saying that "from a physical point of view, this may be taken as the same method, rather than as a distinct one. Most work on mutation since 1926 has made use of this technique."[175] If radium was expensive and hard to get, however, the X-ray machines were often frustratingly cantankerous. As Theodosius Dobzhansky complained in 1934, "My X-rays experiments went on rocks twice. It seems that this Institute does not possess an X-ray tube capable of affecting the flies. One can comment on this only in unprintable language, so I make no comments."[176] Pragmatic considerations of cost, ease of use, or available instrumentation thus largely dictated which of the two sources was employed in a given experiment. For Muller, too, convenience was the name of the game in choosing between radium and X-rays, and he continued to use radium in his experimental work as late as 1943.[177] "X-rays and radium can be used artificially to produce mutations in abundance," he noted.[178]

Both the rays from radium and X-rays could be *effective* in inducing mutations, but whether they could be *equated* in terms of their other biological effects remained a question of lively interest for decades. After all, X-rays were held to produce ionization directly through their β-radiation (the passage of an electron), while γ-rays were held to do so through secondarily induced β-radiation produced when γ-rays passed through matter. It was not to be taken for granted that these different mechanisms produced entirely identical effects. Indeed, Muller's interest in just what sorts of *biological* differences might result from the use of radium as opposed to X-rays was a major reason for his continued use of radium even after his turn to X-rays.

According to Muller, while most radiation work in genetics up to this point had focused on the uses of radiation for "increasing our understanding of genetics," the time had come to begin to explore "the biological action of radiation per se" in order to answer the question of the possible differential effects of the different kinds of radiation in biology. Following up on his X-ray work, Muller therefore drew up a proposal in 1933 to further study the "genetic effects of radium," explicitly comparing the biological effects of β-, γ-, and X-rays. He was interested in "determining definitely what, if any, is the difference between the genetic action of radium radiation and roentgen radiation." Muller had already long ago reduced the meaning of mutation to the level of the gene; the biological effects of radiation, however, existed at varying levels beyond the gene. He proposed, therefore, among other things, a study involving a "quantitative comparison of the relative frequency of gene mutations and of chromosome aberrations of different kinds."[179]

Despite the established physical equivalence of X-rays with the γ-rays from radium, and their functional equivalence for inducing mutations, Muller left open the possibility that there could be real disparities in biological effects between the two forms of radiation and that these disparities might exist for eminently comprehensible reasons.[180] On the assumption that chromosomal aberrations were "caused in a less direct way than the gene mutations, since it has been found that the occurrence of one chromosome break is somehow connected with that of another," Muller speculated whether "changes in the quality of the radiation could affect the former without affecting the latter, and thus change the ratio between the two types." Differing doses could also provide insight.[181]

But others still saw problems in any easy equation of ionizing power with biological effect. These problems lay at the heart of the condition of possibility for the field of radiation genetics, which would later try to use the effects of ionization to gain knowledge about the size and structure of the gene without any idea of the mechanisms by which such ionization might bring about the observed changes.[182]

However, it wasn't until Charles Packard undertook to find "a method by which the biological action of Roentgen and radium radiations can be determined qualitatively" that X-rays and the γ-rays from radium were formally equated in terms of their biological effects in at least one important way beyond inducing mutation.[183] Having irradiated more than 30,000 *Drosophila* eggs, Packard found that "the death rate of radiated cells depends only on the intensity of the beam to which they are exposed and the length of exposure" (that is, on the number of Roentgen units to which the cells were exposed). In other words, there was no demonstrable difference in the biological effects of X-rays and of γ-rays from radium for at least one important measure of biological function. Packard also reworked a novel concept of the half-life based on *Drosophila* eggs, measuring "how long an exposure [to γ-rays of measured intensity] is required to kill half the eggs." Packard also argued that his method could permit the calibration of X-ray machines, which were otherwise still notorious for delivering indefinite doses. (Standard measures of X-ray beams required the use of "rather temperamental galvanometers and ionization chamber[s]" that were not yet "sensitive enough to detect the very weak gamma radiation.") Packard's results thus not only helped establish the equivalence of X-rays and γ-radiation in terms of biological effects, but also provided a new way of calibrating the X-ray machine by means of the "half-life" of *Drosophila* eggs. Faulty X-ray technology could now be corrected by comparison against *biological* standards.

Hanson and Heys similarly praised all the advances made in the biological equation of X-rays with γ-rays: "The cause of variability in fruit flies, and possibly in all organisms, has been tracked down to the ultimate entity known to physics—the electron." And they, like Muller, viewed these successes as the result of a fruitful union of different disciplines: "The fundamental problem of biological variability necessarily waited for its solution upon the most recent developments in physics, again illustrating the interdependence of the arbitrarily delimited fields of science. Genetics has joined general physiology in furnishing a common meeting ground for biology and physics."[184] Radium had become effectively equated with X-rays in physical terms, in mutational terms, and largely in terms of other biological effects. This increasing instrumentalization of varied sources of increasingly identical ionizing radiation—the equation of the rays of radium with X-rays—would contribute to still further transmutations of the once obvious and strong metaphorical, metaphysical, and experimental connections between radium and life.

Muller had also speculated early on about whether other treatments, including poisons and ultraviolet light, could produce genetic effects. If they could, he thought, then it should also be possible to produce mutations by "totally excluding all X-rays."[185] This gesture toward a kind of shielding experiment was similar to an idea that Blaauw and van Heyningen characterized in 1925 as a "deradiation response"—the purported stimulating effect on an organism of the *removal* of radiation.[186] In 1928 Muller called for such an experiment in the context of genic mutation, not simply as a matter of physiological response, but as an interdisciplinary endeavor involving both physicists and biologists:

> Are all mutations ultimately due to rays of short wave-length and to high-speed particles of corresponding energy content? *If* so, biological evolution has been made possible only by the stray radiation present in nature—the beta and gamma rays, and the cosmic rays. This question permits a definite solution, for some organisms at any rate, *if only we can compare the mutation rate in ordinary controls with that in cultures from which a large part of the natural radiation has been artificially excluded*.[187]

Eight years later, Jack Schultz would also call for such deradiation experiments: "If radiations constitute a sine qua non for the mutation process, when radiations are screened off there should be fewer—or no—mutations. This is a most difficult experiment—the spontaneous mutation being low as it is. But it is a critical one."[188] What had been left unexplored by Gager, and perhaps was even considered unattainable in that earlier age of such strong associations of radium with life—investigating the effects of an absence of all radiation—now reemerged with Muller as a novel and interesting line of research. And while the unnoticed inconsistency in Gager's case (his failing to call for the study of the stimulative effects of the *removal* of radiation as a change in stimulative tonus parallel to irradiation) serves as a clear indicator of the power of the associations of radium with life as a constitutive element of his thought, the remarks of these later investigators concerning the possibility of such shielding experiments serve as a clear indicator of the growing instrumentalization and technologization of those associations following Muller's use of X-rays in the late 1920s.

In a multitude of different ways through the late 1920s and into the 1930s, then, the association of radium with life was being reworked and instantiated in new practices with X-rays and with genes. And so, a variety of glowing radioactive paths can be traced from the earlier uses of radium to Muller's Promised Land and the terrain beyond. Powerful and long-standing associations of radium with life had done much to condition both Muller's initial encounters with radium and his later efforts to equate radium's mutagenic effects with those of X-rays. As this link between the half-living atoms of radium and the secret of life transmuted into a powerful new form, becoming even more provocatively established through his work on the artificial transmutation of the gene and the dramatic shift in the meaning of mutation it caused, X-rays came to be understood as having been the first and only source of mutation-inducing irradiation of the gene. As radium gave way to X-rays, as chromosomes gave way to genes, and as the meaning of mutation became restricted and mutation became genic, *mutatis mutandis*, Muller became the first to induce mutation artificially by means of radiation. As his obituary in the *New York Times* read in 1967: "When some future historian contrasts our barbaric 20th century with his own happy era he will not stint himself in praising Muller. 'To his monstrous fruit flies were traced the first, deliberate successful scientific interference with the processes of heredity by external agencies' he will say of the professor."[189] Thus the entire broad and diverse history of earlier successful efforts to induce mutation with radiation came to be obscured. Here, then, in this scintillating mix, the nature of radiation, the meaning of mutation, and the rewriting of history all came to be inextricably intertwined. Further transmutations were still to come.

6 Transmutations and Disintegrations

Creation of life, alteration of species, mutation of genes: the connections between radium and life underwent profound transformations over the course of the first half of the twentieth century, with clear consequences in both the conceptual and experimental realms (shifts in the meaning of mutation, the forgetting of powerful discursive roots, and the shift from radium to other forms of ionizing radiation such as X-rays, among others). At first seemingly steady in the power of its metaphysical and metaphorical associations to produce new experimental systems and to condition the interpretation of their results, radium's connections with life seem to have transmuted with every passing decade.

Transmutation, decay, disintegration—each of these metaphors seems apt in describing these ongoing reworkings of radium's associations with life. Moreover, such transformations are arguably a consequence of the same historical processes at the intersection of metaphor, metaphysics, and experimental hermeneutics that permitted radium and life to become so closely and productively associated in the first place. Tracing these sets of transformations beyond Muller's 1927 work, and beyond the shift in the meaning of mutation from organism to chromosome to gene, understandably leads us into an ever more complex muddle with each subsequent decade. This chapter explores the afterlife of these associations, as the ties that had once bound metaphor and experimental practice so tightly together transmuted into productive new experimental systems and approaches that had seemingly little to do directly with radium—decaying to mere discursive residues or nearly disappearing altogether. The recounting of such an increasingly refractory story seems to test the very limits of historical narrative. In other words, given a history of constant transmutation and ongoing decay that never quite reaches total disintegration, how does a story that defies endings end?

Rather than simply tracing a complicated story across experimental systems, investigators, and decades and wrapping it all up with a compelling and coherent empirical ending, a more reflexive approach would seek to use "radium" as a narrative conceit, a powerful metaphor, and an epistemological tool for the historian as it was for the scientist. Such an approach would take the metaphor of the half-life of the transmutations and disintegrations of the association between radium and life in a *performative*, and not merely descriptive, sense.[1] It might even seek to question in the course of its own telling just how far any such narrative of a decay chain might reasonably extend until leadened with unworkable examples.

By tracing this asymptotic process of decay, and in coming to some points where one is no longer sure whether the historical evidence speaks to a still-extant connection—does radium really have anything to do with radiation genetics, with Max Delbrück, with Niels Bohr, or with any of the other scintillations touched on briefly in this chapter?—I hope to parallel in this chapter's deconstruction of my historical narrative the same epistemological dynamics in the metaphysics of metaphor that were so profoundly at play in the construction of scientific knowledge detailed in the preceding chapters.

As the once pronounced clicking of the Geiger counter of historical narrative (Soddy! Burke! MacDougal! Gager! Blakeslee! Muller!) slowly merges back into background static, both the writing and the *re*writing of the history of radium and the secret of life might thus prove to be a further unexpected transmutation, decay, and perhaps even disintegration of this once all-powerful association. Perhaps, then, as we trace here the shift away from radium and toward a more generalized radiobiology; Muller's surprising later work with radium and the overlap he saw between his theories of the gene and other discourses of "organic radiation"; and the larger physicalist turn in the study of heredity in the 1930s and 1940s with the emergence of "radiation genetics" and "phage genetics," we might more consciously be aware of the curious persistence of links between the realms of the radioactive and the living even as radium itself began to exit the historical scene. We might more consciously be aware, as well, of "the artificial transmutation of the meme."

From Radium to Radiobiology

From its birth in a drafty laboratory at the hands of Marie Curie to worldwide acclaim, and from the foundation of radium institutes in various countries to an intense focus on the medicinal applications of radium, and up to the increasing realization that radium was not "just like fire," but potentially even more dangerous, the role of radium shifted in later years from wonder element that could do no wrong to both tool and mixed blessing.[2]

Radium's penetrating power was undoubtedly formidable, and questions about risks associated with its use grew over the decades with the publication of ever more stories of vials being lost on the Paris Métro or in the snow of Saskatoon (more than one hundred cases of lost radium were known by 1944, and only two-thirds of these were ever recovered),[3] of the accidental ingestion of radium capsules by a woman in Philadelphia who needed an operation to remove them in 1929, and of the deaths of several of radium's greatest promoters from overexposure (including Sabin Arnold von Sochocky and the millionaire playboy and radium tonic enthusiast Eben Byers). As historian Matthew Lavine has noted:

> By the late 1920s, roughly the peak of the element's availability on the medical and consumer marketplaces . . . patients had finally gained enough experience with the substance on an experiential basis for that familiarity to begin to breed contempt, or at least potential disillusionment . . . expectations could not be maintained in the face of the underwhelming reality of radium nostrums. Only because the laity had had actual contact with radioactive substances (or believed they had), could the experiences of Marie Curie, Eben Byers, and the dial painters start to gain a real foothold in the discourse.[4]

While radium "seeds" were still readily used in cancer treatments as late as 1924, in January 1925 one newspaper made reference to the increasing "roll of martyrs to science" killed by radium overexposure. The end of May brought news of the fate of the New Jersey dial painters: "New

Radium Disease Found: Has Killed 5; Women Painting Watch Dials in a Jersey Factory the Victims, Doctor Says . . . Cancer Called Incurable; Trust in Radium Is Unjustified, New York Physician Asserts." By 1932 one headline ran, "Death Stirs Action on Radium 'Cures.'"[5] While radium had been touted as a cure for cancer in the early years, signs of an ongoing dissociation of radium from life were already apparent to some observers by 1920:

> Great attempts have, as we know, been made *to cure Cancer with Radium*; but as far as can be gathered up to the present Radium is certainly not a cure for the disease. All manner of different methods have been tried for curing it, but so far a practical and satisfactory cure for Cancer is not yet known. . . . The author is himself aware of two or three deaths which have been caused by the "Radium treatment," and, furthermore, there have been indications that if "Radium" had not been resorted to life would probably have been prolonged.[6]

Moreover, failures that had previously been documented but generally overlooked—or interpreted as signs of radium's life-giving power, like Gager's own induction of a slew of morphological defects in his radium-treated plants—were becoming increasingly recognized with each passing decade: seeds whose vitality were "destroyed," leaves losing chlorophyll, and guinea pigs succumbing to radium's damaging rays.[7]

Radium continued to be used in laboratory work well into the late 1930s,[8] but with other options becoming available, radium stocks seem to have been perceived as more and more risky with every passing year. As one letter Muller received noted, "A recent event has suggested that there may be some misapprehension on the part of those responsible for holding National Radium on loan from the Medical Research Council for research purposes, as to the correct procedure in cases where there is damage to the containers, with or without suspected leakage of radium."[9]

Gager, however, clearly remained interested in using radium even as others had begun to move on to other sources of ionizing radiation. He wrote to a colleague, W. C. Curtis, at the University of Missouri, requesting further support for investigations with radium: "I would be very much interested in it if you will include *radium* with X-rays, so that the funds shall be in support of investigations of the effect of X-rays and rays of *radium* on plants and animals." As Gager described his plan, he emphasized the central role radium had in his studies, in explicit distinction from X-rays: "My problem for future work is to continue the same kind of investigations as were reported in my 1927 paper (Gager & Blakeslee), which is an investigation of the effect of *radium* rays in modifying heredity and to study the cytology of egg and sperm cells that have been exposed to the rays."[10] Curtis responded, reassuring Gager that radium had been excluded only as an oversight: "I hasten to say that it has been our intention all along to include all radiations in our program. Perhaps my own interest in the X-rays has led me inadvertently to use language indicating such a limitation."[11] It was even specified in "Communication No. 2," issued on March 8, 1928, "that ultra-violet, x-rays, and work with radium are all included" in the program designed to fund research on the "Effects of Radiations upon Organisms."[12]

It was also at this time that the International X-Ray and Radium Protection Committee came to be organized, in July 1928. Although the equation of the biological effects of X-rays with those of radium was long-lived, it was a matter of decades before X-rays came to completely supplant radium in discussions of radiological protection: the committee was renamed the International Commission on Radiological Protection only in 1946. (This was the same year that the American Roentgen Ray Society and the Radiological Society of North America "combined their protection activities into a single committee.") Some integration of the two forms of ionizing radiation had taken place earlier, however, as with the first institutionalization of the Advisory Committee on X-Ray and Radium Protection in early 1929, which published guides on X-ray protection in 1931.[13]

And even with his intense focus on and interest in the effects of X-rays, Muller himself had never really left radium behind (see chap. 5). Despite the loss of his radium in 1924, Muller was again in possession of some by early 1932, this time rented from the Radium Emanation Corporation. In the intervening years he had even suggested to other researchers that *they* investigate various genetic effects of radium.[14] He continued to request and receive more radium sources into the late 1930s,[15] and he continued highlighting the place of radium in his narrative explanations of mutation even years after his more prominent work with X-rays: mutations were "of an ultramicroscopic nature, such as the impact of a minute ray (for instance, from radium) on one of his genes."[16] Muller succeeded in getting radium from the Medical Research Council at least as late as 1938, and even in 1943, he wrote that he continued to value radium "because of the greater ease with which it can be used to give low intensities constantly over a long period, but that there is in fact no quantitative or qualitative difference between x-ray + radium effects, for a given total dose in *r* units."[17] And, as the old became new again, the Science Service reported just after noting Muller's receipt of the Nobel Prize in 1946 that he "has added radium radiations to X-rays as weapons of genetic bombardment."[18]

Others geneticists also continued experimenting with radium—at Woods Hole, at least—until a fund designated for such experiments ran out after 1940.[19] Widespread use of radium was markedly on the downswing by the mid-1930s, however, and cost was a major—though not the only—factor. One source of funding for radiation biologists officially changed its policy to support more explicitly physical work with radiation instead. Within a few short years, for example, Blakeslee and his student John T. Buchholz could get no more funds from the Radiation Committee for their biological work.[20]

The broader trend was readily apparent. Expensive and increasingly more difficult to control than ever-improving X-ray technologies, radium no longer held pride of place. While Gager, Muller, and some others continued to use radium, X-rays began to dominate the scene. Blakeslee himself had begun to use X-rays, in addition to radium, in his collaborations with Buchholz, and he made use of various X-ray tubes while at Cold Spring Harbor throughout the 1930s, alongside Milislav Demerec.[21] And Curtis reported to Gager in early 1928 that "Stadler is just installing an X-ray machine in the building next to ours." The convenience of using X-ray machines over radium was clear to Curtis: "I shall be able to carry on my work with much greater convenience than in the past when I had to depend upon the machine at the hospital."[22] Running X-ray machines was not necessarily any easier or safer than using radium, however. One of the researchers at the Edinburgh Institute of Animal Genetics (where Muller worked for a time) had continued his experimentation with X-rays and radium up until 1938, and one of his coworkers noted that "the difficulty of regulating the x-ray dosage seems the greatest snag of all" in their experimentation. Following the arrival of a powerful new X-ray tube, he reported to Muller that the "workers don't know how to operate it carefully."[23]

In time, newer and ever-improving instrumentation "such as the van de Graaff generator, powerful linear accelerators, betatrons, synchrotrons and microtrons" made possible the production of other ionizing radiations well beyond the strength of ordinary X-rays.[24] "Improved X-rays for Cancer Work," proclaimed one headline: "Harvard Physicists to Use Deeply Penetrating Type in New Laboratory; Hope to Displace Radium." The article beneath noted that "the trouble with the use of X-rays up to this time has been that they are not as penetrating as the so-called 'gamma rays' of radium."[25]

Were X-rays the new radium? By 1928 W. D. Coolidge had commented that "Man, in his effort to equal the power of radium, is locking himself up in a lead-lined room, encaging himself within a cabinet of thick lead and submitting himself to the dangers of high electric currents such as he has never reached before." One journalist noted that Coolidge "has succeeded, so far, in attaining only one-half the power that lies within a fraction of an ounce of radium—nature's most remarkable element."[26] A popular novel written at the time of this transition to X-rays, Rudolf Brunngraber's *Radium* (1936), spoke of one scientist who

> implored electro-technicians and physicists throughout the world to perfect Röntgen apparatus, since Röntgenization was a fairly efficient substitute for irradiation. Surely it would be possible, he went on, to increase the tension of the gamma-rays in Röntgen tubes?

The perfection of the Röntgen tubes even played a significant role in the story's plot:

> Also, thought Francis, Pierre Cynac was the man with whom he himself, Francis, would come into conflict with his Röntgen tubes for contact-therapy and his short-wave-length apparatus for restoring health to diseased cells. Success in these domains would make radium superfluous, and therefore ruin Pierre's schemes as radium king. . . . Life was a horrid muddle.[27]

: : :

But as nuclear physicists claimed the production of "artificial radium" by 1935, even the "most powerful rays" later produced with the invention of a million-volt "giant cancer tube" at Caltech were still said to be "equivalent to [the] entire world['s] radium supply."[28] By 1948 other replacements for radium were on tap: "Atom-Bomb By-Product Promises to Replace Radium as Cancer Aid," reported the *New York Times*, noting that the replacement—irradiated cobalt—was "a 'virtually costless metal,' promised in every way to be as effective as radium in the treatment of cancer, and far easier to use."[29] Indeed, by the 1950s, "the nearly entire focus of gamma rays from radium on plant growth would switch to cobalt-60," especially in "gamma gardens."[30] And in reports on the "atom-smasher extraordinary" Ernest O. Lawrence of the University of California, Berkeley, the newspapers claimed that he had perfected "a new cyclotron which produces the most artificial radium-like rays in the world."[31] These varied novel means of producing powerful ionizing radiations ultimately contributed to a brave new world for radiobiology in the post–World War II context.[32] As Spencer Weart has noted, "Isotopes became an invaluable tool for studying everything from physiology to the way heredity works. Much of the tremendous progress in biology and medicine since the 1950s would have been impossible without radioactivity. Tracer isotopes would unveil the secrets of life itself!" Indeed, as one CBS radio program announced: "When you get deeper and deeper into the secrets of life, you find them so fascinating you sometimes forget that the atom can kill."[33] While the Atomic Energy Commission would later commission Blakeslee to "study the effects of thermal neutrons and radiations from nuclear detonations and from a cyclotron in the production of chromosome and gene mutations by using the *Datura* material," and would later team up with the USDA and over a dozen state agricultural experiment research stations to determine "whether radioactive material does indeed stimulate plant growth," the *Journal of Heredity* was already reporting by 1946 that "there is no reason to believe that a whiff of atomic energy is calculated to improve human germ-plasm."[34] Indeed, the end of the USDA study "marked the end of an era for radium."[35]

Transmutations: The Gene as Atom (of Radium)

Even as X-rays vied with radium as the preferred tool for biological experimentation in later decades, Muller continued to rely on radium not only as a mutagen, but also as an important conceptual tool, seeing radium and life as somehow intimately connected analogically, discursively, evolutionarily, mechanistically, and metaphysically. Even a decade after his epoch-making work, he continued to describe and analyze phenomena in terms that frequently glowed radioactive. Even as he turned to X-rays, agreed with the physicists' equation of X-rays with γ-rays from radium, helped to establish the equivalence of these rays in their biological effect, and displaced the radium-based artificial transmutations of Gager and Blakeslee in the historiography, Muller went further than most in continuing to approach the questions of genetics through the language and frame of radioactivity—a testament to the endurance of the powerful associations between radium and life that had long served as his source of inspiration. Two particular areas of his research agenda following his 1927 work serve as good illustrations of this approach: his claims for a possible role for radium as an internal organic mutagen, and his proposed physicalist analysis of the "auto-attraction" of genes.

Following the artificial transmutation of the gene, Muller sought to find an explanation of observed natural mutation rates with reference to natural sources of radiation. As it turned out, ambient sources of radiation were seen to be insufficient (with the physicist L. Mott-Smith, Muller estimated that the amount of natural radiation was some 1,333 times too low; Nikolai Timoféeff-Ressovsky would later estimate that it was 462 times too low). After having discarded nearly every other possible environmental and cosmic source, Muller thus returned to his favored element to explain the discrepancy: "Practically, we should have left only highly radioactive substances like radium as possible sources of radiation competent to give the observed natural mutation rate."[36] And as he moved from a consideration of the mutation *rate* of genes—his atoms of life—to their very state of *mutability*, the term "half-life" crossed back again from radium to the realm of the living as he referred to the "half-life of the individual self-duplicating gene in *Drosophila*."[37] Atoms of life and living atoms now both had half-lives.

Similarly, while looking for a mechanism to explain purported cases of mass mutation and other mysteries of altered mutation rates—precisely the same sorts of issues that intrigued many *Oenothero*logists—Muller seriously considered speculations by Vladimir Vernadsky and others as to whether organisms might have evolved so as to be able to *store* radium and thereby preserve the ability to mutate. Organisms, in other words, might be viewed as *condensers* of radium.[38] (Others at the time had similarly wondered "whether there is any relation between the power of radium concentration and the variation or evolution of the organism."[39]) Muller explicitly excluded other radioactive elements such as uranium and thorium from consideration: only radium was powerful enough to begin to account for the effect.[40]

Muller thus continued to call on radium to explain some of the unaccounted-for phenomena of life not only in the individual organism (in inducing its own particular mutations), and not only in intriguing explanations of the stability and length of life of its genes, but also in considerations of an individual organism's very *capacity* for mutation and evolution in the first place.[41] Muller's serious—if brief—consideration of whether organisms are condensers of radium and how this might explain evolutionary processes is an interesting transmutation of Burke's earliest claims to have shown "that there is an element, a bio-element, possessing a vast store of potential properties, and of potential energy equivalent to biotic."[42] Recall that even Becquerel had made similar claims in 1925.

Time and again, Muller also insisted that advances in the study of heredity depended on cooperation between physicists and biologists, as it was "in the tiny particles of heredity—the genes—that the chief secrets of living matter as distinguished from lifeless are contained," and it was by "understanding of the properties of the genes," which were "most unique from the standpoint of physics," that biologists and physicists together could "bridge the main gap between inanimate and animate." Genes, these most remarkable entities on the border of life and nonlife—"so peculiar are these properties that physicists, when first confronted with them, often deny the possibility of their existence"—were the new radiobes. Understanding them might "throw light not only on the most fundamental questions of biology, but even on fundamental questions of physics as well."[43]

Genes, as "the ultimate particles of heredity"—which even "probably constitute the ultimate particles of life itself"—also presented Muller with other mysteries he was keen to solve from a physical standpoint. Their manner of replication (or the mystery of their "autosynthesis," as he termed it), was of interest, but first and foremost in Muller's mind was solving the mystery of *the nature of gene attraction* in terms of what could be observed at the cytological level: the lining up and drawing near of homologous chromosomes during the process known as karyokinesis (the nature of nuclear fission was as of much interest to Muller as it had been to Burke).[44] By 1936 Muller sought to explain this phenomenon of "auto-attraction" in terms of *radiation*. Not only did he theorize that the genes themselves "emanated" some kind of "radiation," he argued that "under certain conditions, it becomes evident that each gene forms the center of a specific field of attractive force."[45] Though Muller claimed to "use the word radiation here only in the most general sense," the discursive imprint of radium seems clear: Muller was literally talking about organic "radiations resulting in genic attractions," about physical radiations as somehow emanating *from the genes*, in order to account for auto-attraction of homologous chromosomes during karyokinesis.[46] He thought that the solution of this problem by physical means would do much to enlighten biologists as to the nature of the gene. But he reported that it had been difficult "to make quantitative studies, after the physicist's fashion, of the nature of the force of gene attraction; studies of its variation of intensity with distance; of the effect of varying conditions upon it; of its direction; of its speed of propagation; of the possible interference with one another of the forces emanating from different genes; of its possible polarization, etc."[47] Muller even wondered whether a "Heisenbergian 'principle of uncertainty'" was at the root of how "one tiny gene" with its own "ultramicroscopic determinism" could produce through "growth and development . . . a molar indeterminism" on which natural selection could act.[48] Muller seemed open to the possibility that organisms could, in effect, *mutate themselves.*

Life spans, half-lives, and disintegrations, organisms as condensers of radium, radioactive auto-attraction, and references to Heisenbergian questions of determinism—Muller's work suggests that a link between the realms of the radioactive and the living persisted in some respects at least well into the 1940s.[49] But Muller was far from the only one whose work suggested that further transmutations were afoot. By 1933 some researchers even classified mutants along a "mutation spectrum" (using a word from the study of radiation) using Greek labels—α, β, γ—that exactly paralleled the three kinds of rays given off by radium.[50]

: : :

The novelist Rudolf Brunngraber had framed the matter aptly in his *Radium*: "What objection is there to the hypothesis that the tissues, live, dead, or dying, may emit such radiations? Heatless radiation accounts for the light of the glow-worm, the firefly, and the luminous deep-sea fishes. Why, then, should not animal cells give off other kinds of radiation?" After Rutherford and Bohr's proof of elemental transmutation and the ways in which "we had come to conceive a universe in which matter was a figment of the imagination, in which matter was resolved into a gamut of undulations," Brunngraber noted, "there was no longer any difficulty in conjecturing that the organic cell likewise must be an electrical system which emits and receives radiation."[51]

Chief among these ideas of cells radiating rays in the 1930s was the purported discovery of "mitogenetic radiation"—a discovery that was entirely consonant with Muller's other explorations into the idea of organisms as radium storage units, and which fascinated him. While N-rays had proved crucial on Burke's path to radium and ultimately to radiobes, it was these later "M-rays" that most closely paralleled Muller's own querying of physical radiations emanating from the genes in the 1930s, and which might themselves represent a further transmutation of the associations between radium and life.

First described in 1923 by Alexander Gurwitsch, a Russian cytologist, M-rays were thought to be especially noticeable in actively fissioning living tissue. Characterized first as "mitosis-stimulating radiation," these oscillatory rays were not only produced by actively dividing tissues, but could also stimulate other tissues.[52] A subcommittee of the National Research Council dedicated to mitogenetic radiation was established in 1928, just a year after Muller's artificial transmutation of the gene, and Muller himself was intrigued by the phenomenon. Reflecting on some work by Altenburg, Muller had even speculated that "natural ultraviolet rays ('mitogenetic rays') produced by the chemical reactions occurring in organisms are responsible for some of the mutations that occur naturally."[53] Perhaps, he noted elsewhere, even the "structure of the radiation would be some sort of geometrical resemblance between the arrangement of parts in the gene and the arrangement of parts in a bundle of the radiation itself."[54] Radiation, the gene, and the organism were all held together in one atomic whirlpool by M-rays, just as an earlier generation had held life and light together with N-rays, and just as Crile and Lakhovsky in the wake of Burke had theorized about radiation produced by and emanating from living things.

Over the next few years, some six hundred papers confirming M-rays' existence by two hundred authors from American and European laboratories were published in well-respected journals; reported observations reached a record level in 1935 before tapering off.[55] One commentator in 1933 even directly associated mitogenetic rays with radioactivity:

> In the face of such experimental evidence it is extraordinary that the existence of the Gurwitsch rays should be questioned. The remarkable thing about them is not that they have been discovered but that their presence was not suspected long ago. . . . The complexity of protoplasm is not in itself sufficient evidence of radioactivity, but it does leave one more ready to suspect it of being thus active.[56]

Despite its many defenders, however, mitogenetic radiation was one more radiation consigned to the dustbin of history, and the NRC subcommittee formed to study it disbanded by 1936.

Is the sputtering arrival of mitogenetic radiation another potential transmutation of the radium-life association? The answer to this question returns us to the full spectrum of possibilities relating metaphysics and metaphor to experiment. No longer clearly demonstrating an ontological connection, as Muller seemed to wish for, countless further provocative examples of an ongoing association of radioactivity with life can nevertheless be readily identified, suggesting that something more than mere metaphor is going on. One 1947 book, discussing the nature of radioactivity, described the protons and neutrons in the nucleus of the atom as "cells."[57] The authors of the Smyth Report on Atomic Energy for Military Purposes, published in 1945, concluded that the term "nuclear" still had "primarily a biological flavor" and opted for the term "atomic" instead (which was also "less likely to frighten off readers").[58] Even in 1957 one author referred to cells as "the transmuters of molecules" and described mitosis in terms of tissue having "the ability to make use of the chain reaction principle. . . . It may be said that normal tissues grow by means of a controlled chain reaction"—thus using Niels Bohr's concept of the chain reaction in nuclear physics to describe the original biological process of fission.[59] ("Fission" had itself been a biological term before it became a nuclear one.[60]) Even the association of cells with one another was sometimes described in atomic terms: "Far from being isolated, the cells live in close integration and create an atomic whirlpool."[61]

Even as late as 1947, in a lecture he delivered at Oak Ridge, in which he also acknowledged having read Soddy in his youth, Muller remained explicit about the connections he saw between the world of atomic energy and the world of the gene:

> It may also seem strange that people in my line of study, who have been concerned with the slowest moving and in a sense the most insidious forces in the world, should have anything to contribute which might be of interest in connection with the line of work dealt with here, which concerns the quickest, the most violent and spectacular of forces. . . . I [shall] try to give you an inkling of some of the ways of working of these peculiar, slow-moving forces which, unlike the violent destructive energies of the atomic chain reaction, have very gradually worked *constructively* and themselves, too, in a kind of chain reaction system, on the chemical level, so as to have finally brought into being, not dissolution, but the ultra-complicated organizations found in our own bodies and in those of all higher animals and plants.[62]

In short, the sheer conceptual productivity provided by this metaphorical overlap between the realms of the radioactive and the living clearly remained a major driver for Muller and for others for some years. In other instances, however, it seems evident that the terminology was simply convenient to retain, or somehow remained a compelling usage despite important points of disanalogy. As with the initial formation of a biologically inflected radioactive discourse in the early part of the century (see chap. 1), these later parallels were the results of deliberate choice, mere coincidence, and overdetermination alike.

No matter their individual plausibility, the sheer number of such examples and their widespread occurrence—even as experimental interests and concerns increasingly led many investigators farther afield from radium itself—might well be further evidence of the ongoing transmutations of the radium-life connection not only toward radioactive discursive residues, but also toward the continued use of radiation in understanding the atomic physics of the gene. "Already it is evident that the problem of the gene is the problem of the atom," noted one science journalist in 1945, parroting Muller.[63] Even cytogeneticist Cyril Darlington would write in 1933 that the gene is the "atom of inheritance" and that "we can assume without hesitation that an intra-molecular and therefore intragenic change precedes and conditions all more complicated kinds of change."[64] Others would soon pick up on this call for a more physicalist treatment of the gene.

Decay: The Target Theory, Light and Life, and the Atoms of Biology

Although Gager, Blakeslee, and even Muller for a time continued to use radium in their experiments and in their thoughts, the increasing instrumentalization of radiation—the movement away from a singular focus on the lifelike and life-relevant properties of radium and toward the use of X-rays as tools in the study of life—unquestionably moved many subsequent experimental systems further away from what had once been an all-encompassing association of radium with life. With the gene becoming the natural target and X-rays and other forms of ionizing radiation the increasingly obvious means for studying mutation, Muller's success also inadvertently aided in the growth of the new field of "radiation genetics," which sought to bring radiation ever closer to the secret of life in the genes.[65] As mutation became a physicalized process, paralleling the materialization of the gene, and as new kinds of radiations proved easier or cheaper to use than radium itself, the field of radiation genetics, by drawing on earlier tropes that analogized the hereditary substance to an unstable element, signaled a shift from what was often a population genetic approach (dealing with the half-lives of genes in a population) to a more strictly biophysical and molecular approach. No longer merely genic, mutation was becoming *molecular*. And what had begun as the use of radiation to study the *gene* was transformed, in the hands of radiation geneticists, into the use of the gene to better understand the *molecular effects of radiation*.[66]

One outgrowth of this use of radiation and its effects on molecules to study genetics came to be known as the *Treffertheorie*, or "target theory"—a series of attempts to establish with ever greater precision the character of the genetic material and the size and nature of the genes, and to establish a quantitative relationship between the amount of ionizing radiation deployed and the amount of mutation produced.[67] The target theory held that in many respects, the gene could be understood by analogizing it to an unstable element. In fact, the target theory's attempts to ascertain the size of the gene by atomic bombardment bear at both first and second glances an uncanny resemblance to Rutherford's earlier search for the atomic nucleus. The experimental practices of target theorists are thus arguably also among key further transmutations of the associations between radium and life.

The biophysicist Max Delbrück agreed with Muller that biological problems could be attacked most fruitfully with the tools of physics: "As we enter this new territory, we are rewarded at every step with new insights into the wonderful mechanics of the hereditary mechanism, for the exploration of which radiation has furnished a powerful tool."[68] According to William Summers, Delbrück "used evidence and concepts from target theory experiments to construct a model of mutation and then a theory of gene mutation and structure." Angela Creager has also viewed Delbrück's contributions as primarily theoretical: he "drew on quantum mechanics to interpret mutations in terms of shifts in atomic configuration from one stable energy state to another."[69] (Gunther Stent would later characterize this as a "quantum mechanical" model of the gene—a clear echo of Muller's quantum understanding of evolution.[70]) The biophysicist Delbrück, geneticist Nikolai Timoféeff-Ressovsky, and physicist Karl Zimmer teamed up to explore how radiations might be used to better characterize the "gene molecule" and understand it from a quantum mechanical point of view. The end result of their collaboration was an important "green pamphlet" known familiarly among those in the new field as the *Dreimännerwerk*, or "Three-Man Paper," of 1935.[71]

As Delbrück later recalled, "The major paper got a funeral first class. That means it was published in the *Nachrichten der gelehrten Gesell-*

schaft der Wissenschaften in Göttingen, which is read by absolutely nobody except when you send them a reprint." The three men circulated reprints among the small community of interested geneticists and physicists, however.[72] They concluded that

> mutation was a "one-hit" process, a single ionization produced by a quantum of radiation within a certain sensitive region. They calculated the size of the sensitive region to be on the order of a large organic molecule. Although they were cautious about identifying this sensitive region with the gene itself, they argued from the target-theoretical analysis that the gene could be understood as a "group of atoms."[73]

One of the (few) longer-term accomplishments of the target theory was thus the literal equation, long in the making, of genes with the atoms of physics.[74]

The target theory—and the rapid growth of radiation genetics as a whole—has often been viewed as a sort of imposition of physicalist methods on biology, or at least of physicists on biological questions, a topic with its own large literature.[75] More specifically, Summers has carefully traced the ways in which specific ideas from "the atomic physics of Thomson and Rutherford" were applied to the study of the gene and how they "depended in the first instance on a conjunction of events and an individual with specific interests and knowledge"—reflecting a classic historiographical interest in compensating factors, interests, and the play of contingency.[76] It may be equally useful, however, to view at least some of the development of radiation genetics and the target theory as later curious decay products of the once powerful associations between radium and life—later, having largely happened after Muller's most compelling work, and curious, because while the constant reiterations of "physics" and "biology" (and the atoms of each) owe much to discursive modes established earlier, radium *itself* had by and large disappeared from consideration in the new radiation genetics. The initially powerful associations between radium and life not only transmuted, one might say, but also decayed.

: : :

Viewed in this light, even Niels Bohr's famous lecture "Light and Life" of 1932—delivered as the inaugural lecture of the International Congress of Light Therapists in the Rigsdag in Copenhagen—might be viewed as a further decay product.[77] Bohr's claim that perhaps there were new laws of physics to be discovered in the biological realm paralleled older, familiar claims that the discovery of radium necessitated the rethinking of the laws of physics.[78]

Moreover, his association of "light and life" was one that dated back well before Burke. Some commentators at the time even drew direct links from Muller's work to Bohr's lecture, stating that "there is a connection" between "fruit-fly eggs which have been genetically jolted by radiations" and Bohr's "Light and Life."[79] (In 1935 Timoféeff-Ressovsky, Delbrück, and Muller met with Bohr at the Carlsberg Laboratory in Copenhagen specifically to discuss the nature of mutagenesis.[80])

But while Burke's work proved to be sensational and Muller's bent was experimental, Bohr's reworking was more thoroughly conceptual: "We are not dealing here with more or less vague analogies," between light and life, Bohr said, echoing his many predecessors in sounding a note of caution about metaphors. Rather, the concern was "with an investigation of the conditions for the proper use of our conceptual means of expression."[81] And because living things are constantly in flux, Bohr argued, the application of mechanical or quantum ideas from physics to the analysis of life is difficult: "This fundamental difference between physical and biological research implies that no well-defined limit can be drawn for the applicability of physical ideas to the problem of life. . . . This apparent limitation of the analogy in question is rooted in the very definitions of the words . . . which are ultimately a matter of convenience."[82]

Bohr held that the analysis of words was important to understanding the nature of the claim that could be made for a particular association between physics and biology. In so doing, he was only the most recent exemplar of investigators into the relationship between radiation and life for whom words were central, including Soddy ("words would not come . . . as though propelled by some outside force I heard myself utter unbelievable words"), one of Burke's critics ("Put in this way the whole matter resolves itself into a question of words"), de Vries ("the instability seems to be a constant quality, although the words themselves are at first sight, contradictory"), Blakeslee ("We all feel a difference in meaning between the words mutant and mutation"), and Muller ("Mutation and Transmutation—the two key ~~words processes~~ stones of our rainbow bridges to power!") When dealing with the borderlands between radiation and life, words mattered as much as things for Bohr, as they did for his predecessors. Perhaps Muller's emendations to his description of mutation and transmutation say it all: beginning as mere words and metaphors, then functioning as processes and experiments, they end as "stones" in a metaphor of another kind.

Bohr also suggested that biology, like physics, has an irreducible aspect known as the quantum, but that this quantum is not the mutant, the cell, or the gene (as Muller would have it), but the very "secret of life" itself. This "secret of life" was always inherent in our knowledge of the living world, a residue that could never be "explained away." The quantum of life, Bohr suggested, was not waiting to be found inherent in some entity, but lay rather in our considerations of *what would make it so*. Was radium alive? Were radiobes examples of primitive life? Were chromosomes truly the determiners of heredity? Were genes more properly conceived of as the ultimate basis of life? Debates over the precise status of each of these entities in the field of life could now be superseded through better consideration of the meaning of "life." Framed in such reflexive ways, this epistemological twist in Bohr's "Light and Life" might be productively viewed less as an intrusion of physicalist thought into biology than as an ever more distant decaying residue of radium and life. Bohr's twist suggests that the "secret of life" is not "out there," waiting to be empirically discovered using the radium of the moment. Rather, a new and different mode of investigation and analysis—suggesting a new way of narrating and analyzing this complex history of transmutations—would call less for a "just the facts" narrative with an unproblematic recounting from empirical sources than for a reflexive narrative reconceptualizing what counts as "proper" evidence and reasoning, and challenging the very presumptions of historical narrative itself. These sorts of deliberately artificial transmutations of the meme might be central to alternative ways of telling the history of the complex reworkings of "radium and life." Perhaps, just as Morgan and Muller had to doubt radium's mutagenicity in order to make their own advances into the nature of induced mutation, the historian should reflexively seek to doubt his own compelling narrative and deconstruct its own interweavings of metaphors, metaphysics, and modes of historical reasoning (a possibility explored further in the Conclusion).

: : :

Bohr's contemporaries found more concrete insights in his lecture. Sitting in the audience in Copenhagen was the young Max Delbrück: "I was interested—well, anybody who was *at all* interested in quantum mechanics couldn't help but be fascinated."[83] Delbrück found inspiration in Bohr's lecture to search for an analog of the physicist's notion of complementarity in the biological realm—even if that meant a new principle that was in some way unpredictable. As Lily Kay has noted, "Bohr inspired Delbrück to explore biology—the 'secret of life' as he put it."[84] Once an assistant to Lise Meitner in the laboratory that had produced the first artificial fission, Delbrück envisioned numerous interconnections between biology and atomic physics—indeed, he chose to work with Meitner from 1932 to 1937 largely because of the proximity of her laboratory to the Kaiser Wilhelm Institutes for Biology ("I thought it would be a good opportunity for me to pal around with biologists"). Bohr's speech, Delbrück noted, "was sufficiently intriguing for me . . . to decide to look more deeply specifically into the relation of atomic physics and biology."[85]

As Delbrück would later remark in 1944, "Perhaps we are approaching a similar phase in biology" to that in "physics around 1890," just before "the discoveries of radioactivity, of X-rays, and of the electron." He continued, "It would seem that the principles of atomic physics will have a large share in the construction of this 'modern biology.'" He even referred to "the atomic theory of biology, i.e., genetics." This reference was intended as more than mere rhetoric or analogy: where Bohr was vague, Delbrück wanted "to find out just how far atomic physics does carry us in the understanding of the phenomena of the living cell." In so doing, he was asking new versions of some very old questions: Were there atoms of life?[86] And just what was the relationship between the half-living atom and those atoms of life?

While Muller had tentatively proposed in 1922 that bacteriophage—a virus that infects bacteria—"would give an utterly new angle from which to attack the gene problem," it was up to Delbrück to lead the field of "phage genetics." (Delbrück, along with Salvador Luria and Alfred Hershey, "dominated this nascent phase of molecular genetics.")[87] Delbrück recalled his first discovery of bacteriophage-induced plaques in the lawn of bacteria on Carl Lindegren's petri dishes as a "simple experiment on something like atoms in biology."[88] But a colleague remembered Delbrück exclaiming upon seeing the plaques, "Oh, my God, you have atoms in biology. I'm going to work on that."[89]

No mere passing fancy or simple descriptive technique, these "atoms in biology" inspired Delbrück to draw connections between the atomic theories of physics and a quantum understanding of genetics (the so-called "atomic theory of biology") as he used ionizing radiation to produce mutations to better study the physical nature of the gene. Erwin Schrödinger, the son of a botanist, drew further on Delbrück's work with Timoféeff-Ressovsky to happily equate de Vries's mutation theory with the quantum. Schrödinger's famous *What is Life?* (1944) is a paean to the idea of the hereditary unit as an atomic structure; he claimed in 1945 that "in the light of present knowledge the mechanism of heredity is closely related to, nay, founded on, the very basis of quantum theory."[90] But as Schrödinger suggested, the reduction of genetic fundamentals either to the phage system or to a basic code-script—rather than the more complicated systems of heredity in *Oenothera*, *Datura*, and other higher organisms with admittedly much more complex hereditary mechanisms at levels above the gene—also meant further reductions in the meaning and phenomena of mutation.

But even Luria's later discovery that phage could be both inactivated and *re*activated by irradiation—with radiation now not only causing damaging gene mutations, as Muller had established, but also able to re-induce some of the characteristics of life—may serve as yet another far-removed residue of the association between radium and life. And Delbrück's later characterization of the discovery of the structure of DNA also suggests ongoing resonances: "Very remarkable things are happening in biology," he wrote to Bohr, after learning about the discovery in a letter from James Watson. "I think that Jim Watson has made a discovery which may rival that of Rutherford in 1911."[91] Perhaps, then, even the ascription of the "secret of life" to DNA—refracted through Delbrück's "riddle of life"—is another product in the decay chain of radium and life.[92]

Orthogonal to traditional narrative progressions from target theory and radiation genetics to Bohr's "Light and Life," and from Delbrück's talk of "atoms in biology" to the "secret of life," these brief scintillations suggest an alternative historical narrative of ever-sporting associations between radium and life over decades and across subfields.[93] But if these varied cases seem increasingly less compelling or tangential than the chapters that preceded them—if something has seemingly decayed in the association of radium and life from Muller to Bohr and Delbrück—still further disintegrations are yet to come.

Disintegration

In the aftermath of Muller's work, genes continued to be strongly associated with the atoms of biology. Gager's *General Botany* of 1926, for example, reported that "most mutants are exceedingly stable . . . the genes of the vinegar fly, *Drosophila*, indicate a minimum stability on the average for each gene comparable to that of radium atoms, which have a so-called 'mean life' of about 2000 years."[94] Yet points of disintegration began to emerge. By 1936 Muller had calculated the life span of a gene to be of a different order of magnitude than that of an atom of radium—on the order of 100,000 years.[95] And despite the curiosities of the background levels of radioactivity in living matter, by 1959 Muller was able to remark that "the genetic material, unlike protoplasmic constituents, is not subject to flux: that is, the atoms within it remain there permanently, without turnover."[96] What had been an intra-atomic connection to life was rapidly becoming merely extra-atomic chemistry. Even radium's use as a central metaphor was eclipsed by Muller's increasing use of alternative tropes, such as the older trope of the fire of metabolism.[97]

Nor was radium, or radiation in any form, still necessary for the production of mutants. Charlotte Auerbach discovered alternative mutagens in mustard gas in 1943, and by 1937 Blakeslee and others had already begun to use the chemical colchicine in efforts at what they called "genetics engineering."[98] In light of the chromosomal evolutionary engineering first pioneered by Blakeslee, Barbara McClintock began to construct maize stocks that could, through their own dynamics, produce random mutations. The discovery was as shocking to McClintock and her contemporaries as Muller's and Blakeslee's had been to theirs. (McClintock reported she was "astounded. . . . It had gone wild. The genome had gone wild."[99]) As Nathaniel Comfort has noted, not only did McClintock's technique involve a thoroughgoing and contested reconsideration of the nature of mutation,[100] but her use of the breakage-fusion-bridge cycle for producing mutation was positioned to replace the use of X-rays, casting them aside as "expensive and dangerous"—just like radium before them.[101]

As the effects of radiation increasingly came to be seen as fundamentally different from (and more damaging than) the diverse new ways in which mutations could be spontaneously induced in organisms, the role of ionizing radiations in the study of heredity became much less clear. Even as X-rays replaced radium on many fronts as tools of "extraordinary nicety," as Muller had put it, a fuller understanding of mutational processes meant that they were increasingly viewed as potentially *misleading* tools: Muller and Stadler disagreed repeatedly over whether the mutations induced by X-rays were the same as those occurring naturally.[102] And as X-rays, unlike radium, were without a "natural" analogy to life to fall back on for discursive comfort, more such contestations rapidly emerged. Even Delbrück would note by 1949 that "it may turn out that certain features of the living cell, including perhaps even replication, stand in a mutually exclusive relationship to the strict application of quantum mechanics, and that a new conceptual language has to be developed to embrace this situation."[103]

The connections between radium and life were thus, in some respects, disintegrating. Disanalogies between radiation and life had, of course, always been present. But with the development of novel experimental systems that increasingly technologized and instrumentalized the uses of radiation, and with the reduction of the basis of life to the gene in the view of many geneticists, the significance of these disanalogies became easier to discern. And these disanalogies were combined with changes in understandings of the nature of various forms of radiation, advances in X-ray technologies, and fully instrumentalized experimental techniques in radiation genetics that needed no further metaphorical justifications, thereby challenging the earlier connections once so obvious to Soddy, Burke, and their contemporaries, and even to Muller. The once obvious truth that radium had curious properties reminiscent of, ontologically similar to, and perhaps even generative of life became increasingly difficult to see by midcentury as ever more complicated understandings of the genic nature of mutation dovetailed with rising concerns about the uses of radium and popular notions of radiation as life-stealing rather than life-bearing.

In fact, germs of decay—such as the idea that radium could be detrimental to life, rather than stimulating or otherwise positive in its effects—were already present in the earliest literature on radium, from early visions of the potential inherent dangers of its untapped but unlimited power for misuse to understandings of radioactive decay as a kind of backward evolution from the more complex to the less complex.[104] Disanalogies had always been present: radium didn't *really* reproduce, and even the view of its daughter elements (decay products that were not the same as the original radium) as "mutants" seems to have strained the analogy too much for contemporaries to offer more than the first outlines of such an account. Soddy eventually came to blame radium for his infertility. And even Muller's search for the sterile products of irradiation can be seen as marking an important disintegration of the association of radium and life. Unlike Morgan's avoidance of sterile mutants (seen as noise obscuring the signal he was trying to detect, and therefore not a factor in his discoveries), and unlike Gager and Blakeslee's search for possible new species of *Datura* (which they clearly found), Muller's complicated techniques for calculating mutation frequency in *Drosophila* relied centrally on the identification of lethality itself—dead flies—as revealers of mutations, the observable evidence for his carefully designed tests. Neither dead flies nor X-rays (recall Frau Roentgen) would be the most obvious candidates for an ongoing association with life. Even as Muller's radioactive metaphysics increasingly construed the gene as something akin to an atom of radium and placed it at the center of all evolutionary change, the seeds of decay were already present.

Finally, as the "secret of matter" was to be found in the atomic nucleus, it stood to reason for Muller, as for many others, that the "secret of life" might reside in the biological nucleus, and that radium and its daughters might be the means to get there. "Whatever the secret of the gene's ability to reproduce itself and its mutation may consist in," Muller noted in 1950, "it seems today clearer than ever . . . that this is also the most fundamental secret of life itself."[105] But just as the biological claimants for the role of "secret of life" continued to shift over the years—from animalcule, organic molecule, or monad in earlier times to cell, chromosome, and gene, as told in these pages—so, too, did the physical claimants to act most intensely or instructively on them. Therefore, even as the shift from radium to other forms of ionizing radiation provided for multitudes of new experimental possibilities and for the emergence of the new fields of radiation genetics and radiobiology, the once familiar associations of radium with life continued to disintegrate as experimental setups and tools strained any immediately obvious connection between the two. As experimental productivity and epistemically helpful framings parted ways, such associations became increasingly unrecognizable by midcentury, and radium came to seem to have almost nothing to do with the secret of life at all.

Various other developments contributed to this destabilization, of course, and further transmutations, processes of decay, and disintegrations continued apace over the span of decades and across contexts. A general belief in radiation hormesis (the stimulating effects of radiation), so much a part of the early twentieth-century association between radium and life, began to go out of fashion by the 1940s.[106] Moreover, the rise of an instrumentalized radiation genetics, as well as larger cultural contexts in which the hydrogen bomb became the ultimate symbol of radioactive contamination and concern, led to a dawning "radiophobia."[107] The association of radiation and life, brought into being at a certain moment, the product of a particular reach and lifetime, was now also accompanied by world-historical events. By 1951 one book—*Our Atomic Heritage*—could even declare plainly in one chapter subtitle, "Radiations equal mutations," and in the next, "Mutations are not good."[108]

Dovetailing with the multitude of other concerns and events by the 1950s, radium's associations with life continued to disintegrate. Muller's own mutagenic studies with radium and X-rays contributed prominently to the concept of radiation-induced hereditary damage; his work brought concerns regarding the lack of a minimum threshold below which no mutational damage could be expected and introduced the concept of "our load of mutations" to a wider audience in 1949.[109] This view of the effects of radiation as lethal, damaging, and generally "bad"—Muller's rallying cry throughout the 1940s and beyond as he conducted further experiments—was the opposite of the widespread conception of mutation as profitable and life-enhancing held by most biologists in the first decade of the century: the shift from "Our Lady of Radium" to "Our Load of Mutations" was clear. In a Cold War battle for the planet in which the threat of exposure and of radioactive fallout were ever present, such concerns were more than merely biological. And so mutations themselves—once the high goal of experimental evolutionary efforts at Cold Spring Harbor and elsewhere—came to be routinely seen by geneticists as *detrimental* in nature, rather than the desirable new means for the production of agricultural superstars they had once been.

Although echoes of the radium-life connection continue to turn up in the most unexpected of places,[110] as is only appropriate for an association with a half-life, the overall trend was clear: by midcentury, radium—and by extension, radioactivity and ionizing radiation more broadly—had by and large transmuted. Now rarely seen to be helping to unveil the secret of life, radiation became increasingly associated with fears of cumulative and irreversible genetic damage, contamination, and death.[111]

Transmutations, decay, and disintegrations! Are radium and life the same thing? Or were they taken to be such, and how did that change over time? Were genes the atoms of life, the radium of the cell—or was it chromosomes, or viruses? Or "the quantum"? And why are these particular cases mentioned and others not? Why weaken the richly detailed and coherent narratives of the case studies in the previous three chapters with such a seeming smorgasbord of increasingly less compelling cases? Just what is going on here?

By midcentury, this proliferation of narratives, of possibilities, of likely descendants and dubiously relevant cases—only a few of which have been traced here—all becomes terribly confusing within the confines of one synoptic historical narrative. Just so much is to be expected in a historical narrative that takes transmutation and decay seriously as immanent tools of analysis and seeks to *perform* the same sorts of multiplicities and confusions and to raise the same sorts of hints and doubts that its actors themselves clearly grappled with by midcentury. Perhaps Brunngraber captured it best in his novel *Radium* (1936):

> "Radioactivity and life are one and the same thing."
> "Maybe, maybe," replied George in a dubious tone. . . . "I'm sorry," he said, "but my mind is growing somewhat confused."[112]

(That, indeed, is somewhat the point.)

Conclusion: The Secret of Life

"Even as to the fact, science disputed," Henry Adams remarked, "but radium happened to radiate something that seemed to explode the scientific magazine, bringing thought, for the time, to a standstill." The discovery of radium had, for Adams, brought a fundamental "snap" in the continuously swerving path of history. "Only in 1900," he said, did history experience such a profound discontinuity.[1] The insertion of radium into the broader discourse was for most historical intents and purposes an originary moment, albeit one reworking, refracting, and perpetuating a set of earlier traditions in novel form.

Radium thoroughly captured the public imagination in the first decade of the twentieth century, and the ways in which it was conceptualized and described in the popular realm played vital roles in biological theorizing and experimentation for decades to come. More than a mere "trickle-up" of scientific popularization, these "popular" and "scientific" understandings of radium and of life were mutually conditioning. At a first level, this book has aimed to reconstruct the experimental and discursive half-life of that initial moment of crystallization when radium and life first became interrelated and when talk of "life" was found to be suitable for the properties of the radioactive. *Radium and the Secret of Life* has sought to address how popular interest in radium and in new modes of conceptualizing radium's relationship to life surged at the same time, at a moment when "radioactivity somehow reminded people irresistibly of life." And I have tried to show how in the years following the radium craze, and even decades later, *life* itself in many ways somehow irresistibly reminded people—even biologists—of radioactivity.

This irresistible reminding took place largely through the application of metaphor. This, then, has also been the story of the generative power of metaphor across a field of experimental systems, a story of the metaphysics of metaphors, of metaphors made real in experiments, and of metaphors and experiments forgotten—a diachronics of performative metaphor. Moreover, we have seen how a series of main figures in the history—from Soddy, de Vries, Darwin, and MacDougal to Gager, Blakeslee, and Muller, among others—each suffused with the glow of the associations of radium and life, drew on metaphorical and metaphysical modes of description relating radium to life and transmutation to mutation. Over the course of the first half of the twentieth century, they refashioned these provocative resonances in ways that produced both novel conceptual understandings of the phenomena of life and new experimental techniques suited for further investigations.

The transmutation of radium's association with life into various and perhaps countless traces through thrown-off experimental systems and scientific practices is a remarkable phenomenon (albeit one that has drawn far less popular, scientific, or historical attention than have radium's own sensational properties or the remarkable circumstances of its discovery). From Burke's half-living radiobes to MacDougal's half-life calculations of mutation frequency, to Gager and Blakeslee's radium-based experiments to Muller's irradiation of the gene and his production of "half-alive" flies that enabled him to statistically measure the half-life of a gene, among the many other aspects of his radium-infused approach—in time, these transmutations not only transformed various key fields of research in the life sciences, but even came to obscure the once entirely obvious and all-powerful association of radium and life. And so, even the disintegration and dissipation of this association between radium and life reveals something of the deeper dynamics of history.

As the living atom shifted from radium itself toward radiobes, and the atoms of life from cells and chromosomes finally toward genes, the secret of life moved ever inward—though not without much multiplicitous veering and "throwing off"—until it reached deoxyribonucleic acid (DNA), where the application of this trope to the structure of the genetic material finally stuck. Although many other discoveries in biology have since laid claim to the title, DNA has remained intimately associated with the "secret of life" in ways that no other biological discovery to date has yet equaled. Radioactive traces, though faint, remained: one newspaper account of the discovery of the structure of DNA reported that Watson and Crick had discovered "the structural pattern of a substance as important to biologists as uranium is to nuclear physicists." Not only was DNA "the vital constituent of cells, [and] the carrier of inherited characters," the article concluded, it was "the fluid that links life with inorganic matter."[2] DNA was the new radium.

And so, just as investigators and commentators at the turn of the century once held that the discovery of radioactivity entailed the unveiling of the "secret of matter," the ascription of the "secret of life" to DNA seems to be one of the last and most powerful *discursive* remnants of the disintegrating association between radium and life, some fifty years after its inception.[3] Much more than a congenial and now familiar phrase with potent and seductive promise, the "secret of life" was the residual fallout from the bomb that radium had dropped on biology.

But there are any number of glowing radioactive paths one might trace from Muller's claim that the secret of life resided in the genes through the later disintegrating history of radium and life. For a generation already, scholars have rightfully warned against overly simplistic reductions, such as that of "the origin" of molecular biology, and against seeking to find the roots of a large, diverse, and multinational field in the work of some particular group of scientists (such as Watson and Crick), in some particular text (such as Schrödinger's *What is Life?*), or in the funding techniques of an important foundation (such as the Rockefeller Foundation). Angela Creager, for instance, has cautioned that a "linear narrative from Delbrück's experimentation with bacteriophage in the 1930s to the identification of DNA as the hereditary material in the 1950s overlooks the way the virus-gene analogy veered in the 1940s."[4] Bearing such warnings against Whiggery in mind, then, seeing DNA as the inheritor of the title of "the secret of life" is not an attempt to glorify that midcentury moment above others, nor to suggest that Watson and Crick's discovery was somehow the ultimate "outcome" of these associations between radium and life, any more than one of the γ-rays emitted in the decay of radium can stand supreme over any other. (The asymptotic reasoning employed in the very narrative device of tracing the half-life of these connections prohibits any such interpretation.)[5] But it is to follow a trace. And when a distinct and curious click emerges from the noise in the Geiger counter of history, and when this click can be directly related to an earlier and furiously radioactive period some ages past, it perhaps rightfully draws the historian's attention. If nothing else, this resonance provides a useful, familiar, and convenient ending for a story that, in approaching an asymptote, otherwise has no neat and easy narrative conclusion.

But perhaps there is a more theoretical insight to be earned. Perhaps this pervasive trope of the "secret of life" offers an important and powerful place to pause and reflect on just what it means to narrate the history of a radium-life nexus. Is radium's role in the history of biology an important one whose many uses should be detailed and explored in straightforward fashion? Or is it a wonderfully clever metaphor that usefully ties together disparate case studies into a coherent and wide-ranging narrative?[6] Or is it something still more, and might our own struggles to trace how far this connection between radioactivity and life persisted over later years provoke still further questions about how to best write a history of that which never quite disappears?

: : :

The connections between radium and life certainly proved much more than merely metaphorical and airily metaphysical: radium not only had everything to do with the origin of life, the origin of species, and many of the earliest experimental efforts to induce mutations artificially, but was centrally involved in the study of the nature of heredity contained in chromosomes and—following the further work of Muller—the study of mutation at the level of the gene. But in this breeder reactor of a history, the living atom and the atom of life have never been far apart. For several decades already, for example, the gene has often been described in ways that made it seem as if it were a master molecule radiating powerful controlling forces throughout its biological surroundings. Ever since Muller, genes have glowed. But our "play with words" has not led us "down a thorny path of 'merely dialectic exercises,'" to quote one of Burke's critics, nor have we "cantered off on a metaphor." Rather, we have seen how the interplay between metaphors and evidence has both constituted and transformed our understandings of radioactivity, of heredity, and even of the gene itself.

While many scholars engaging with the history of radium have dealt with technical aspects of radium's discovery, industrialization, and medicinal and therapeutic uses, this book has therefore attempted to offer a more intellectual and cultural history of radium in scientific practice (and in biological experimentation in particular), presenting the changing particulars of the associations between radium and life over the years and across experimental systems in order to illustrate how the experimental and discursive productivity of these associations eventually outpaced their ability to maintain a unified coherence. Indeed, as discursive tropes and material agents alike continued to transition, the initial association began to disintegrate.

Even successfully identifying midcentury historical examples of the ongoing association between radium and life (much less relating them to one coherent narrative) can prove confusing, if not maddening. But this effort, too, can be instructive—if not about radium or genetics, then about writing their interlinked history. Drawing on the important work of many other scholars in the history of biology, *Radium and the Secret of Life* has sought not to encapsulate the history of any particular subfield—it is neither a comprehensive history of any of the sciences of heredity or their approaches in the twentieth century, nor a synthetic account of early twentieth-century investigations into the origin of life, or of mutation. It has aimed to be neither a biographical account of the main historical figures nor an institutional history (though it has made use of all these accounts). And as a first attempt to contribute to the "prehistory" of radiobiology, it has certainly attempted to cut across traditional histories of radioactivity and biology. But more than this, as an empirically rich meditation on the experimentally productive uses of metaphor, it has tried to investigate what happened when analogies and metaphors in early twentieth-century biology became fruitful producers of ontological novelty. And by studying the changing interactions of metaphorical and metaphysical modes of conceptualization with a series of key experimental practices over the first half of the twentieth century, it has not only attempted to examine in detail the multifold connections between radium and life, but has also sought to trace the transmutations and ultimately the *disintegrations* of these performative metaphors and the metaphysics they brought with them. In so doing, it has followed sometimes slippery recognitions of ghostly radiances across time and experimental systems, and it has explored the ways in which these ever further transmutations might eventually lead even to the historical near-erasure of our very ability to recognize that such associations once had the generative power they did. Rather than merely rewriting the history, we see instead what makes it possible for the same events to be narrated and incorporated in different ways.

Radium and the Secret of Life has thus sought to tell its story in a reflexive manner: by performing ongoing acts of creative historical interpretation with evidence that has always been available to us, and—more importantly—by seeing this very narrative as part of the same half-life of ongoing transmutations and disintegrations of the association of radium and life as the science itself. More than a productive reflection on certain themes in the larger history of biology, and far from being "stark mad in metaphysics," this account is thus a provocation of the possibilities and consequences of writing history. The powerful associations between radium and life both reflected and made possible new modes of doing and talking about biology—and perhaps new modes of doing and talking about the history of biology as well. When the ever-scintillating world of radium meets the ever-sporting world of mutation, there are always more, and different, stories to tell: the historical is always already the historial.

: : :

In what Hans-Jörg Rheinberger has explored as the *historial*, there is "no logic of development that is ontologically or methodologically grounded." Rather, the "differential reproduction" of "research systems" involving "the generation of the unknown becomes the reproductive driving force of the whole machinery" and extends far beyond any particular technique, model organism, or individual:

> An experimental system has *more* stories to tell than the experimenter at a given moment is trying to tell with it. It not only contains submerged narratives, the story of its repressions and displacements; as long as it remains a research system, it also has not played out its excess. Experimental systems contain remnants of older narratives as well as fragments of narratives that have not yet been told. Grasping at the unknown is a process of tinkering; it proceeds not so much by completely doing away with old elements or introducing new ones but rather by *re-moving them*, by an unprecedented concatenation of the possible(s). It differs/defers. . . . The historical, without realizing it, obeys and discloses the figure and the signature of the historial.

And when a set of experimental systems relating radium to life is narratively assembled by a historian, this whole "*field* of systems," differing and deferring, may be said to be ultimately linked "not by stable connections but rather by possibilities of contacts generated by the differential reproduction of the systems." Such a "historial ensemble," in Rheinberger's terms, constitutes a "field of the possible."[7]

And so, rather than trying to delimit exact contours for its cases or its traces, *Radium and the Secret of Life* has accordingly attempted to move beyond the constraints of traditional historical narratives. Rather than using simple biographical tropes—telling the "life and death" of the association between radium and life—it has sought instead to narrate the emergence and ongoing decay of this association in and across experimental systems as a phenomenon of "the trace" within a larger field of possible realities, and of possible historical narratives. Deliberately reveling in the instabilities of the ways in which radium and life served as metaphors and experimental realities for each other, we come to acknowledge that these instabilities are as much in play for the historian as they were for the scientists. The "analytical moves" here are less "insights" than they are "the very elements of the analysis that one seeks to describe."[8]

Moreover, in following various historical traces in the previous chapters from the most obvious of connections to the more tenuous, I have attempted to extend the narrative as far as possible by following its contours until its traces began to slip through my fingers—to go beyond the plainly evident to the point where it becomes radically unclear whether or not a given datum is "actually" a residue or not. For only by narrating a story that comes from traces and ultimately returns to them—as Rheinberger has noted, "It is only the trace that will remain which creates, through its action, the origin of its nonorigin"—will the historical narrative itself enact the same processes of blindness and insight, of provocative inspiration and periodic forgetting, that were at play for its scientific actors.[9]

Indeed, it is only in the very "disintegration" of this historical narrative into traces by midcentury that one can begin to see the possibilities for a more "deconstructive" history of science. In short, the very dissolution of a compelling narrative association between radium and life as we near midcentury is not only seemingly inevitable, but *intentional.* The point here is historiological.[10] One need only look at how MacDougal, Gager, Blakeslee, and Muller each praised and later doubted the mutagenic efficacy of radium on their path to gaining clearer insights into the real nature of mutation to envision how deconstructing my own narrative might be the most promising way for historians to gain greater clarity on this history. And so, rather than declaiming the limits of an association between radium and life, and rather than hypostatizing or reifying it into concrete existence, *Radium and the Secret of Life* has sought instead, *pace* Henry Adams, to use radium as a metaphysical bomb for history just as it was for science.

It seems only appropriate to the study of the historical transmutations of radium's association with life over decades and across experimental systems that just as radium transmutes, transforms, decays, and disintegrates into its daughter elements, so, too, has this historical narrative been concerned to trace the disarticulation of experimental and discursive dimensions, and even to set up the conditions for its own dissolution. And while no ultimate conflation between radium and life is quite as possible for the historian as it was for various historical actors—we deal here with "radium," not radium—*Radium and the Secret of Life* has suggested the power inherent in a hermeneutics of *transmutation*: how the constant dissemination of meaning, the stimulating multidirectional decay of metaphors and metaphysics over decades and across experimental systems, is not only a phenomenon of history "back then," but even of history "right now" as it is being written (and read).

Indeed, in a sense, this is what historians *always already* do when they say they are "uncovering" new truths about the histories they thought they already knew so well. Starting with solid and definable case studies and then adding in ever-proliferating examples of increasingly "suspect" historical evidence—the work of chapter 6—is thus one way of highlighting the interpretive work that is always done in arranging historical sources into coherent narratives—or even in finding such resonant associations in the first place. Although we historians may not be as close to radium as our scientific actors, perhaps the creative processes of historical discovery are not so different from theirs after all.

But some will still seek to stub their toe on leaden residues and will want to ask, just how solid is this association between radium and life? Was it ever really there? Or has the historian perhaps just richly imagined this association and construed the evidence accordingly? In this intentional pushing of the limits of historical narrative, there is indeed a reason why, compared with Burke or Gager or Muller, it becomes more difficult to view aspects of the biology of the 1940s and 1950s as part of the narrative—because they very nearly aren't. Indeed, any given example may prove to be as powerful, contestable, or unstable as radium. And this is precisely as it should be. That any given case may always already be disappearing, transmuting into a sign of significance of some other potentially better and more accurate historical truth, or moving toward some asymptote of ultimate decay, is (in both senses) my claim's *end*.[11] Any attempt at a more consciously reflexive history must accordingly seek not only to recount, but to *perform*, these associations. In so doing, in testing the limits of evidence and argument, it will contain within itself the seeds of its own narrative decay. Thus that metaphysical bomb radium, that ultimate disseminator, has not only proved to be a wonderfully appropriate element for exploring the secret of life—it has also proved to be the central epistemic thing for this book.

Like a "Curie of the laboratory of vocabulary," we have, as the poet Mina Loy once wrote, "crushed the tonnage of consciousness, congealed to phrases, to extract a radium of the word."[12] The use throughout this book of terms such as "transmutation," "decay," "disintegration," and "throwing off" are accordingly far from mere wordplay. As we have seen, these were always already "contaminated" words, and they held simultaneously alchemical, radioactive, physical, biological, conceptual, and analytical valences for the historical actors, as they must for us. As there is no way to escape from the plasticity and constraints of language and the inherent and constitutive power of metaphor, the significance of this "contaminated" language is always up for consideration and reconsideration. The methodological aim of this history has been to revel in this instability and to let the radioactivity of language pass through from the actors' categories to our own, mutating and illuminating with a soft (but perhaps deadly) glow what it may along the way.[13]

: : :

But let's leave the glow behind for the moment and fall out back down to earth. *Radium and the Secret of Life* has attempted to reveal the numerous and powerful interconnections and ramifications of the associations between radium and life across a number of realms of the biological sciences in the first half of the twentieth century. These interconnections have to date gone largely unrecognized, and yet they were clearly central and highly visible in the early history of the physics and chemistry of radioactivity, in the vitalistic metaphors and metaphysics of the popular radium craze, in debates over how to experimentally investigate the historical origin of life, and through the work of major figures in the history of experimental evolution and classical genetics who were attempting to tease out fundamental facts of heredity. At its most basic level, my aim has been to recover this forgotten story with its turn-of-the-century roots and to examine some of the remarkably productive connections between radioactivity and the life sciences long before the dawn of molecular biology made such connections seem commonplace.

By detailing the emergence of this ever-sporting connection between radium and life and by tracing some of the ways in which it transmuted over the decades, I hope to have gone beyond a standard history of radioactivity and mutation research not only by uncovering the important "prehistory" preceding the later and better-known interwar and postwar fascination with experimental radiobiology and mutation studies, but also by showing how some of the guiding tropes linking radioactivity and life in time themselves became constitutive of new scientific theories and practices, especially in the heyday of classical genetics.[14] This swapping of metaphors and metaphysics, terminology and technique, between the realms of the radioactive and the living (and those who studied both) proved enormously fruitful and, as we have seen, not only conditioned the continuing use of radium in evolutionary and genetic research for years to come, but even involved the forgetting and rewriting of this very history. From the secret of matter to the secret of life, and from the secret of life to the secret of death: time after time the associations between radium and life have been, and are being, transmuted.

Acknowledgments

Researching and writing a book whose recurrent theme is that of the half-life accurately suggests that my gratitude knows no definite end. Primary thanks go to my intellectual mentors Mario Biagioli and Hans-Jörg Rheinberger for inspiring me with their scholarship and for guiding me from afar. Gratitude is due as well to Stephen Greenblatt for his provocative injunction to "find a resonant text!" which led me to stumble onto those first few passages relating radium to life in exceedingly curious ways many years ago. Many transmutations later, I see now how I have somehow arrived nearer the secret—if not yet the meaning—of life.

There is no warmer community of scholars than historians of biology, and my sincerest gratitude goes to Garland Allen, Peder Anker, Robert Brain, Soraya de Chadarevian, Nathaniel Comfort, Helen Curry, William deJong-Lambert, Evelyn Fox Keller, Dan Kevles, Barbara Kimmelman, Kim Kleinman, Sharon Kingsland, Robert Kohler, Matt Lavine, Susan Lindee, Elisabeth Lloyd, Jane Maienschein, Erika Milam, Staffan Müller-Wille, Lisa Onaga, Diane Paul, Karen Rader, Joanna Radin, Marsha Richmond, Neeraja Sankaran, Alexander Schwerin, Betty Smocovitis, Jim Strick, William Summers, and Mary Terrall, among many others. I would especially like to thank Angela Creager, who shares my passion for the life atomic. I am also grateful for wisdom from Gillian Beer, David Bloor, Sydney Brenner, I. B. Cohen, Diane Destiny, John T. Edsall, Donald H. Fleming, Richard C. Lewontin, Everett I. Mendelsohn, and John Stilgoe. I am deeply grateful to Karen Merikangas Darling and the University of Chicago Press, to my copyeditor Norma Sims Roche, and to three anonymous reviewers for their keen comments and queries.

This work would not have been possible without the support of the many caretakers of our cultural and scientific past. I am deeply indebted to the librarians, archivists, and assistants who have made the archival research on which this project depends both possible and thoroughly enjoyable: Rob Cox, Roy Goodman, Charles Greifenstein and Valerie-Anne Lutz at the American Philosophical Society; Mae Pan and Kathy Crosby at the Brooklyn Botanic Garden; Shelley Erwin and Loma Karklins at the California Institute of Technology; Clare Clark, John Zarrillo, Judy Wieber, and Mila Pollock at the Cold Spring Harbor Laboratory Library and Archives; and the hardworking staff of the Library of Congress, the University of Missouri at Columbia, and the Lilly Library of Indiana University. I would also like to thank each of these archives for their permission to quote from documents in their care and in some instances to reproduce photographs from their collections. In addition, I would like to express my gratitude to Fred Burchsted and the staff of Widener Library, the Ernst Mayr Library, and the library of the Harvard Herbaria for their unfailing help in tracking down sometimes difficult-to-locate materials. Thanks as well to Patrick McDermott, Paul Frame, Amy Viars, Joel Lubenau, Oak Ridge Associated Universities, and the University of Texas for locating and providing images, and to Eileen Price, pictorial archivist of the University of New Mexico, for help with securing reproductions. Just before this book went into production I had the serendipitous pleasure of meeting Helen Muller, whom I am delighted to have as a colleague at the University of New Mexico, and to whom I am indebted for her permission to quote from her father's papers and to reproduce images in her care.

Early portions of this book have appeared in print as "The Birth of Living Radium," *Representations* 97 (2006): 1–27, and I gratefully acknowledge the University of California for permitting republication here. Financial assistance has been scarcer than radium, but some small support was provided by Harvard University, the National Science Foundation, the Dibner Institute for the History of Science and Technology, Mel and Rita Wallerstein, and the American Philosophical Society.

On a more personal note, a shout-out goes to Glenn Adelson, who is everything a teacher and an intellectual role model should be; Diana

Eck, Dorothy Austin, Tom Batchelder, Beth Terry, and my fellow tutors of Lowell House, who made our great House my beloved home "from the age that is past, / To the age that is waiting before." My gratitude as well to Livingston Taylor and to Kim and Karen Serota for unparalleled writing and revising retreats on Martha's Vineyard and in Scottsdale, respectively. And unanticipated acknowledgment even to those sun-dewed lilies amid thorny roses and burnt stones who drew me from green-walled wise books and readied me to head west into the high desert most radiant and fair.

Heaps of gratitude most deservingly go to my parents, Therese and Carlos; my sister Ana and her husband Dave; and to my extended family of friends past and present, whose support at various points over the past decade has stood me in good stead: Theresa Choe, Erika Evasdottir, Jamie Jones, Matthew L. Jones, Elizabeth Lee, Johannes Lenz, Heather Lynch, David Munns, Carolyn Parrott, Aaron Pedinotti, David Pollack, Tonya Putnam, Anthony Sagnella, Anna Skubikowski, Cheoma and Derek Smith, Mark Solovey, Benjamin Tittmann, Lisa Vogt, Tina Warinner, Audra Wolfe, and Elizabeth Zacharias. I am most especially indebted to Marga Vicedo and Colin Milburn for untold aid, and to KC, who makes me glow. To each of you, thank you, and may you all find the book as stimulating to read as I have found it stimulating to write.

: : :

This book is dedicated to Lily Kay, whom I met only briefly many years ago, but whose magisterial work in the history of biology has been a constant stimulus to my own. She once asked about the history of the "secret of life." I hope here to have honored both her query and her memory.

Notes

INTRODUCTION

1. Adams, *Education*, 355.

2. James, *Pragmatism*, 62.

3. Adams, *Education*, 355.

4. Ibid., 381–82.

5. "Radium and Its Lessons" (*Lancet*); James, *Meaning of Truth*, 88: "Radium, discovered only yesterday, must always have existed, or its analogy with other natural elements, which are permanent, fails. In all this, it is but one portion of our beliefs reacting on another so as to yield the most satisfactory total state of mind. That state of mind, we say, sees truth, and the content of its deliverances we believe."

6. Blakeslee and Avery, "Methods of Inducing Doubling," 404, 408.

7. Recent scholarship on metaphor in science has emphasized that metaphor is not just representative, but performative, and that it "operates on the level of both scientific discourse *and* practice," according to James J. Bono. Recent scholarship has also repeatedly highlighted that the literature on metaphor in science is vast and unwieldy, and it has accordingly become routine for scholars to apologize for not trying to define "metaphor" or to precisely delimit its role in science, while moving on to analyze the fascinating use of metaphors in empirical cases. See, for example, Mirowski, *Natural Images in Economic Thought*, 451; Dörries, *Experimenting in Tongues*, 3; Keller, "Language in Action," 87; and Kay, *Who Wrote the Book of Life?*, 36. Rather than worrying about it, Bono suggests instead that a recognition of "the complexities

and rich variations of embodiment leads us away from an account of metaphor that stresses its universal features and foundations . . . [and] leads us to acknowledge that schemas, metaphors, and metaphoric systems of meaning are themselves subject to and situated in the particularities and specificities of history, culture, discourse, and all sorts of webs of relations." Bono, "Why Metaphor?," 216. Classic sources specifically on metaphor in science include Hesse, "Models, Metaphors, and Truth"; Lakoff and Johnson, *Metaphors We Live By*; and Bono, "Science, Discourse, and Literature." Finally, as Jacques Derrida has noted, "This epistemological ambivalence of metaphor, which always provokes, retards, follows the movement of the concept, perhaps finds its chosen field in the life sciences. . . . Where else might one be so tempted to take the metaphor for the concept?" *Margins of Philosophy*, 261.

8. For all the historiographical contributions this study seeks to make to the history of genetics and of radioactivity, therefore, it also seeks to make a *historiological* contribution (Novick, *That Noble Dream*, 8).

9. Rheinberger, "Experimental Systems."

10. Experimental systems are "systems of manipulation designed to give unknown answers to questions that the experimenters themselves are not yet able clearly to ask . . . vehicles for materializing questions." Rheinberger, *Toward a History of Epistemic Things*, 28. Or, as he put it elsewhere, an experimental system is "the smallest functional unit of research, designed to give answers to questions which we are not yet able clearly to ask. It is a 'machine for making the future.' It is not only a device that generates answers; at the same time, and as a prerequisite, it shapes the questions that are going to be answered. An experimental system is a device to materialize questions. It co-generates, so to speak, the phenomena of material entities *and* the concepts they embody." Rheinberger, "Experiment, Difference, Writing," 309. For earlier work on "experimental systems," see Kohler, "Systems of Production," and Kohler, "*Drosophila* and Evolutionary Genetics."

CHAPTER 1

1. Rutherford, *Radio-Activity*, 13.

2. As historian of radioactivity Lawrence Badash has noted, "Radioactivity . . . was not regarded as a major discovery." In what he called "an heroic age of radiations," radioactivity, simply enough, "because it did not 'do' anything of particular note, did not stand out among this plethora of rays. . . . Still, the world was not so jaded with new elements to ignore the Curies' claim." Badash, "Radium," 145–46.

3. Trenn, *Self-Splitting Atom*, 42.

4. Howorth, *Pioneer Research*, 83–84, 272–73.

5. Trenn, *Self-Splitting Atom*, 42.

6. Howorth, *Pioneer Research*, 90–91. As Spencer Weart has noted, "The connection between transmutation symbols and radioactivity was not only a matter of ancient traditions revived in modern laboratories; it also came through the deliberate use of imagery." Weart, *Nuclear Fear*, 73.

7. Howorth, *Pioneer Research*, 87, 90. Pierre Curie, however, continued to use the term: "Here we have a veritable theory of the transmutation of simple

bodies, but not as the alchemists understood it. Inorganic matter must have evolved, necessarily, through the ages, and followed immutable laws." Curie, *Madame Curie*, 220.

8. Trenn, *Self-Splitting Atom*, 3; Badash, "'Newer Alchemy,'" 89, 91.

9. Quoted in Howorth, *Pioneer Research*, 90.

10. Eve, *Rutherford*, 41. Rutherford described himself as having a "natural radioactivity." Rutherford to Loeb, January 4, 1908, LOC, folder 13, "Rutherford, Ernest."

11. Howorth, *Pioneer Research*, 98–99; see also Howorth, *Greatest Discovery*, 78. For more on the emerging market for radium as a commodity, see Rentetzi, *Trafficking Materials*.

12. Helium itself had only been discovered in 1895.

13. "Das war die chemische Sensation des Sommers 1903," from Karl Kuhn, *Natur und Kultur*, cited in Howorth, *Pioneer Research*, 104.

14. Howorth, *Pioneer Research*, 149.

15. Badash, "'Newer Alchemy,'" 92–93. Armstrong is quoted in Hampson, *Radium Explained*, 31.

16. Rutherford and Soddy's early description of radioactivity explained only radioactive change *within* the atom, however, as did the Curies' initial definition of "radioactive," which meant spontaneously giving off radiation. The other attendant features of radioactivity—such as whether other particles were thrown off as a result of this internal transformation—had yet to be established (Trenn, *Self-Splitting Atom*, 14). By 1906, when Soddy presented a paper entitled "The Evolution of the Elements" at the BAAS meeting in York, he still found colleagues, such as Armstrong, "astonished" at what he had to report, and many of the older generation still refused to believe what they were hearing (Howorth, *Pioneer Research*, 149). For more on Soddy, see Merrick, *World Made New*; Kauffman, *Frederick Soddy*; Sclove, "From Alchemy to Atomic War," and Freedman, "Frederick Soddy." For more on Rutherford, see Jenkin, "Atomic Energy Is 'Moonshine.'"

17. Thomson, *What Is Physical Life?*, 178–79.

18. For more on the nineteenth-century associations of cosmic and organic evolution, see Schaffer, "Nebular Hypothesis"; Lightman, "Evolution of the Evolutionary Epic"; and Secord, *Victorian Sensation*.

19. Badash has noted that the discovery of the transmutation of elements was not viewed with as much skepticism as one might expect (and as many secondary sources imply): "A careful search through the archives of the time . . . fails to support any such assertion!" Badash, "'Newer Alchemy,'" 92.

20. Ibid., 95. As Rutherford noted early on, however, Lockyer's theory of the evolution of matter was based "on evidence of a spectroscopic examination of the stars, and considers that temperature is the main factor in breaking up matter into its simpler forms. The transformation of matter occurring in the radio-elements is on the other hand spontaneous, and independent of temperature over the range examined." Rutherford, *Radio-Activity*, 500.

21. Howorth, *Greatest Discovery*, 94.

22. Quoted in "Disintegration of the Atom."

23. Soddy, "Evolution of Matter," 12.

24. Soddy, "Radioactivity" (1905), 276.

25. Soddy, *Matter and Energy*, 244.

26. Rutherford to Loeb, February 3, 1907, LOC, folder 13, "Rutherford, Ernest."

27. Cited in Weeks, *Discovery of the Elements*, 800.

28. Soddy, "Evolution of Matter," 10–11.

29. "Mystery of Radium," 10.

30. Soddy, "Evolution of Matter," 41.

31. "Notes" (*Electrician*), 437.

32. Millikan, "Radium," 9–10.

33. The equivalent words chosen as translations for these radioactive terms sometimes tend to obscure the resonant value these terms have in English: German's *halbwertszeit*, for instance, and even French's *vie moyenne*. I am fully aware that there may be different valences to this radium-life connection in different languages and cultures, and more work on these different contexts needs to be done. In the present work, I am primarily concerned with tracing these connections in the Anglophone context.

34. "The radiation of radium was 'contagious'—contagious like a persistent scent or like a disease. It was impossible for an object, a plant, an animal or a person to be left near a tube of radium without immediately acquiring a notable 'activity' which a sensitive apparatus could detect. This contagion, which interfered with the results of precise experiments, was a daily enemy to Pierre and Marie Curie." Marie Curie even referred to this problem as an "evil [that] has reached an acute stage." Curie, *Madame Curie*, 220, 222.

35. Shenstone, "New Chemistry," 521. See also Waters, "Radium and Human Life," 329; Eve, "Infection of Laboratories by Radium," 460–61; and Rona, "Laboratory Contamination"; as well as Burke's observation that "the radio-activity is thus infectious, but the infected body recovers in the course of time," in "Radio-Activity of Matter," 126.

36. Hampson, *Radium Explained*, 15; emphasis added.

37. According to the *Oxford English Dictionary*, the first occurrence of the term "half-life" is with respect to radium, though Rutherford had made determinations of the rate of decay (without such a label) for thorium as early as 1900, according to Pais, *Inward Bound*, 121. Other terms—such as "radioactive periods"—coexisted with "half-life" in the early decades: see, for example, Rafferty, *Introduction to the Science of Radio-Activity*, 81. Soddy referred to "life periods," as did Francis Venable in his *Brief Account of Radio-Activity*.

38. Soddy, "Present Position of Radio-Activity," 53.

39. Rutherford, *Radioactive Transformations*, 148ff.

40. "Facts about Radium," 966.

41. "Says Radium Is Modern Miracle," 30.

42. Whetham, *Recent Development of Physical Science*, 236, 291; emphasis added.

43. Ibid.

44. Chamberlin, "Introduction," 6.

45. "Science and Life," 254.

46. Salomons, "Wonders of Radium Explained," 13. *Science* also reported

on this phenomenon, noting that it took about a month to reach maximum activity: "Radium" (*Science*), 347.

47. One article commenting on Curie's interest in Crookes's new invention, the spinthariscope, noted, "It was as if he had been allowed to assist at the birth of a universe—or at the death of a molecule." Moffett, "Wonders of Radium," 11. Compare this with a French source from a decade later, which declared, "Nous savons comment le radium meurt." Houllevigue, *La Matière*, 119.

48. "Growth of Non-living Matter," 590.

49. B. B. Boltwood commented in 1915 that the comparative ease of encountering and studying radium led to its being "considered and accepted as a standard or typical radioactive substance." Boltwood, "Life of Radium," 852.

50. Hammer, *Radium*, 16.

51. Curie, *Madame Curie*, 265.

52. In the case of all the others, he said, there is just too little evidence of their existence other than their radioactivity. Pais, *Inward Bound*, 116.

53. Soddy, *Chemistry of the Radio-Elements*, 44, 52.

54. Soddy, "Evolution of Matter," 37; emphasis added. For more on such attempts to "grow radium" (including a letter from Rutherford to Soddy on June 20, 1904), see Soddy, "Life-History of Radium"; and Soddy, "Production of Radium from Uranium."

55. Soddy, "Radio-Activity" (*Electrician*), 725.

56. Quoted in Howorth, *Pioneer Research*, 267.

57. "Says Radium Is Modern Miracle," 30.

58. See, for example, Whetham, "Life-History of Radium"; Raveau, "L'origine, la longévité, et la descendance du radium"; Hahn, "Muttersubstanz des Radiums"; "The 'Life' of Radium"; Boltwood, "Life of Radium"; Turner, "Ionium, the Parent of Radium"; and Fajans and Makower, "The Growth of Radium C from Radium B."

59. "For the present it seems preferable to refer to the body simply as the parent of radium." Soddy, "Parent of Radium," 256–74.

60. Howorth, *Greatest Discovery*, 89.

61. Soddy, "Evolution of Matter," 13.

62. Ibid., 21. Trenn notes that λ would have to wait for later work by Einstein before it could become a truly *atomic* constant (*Self-Splitting Atom*, 130, 136, 142), while Pais notes that it was only in 1927 that it was discerned why different radioactive elements have the particular lifetimes they do (*Inward Bound*, 103).

63. Pais has suggested that work done on luminescent phenomena "made the lifetime concept familiar" for those working on questions surrounding "unstable systems of atomic dimensions" (*Inward Bound*, 121).

64. Howorth, *Pioneer Research*, 126.

65. Chamberlin, "Introduction," 4.

66. Mendelsohn, *Heat and Life*, 8, 21.

67. Ibid., 95.

68. Hammer, *Radium*, 3.

69. Curie, *Pierre Curie*, 119.

70. See, for example, "To Make Luminous Drinks from Radium," 2.

71. It was only when he realized that the "invisible phosphorescence" of his sample did *not* decrease over time once the external stimulus (sunlight) had been removed, and that it did not show the phenomenon of a "lifetime" like all other phosphorescent phenomena he had studied, that Becquerel realized he was facing an entirely new phenomenon.

72. Humboldt, *Cosmos*, 135, 202, 309, 342.

73. Otis, *Müller's Lab*, 7.

74. Wood, "Scintillations of Radium," 195.

75. Crookes, "Emanations of Radium," 523.

76. Saleeby, "Radium and Life," 226.

77. He goes on to describe "une hérédité d'éléments," "la généalogie du radium," and the discovery of radioactivity as not only the discovery of transmutation, but also the discovery of "l'élixir de vie." He concludes, "Loin d'être une exception, une monstruosité de la nature, le radium n'est que la représentant le plus éminent d'une propriété, peut-être universelles de la matière." Houllevigue, *La Matière*, xxiii, 123, 130.

78. Wood, "Scintillations of Radium," 195–96.

79. De Kay, "Color Visions of the Kiowas," SM13.

80. Lodge, "Radium and Its Lessons," 85.

81. Bottone, *Radium*, 60.

82. Millikan, "Radium," 9–10.

83. Wells, *World Set Free*, 22–23.

84. Maceroni, quoted in Morus, *Frankenstein's Children*, 130–31.

85. Rutherford, *Radio-Activity*, 492; Rutherford, "Radium," 390–96.

86. Eve, *Rutherford*, 107.

87. "New Rays Discovered," 14.

88. Pais, *Inward Bound*, 105. The first such mention of these doubts followed Marie Curie's discovery of polonium, when its energy proved even more intense than that of thorium.

89. Ibid., 109–11.

90. "Says Radium Is Modern Miracle," 30. Comparisons to perpetual motion are legion in the literature of this period; see, for example, A. Frederick Collins, "Common Sense Applied to Radium," SM4; H. Greinacher, "Ein neues Radium-Perpetuum mobile"; and Henri Poincaré's "Éloge de Curie" delivered at the Académie des Sciences in Paris.

91. The rhyme in *Punch* continues: "Take but a pinch of the same, you'll find it according to experts / Equal for luminous ends to a couple of million candles / Equal for heat to a furnace of heaven knows how many horsepower." Keller, *Infancy of Atomic Physics*, 107.

92. Coues, *Daemon of Darwin*, 30–31.

93. Prior to the theory of atomic disintegration, it was even suggested in some quarters "that the energy of radium might be due to the analogous power of that element to derive its energy from outside sources by sifting out the molecules of different speeds impinging on it." "Bacteria and Radio-Activity," 127. For more on the afterlife of Maxwell's demon in twentieth-century history of biology, see Keller, "Molecules, Messages, and Memory." For

more on the demon itself, see Leff and Rex, *Maxwell's Demon*; and Canales and Krawjewski, "Little Helpers."

94. Spengler, *Decline of the West*, 420, 423.

95. Wells, *Tono-Bungay*, 104, 183, 267.

96. "One also thinks of the 'fatigue' of metals, and the hysteresis or 'memory' of certain materials": Gray, *Advancing Front of Science*, 215. For more on the history of the transition from "life-units" to "life-processes," and on various attempts to characterize inanimate entities as potentially living, see Wilder, *Life*, especially chapter 8, "Does Life Inhere in Matter?" For a very useful scholarly account, see Lehman, *Biology in Transition*, 128, 133. See also Singer, *History of Biology*, 464, 573; Kay, "W. M. Stanley's Crystallization of the Tobacco Mosaic Virus"; Lorch, "Charisma of Crystals in Biology"; and Donna Haraway, *Crystals, Fabrics, and Fields*.

97. Maurice Cornforth, *Dialectical Materialism and Science*, 47.

98. Lionel Beale, *On Life*, 52, 39.

99. Saleeby, "Radium and Life," 226.

100. "*Vitality* acts in living centers upon matter only infinitely near the centre." Beale, *On Life*, 41; cf. Beale's argument against a belief in living atoms as more literally understood, 39.

101. Ibid., 58; emphasis added.

102. Dolbear, "Life From a Physical Standpoint."

103. Even the ways in which radium is described as "constantly and without cessation throwing off from itself, at terrific velocity, particles of matter—Energy Force, Power—call it what you will," owes its discursive roots to this hereditarian branch of the living atom tradition (Degnen, "Radio-Active Solar Pad.") Just before introducing the term "gemmules," Darwin refers in his pangenesis hypothesis, Chapter 27 of his *Variation of Animals and Plants Under Domestication*, to his assumption that "the units [of the body] throw off minute granules which are dispersed throughout the whole system." This talk of body parts "throwing off" hereditary particles returned in de Vries's later discussion of mutation, in which some particularly mutable species were said to "throw off" new varieties and species. This use of language will be explored further in the Conclusion.

104. Lehman quotes Singer's 1965 entry in the *Encyclopedia Britannica*: "Underlying much biological thought of the early twentieth century was a sense that 'the substance of life,' like inert matter, must be resolvable into ultimate particles—'quanta' of life, comparable with 'quanta' of inheritance (Mendel) and 'quanta' of energy (Planck)": *Biology in Transition*, 133. Similarly, the only nonbiological milestone in L. C. Dunn's chronology of major events in the history of genetics is Planck's discovery of the quantum: "Now, with the hindsight provided by discoveries of the last decade, we can see that 'quantum biology' might have been a better designation than 'particle biology' [for the study of heredity since 1900] . . . since the essential feature is the discrete nature of the elements first recognized by segregation and recombination." Dunn, "Genetics in Historical Perspective," 81; for a timeline, see the draft version of Dunn's chapter in APS Dunn, box 7, 122.

105. "Artificial Biogenesis," 705.

106. Jan Sapp has noted, "During the second half of the nineteenth century, many leading theorists postulated that underlying the structure of the cell there existed microscopically invisible living units standing somewhere between the cell and the ultimate molecules of living matter. These living units, or hypothetical 'elementary organisms,' were the starting point of every leading theory of heredity and development." Sapp, *Evolution by Association*, 38.

107. Osborn, *Origin and Evolution of Life*, 6. By the second half of the century, François Jacob noted that the growth of genetics and biochemistry had "changed the centre of gravity of living bodies. Organisms were no longer thought of simply as organs and functions arranged in depth; they no longer appeared as curled round a source of life from which organization radiated." Jacob, *Logic of Life*, 243.

108. "Radium" (*Electrician*), 277.

109. Quoted in Howorth, *Greatest Discovery*, 91.

110. Keller, *Infancy of Atomic Physics*, 103.

111. "A Prophet of Radium."

112. "Mr. Soddy's Views," 5.

113. "Wells's book indeed includes several near-verbatim renditions of Soddy's prose." Sclove, "From Alchemy to Atomic War," 177. For more on Wells, see Seed, "H. G. Wells and the Liberating Atom." Other than a 1906 reference in the *London Magazine* and Anatole France's *L'île des Pingouins* in the following year, Wells's book is among the earliest known references to an atomic explosion or bomb. It was Wells's story that would motivate Leó Szilárd to join the Manhattan Project after reading the book in 1932.

114. Wells, *World Set Free*, 23.

115. Ibid., 25, 27.

116. de la Peña, *Body Electric*, 173–74. For more on radium as a commodity, see Rentetzi, "Packaging Radium, Selling Science."

117. de la Peña, *Body Electric*, 175.

118. "Costly Particles of Radium," 6.

119. de la Peña, *Body Electric*, 178.

120. Hammer, *Radium*, 16.

121. "Radio-Active Substances," BR7.

122. See, for example, Soddy, "Present Position of Radio-Activity," 47; and "Radium and Its Lessons." The equation of matter and electricity was criticized by others.

123. Howorth, *Pioneer Research*, 104–5.

124. Iwan Morus has traced these earlier traditions of popular paid electrical lectures and demonstrations displaying rare electrical phenomena to the public in the British context in his *Frankenstein's Children*.

125. de la Peña, *Body Electric*, 174, 9.

126. Weart, *Nuclear Fear*, 43. Weart traces this connection between radium and electricity even years later, citing a 1935 "movie serial, updating *The Exploits of Elaine*, [that] brought cowboy actor Gene Autry back from the dead in a 'radium reviving room,' although the apparatus also crackled with traditional electric sparks."

127. While in the case of electricity, distinctions were often made between

electrical and galvanic phenomena (the electricity of life), with radium, these two realms—the putatively physical and that more inclined to the biological—were conflated from the very beginning.

128. This was in addition to showing how to make gold, facilitating interplanetary communication, and, perhaps presciently, determining "how the world will ultimately be destroyed." Theodore Waters, "Radium and Human Life," 328.

129. Morus, *Frankenstein's Children*, 71.

130. de la Peña, *Body Electric*, 174–75, 109.

131. *Sacramento Bee*, November 7, 1903, quoted in de la Peña, *Body Electric*, 181.

132. Badash, "Radium," 150.

133. Quoted in Badash, "'Newer Alchemy,'" 91.

134. "Women and Radium," 6.

135. "The Popular Interest in Radium," 6.

136. Keller, *Infancy of Atomic Physics*, 106. The title of this section, "An Indecent Curie-osity," is quoted from Keller, 108.

137. Ibid., 107.

138. Badash, "Radium," 147, 150; Moffet, "Wonders of Radium," cited in Howorth, *Greatest Discovery*, 85.

139. "Radium" (*New York Times*), 6.

140. Ackroyd, "Radium and Its Position in Nature," 859.

141. "A Possible Use for Radium," 338.

142. Hampson, *Radium Explained*, 32–33.

143. Badash, "Radium," 148.

144. "Radium Energised Wool," 389.

145. "Odor and the New Radiation," 677.

146. And yet, the author continued, "it is only reasonable to suppose that instead of killing . . . microbes [radium] would act as a delightful stimulant, and it still remains for any investigator to prove the contrary." Collins, "Common Sense Applied to Radium," SM4; emphasis added. With respect to radium's medical potential, the *Lancet* had cautioned already by 1909: "We utter this word of warning to those who expect too much from radium in the present state of our knowledge. Should it prove true that in radium we possess a method of combating malignant disease anywhere and everywhere we should rejoice indeed, but we urge that hopes should not be unduly raised." "The Radium Institute," 773.

147. "Do Men Radiate Light?," 6.

148. "Radium was the rage." A. S. Eve, Rutherford's biographer, on meeting Rutherford in January 1903. Quoted in Badash, "'Newer Alchemy,'" 91.

149. Hampson, *Radium Explained*, 32–33.

150. Harvie, "Radium Century," 100; see also Landa, *Buried Treasure to Buried Waste*, 35–36.

151. Harvie, "Radium Century," 100–104. For more on radium as commodity, see Rentetzi, "Packaging Radium, Selling Science." For more on radium therapy, see Macklis, "Radithor and the Era of Mild Radium Therapy"; and American Institute of Medicine, *Abstracts of Selected Articles on Radium*

and Radium Therapy. For more on radium and its relationship to the American public, see Lavine, *First Atomic Age*.

152. Allyn, “Costumes Treated with Radium Paint,” 6.

153. Badash, “Radium,” 147.

154. “Radium as a Preservative,” 5.

155. de la Peña, *Body Electric*, 174.

156. Shaw, *The Doctor’s Dilemma*, xxxii.

157. The epithet “Our Lady of Radium” was bestowed on Marie Curie by the American journalist Israel Zangwill in a batch of three short essays in the May 1904 *Reader Magazine*. See Kevles, *Naked to the Bone*, 26; and Badash, “Radium,” 147.

158. Clay, “Radium.”

159. “Religious Notices,” 9. A religious fascination with “spiritual radium” has persisted up to the present day: “There exists a great amount of ‘pitchblende,’ referring to all of those who profess belief in the Lord Jesus Christ. God loves His believers and if they persist in their hope they will be saved in the Day of the Lord . . . there must be found in the mass of believers a quantity of ‘radium’—people who will give their all that God’s will might be performed throughout His creation.” Trumpet Ministries, Inc., *The Word of Righteousness*, accessed June 1, 2014, http://www.wor.org/Books/r/radium.htm.

160. Adams, *Education*, 356.

161. Sharp, “Radium of Romance,” 69.

162. Russell, “Research in State Universities,” 853.

163. “‘Macbeth’ in ‘Pure Radium,’” 30.

164. “From the Lives of Players,” 21.

165. Corelli, *Life Everlasting*, 18.

166. Ibid., 19, 33.

167. “When Silence Is Golden,” 6; “Limit Their Size,” 1. A similar usage can be found in “The Cure,” 1.

168. On the 1950s radioeuphoria, see, for example, Weart, *Nuclear Fear*; and Boyer, *By the Bomb’s Early Light*.

169. “The Rays of Radium,” 6.

170. “Lost Radium Tube,” 5.

171. “Diamond Rays Pierce Paper,” 10.

172. “Crowds Gaze on Radium,” 3; “Popular Interest in Radium,” 6.

173. “Billiard Room Buzzings,” SM10.

174. Harvie, “Radium Century,” 100.

175. “Columbia St. Louis Exhibits,” 6. In fact, “the lectures and the exhibit of radioactive preparations and minerals were considered the outstanding attractions of the fair.” Badash, “Radium,” 148–49.

176. “‘Liquid Sunshine’ on Tap,” 6.

177. “The Roentgen ray has been of immense value in curing cancer,” Morton also noted, “but radium promises to go far ahead of it.” “To Make Luminous Drinks from Radium,” 2.

178. “Man Competing with Radium,” 209.

179. Clark, *Radium Girls*, 52; Harvie, “Radium Century,” 100–104; “Bottling the Sunshine,” 9.

180. Lavine has noted that as radium "began to become more commonly available at the clinic—and, to a much greater extent, when it became part of the spa culture—it acquired the characteristic of being not merely vital, but vital to the processes of life. When, during this period, spa-water sellers compared water without radium emanation to air without oxygen, they meant it literally." Lavine, *First Atomic Age*, chap. 4.

181. "I think it is highly probable that there is radium in the sun. . . . The electrons given out by the sun sometimes strike our atmosphere and make a rare gas, called krypton." Saleeby, "Radium the Revealer," 88.

182. Wickham, *Radiumtherapy*, 15. When she saw an X-ray of her hand, one of the first X-rays ever made of a human being, Frau Roentgen spoke of having seen her own death: Kevles, *Naked to the Bone*, 38.

183. Kevles, *Naked to the Bone*, 70.

184. Field, "Radium and Research," 764.

185. Saleeby, "Radium the Revealer," 86.

186. "Revelations of Radium," 490.

187. Le Bon, *Evolution of Matter*.

188. Weart, *Nuclear Fear*, 36–37. I am indebted to Weart's research and analysis here; all publications mentioned in this paragraph are cited in *Nuclear Fear*, 37.

189. "Does the Dead Body Possess Properties Akin to Radio-Activity?," 1201.

190. For more on the history of the N-ray controversy, see Klotz, "The N-Ray Affair"; and Nye, "N-rays." The later mitogenetic ray controversy of the early 1920s can be viewed as a further transmutation of similar resonances between radiation and life; see chapter 6.

191. Nye, "N-rays," 133.

192. Wilson, "Is Radium an Element?"

193. Proumen, *Les Rayons X, Le Radium, Les Rayons N*, 67; Clay, "Radium," 234.

194. Nye, "N-rays," 130, 132.

195. "Radio-Activity of the Animate," 104.

196. Nye, "N-rays," 125.

197. Klotz, "The N-Ray Affair," 170.

198. Wood, "The *n*-Rays," 530–31.

199. Burke, "The Blondlot *n*-Rays" (*Nature*, February 8, 1904), 365; Burke, "The Blondlot *n*-Rays" (*Nature*, June 30, 1904), 198.

200. Howorth, *Pioneer Research*, 146.

201. Ibid.

202. Ibid., 92.

203. Howorth, *Greatest Discovery*, 88.

204. Soddy, "Evolution of Matter," 42.

CHAPTER 2

1. "Scientist's Great Discovery," 1.

2. "A Radium Product That Seems to Live," 1; Hale, "Has Radium Revealed the Secret of Life?," 7.

3. "Origin of Life: Mr. Burke Describes His Experiments," 5.

4. Hale, "Has Radium Revealed the Secret of Life?," 7.

5. "They are introduced to admiring physicists as radiobes, the discoverer being too modest to name them after himself." "Cambridge Radiobes," 11.

6. As noted in chapter 1, for more on the nineteenth-century associations of cosmic and organic evolution, see Schaffer, "Nebular Hypothesis"; Lightman, "Evolution of the Evolutionary Epic"; and Secord, *Victorian Sensation*.

7. Burke, "Physics and Biology." Burke's planetary model of the atom, explicitly comparing the electron corpuscle to a planet and the positive nucleus to a sun, derived directly from his interlinking of the discourses of cosmic and organic evolution. Burke's theories might sound like those of a crackpot, but Rutherford would develop his own planetary account of the atom four years later.

8. Strick, *Sparks of Life*, 192. Strick has masterfully explored the complexities of the debate in the late nineteenth century, including the relationship of spontaneous generation controversies to larger issues of "Darwinism" in Britain: Strick, *Sparks of Life*, 12, 191. The canonical source for a history of the spontaneous generation debates is Farley, *Spontaneous Generation Controversy*. See also Kamminga, "Studies," and Kamminga, "Historical Perspective," as well as Fry, *Emergence of Life on Earth*.

9. See Geison, "Protoplasmic Theory of Life."

10. Strick, "From Aristotle to Darwin," 54.

11. On Buffon, see Sloan, "Organic Molecules Revisited." On Maupertuis, see Terrall, "Salon, Academy and Boudoir"; and Wolfe, "Endowed Molecules and Emergent Organization." On Haeckel, see Haeckel's *Kristallseelen* and Snelders's "Zijn vloeibare kristallen levende organismen?"

12. For a list of examples, see Moriz Benedikt, *Krystallisation und Morphogenesis: Biomechanische Studie* (Vienna: Perles, 1904), 65, quoted in Thomas Brandstetter, "Imagining Inorganic Life." For more on this early history of "synthetic biology," see Campos, "That Was the Synthetic Biology That Was."

13. Brandstetter, "Imagining Inorganic Life."

14. Burke, *Origin of Life*.

15. Pycraft, "What Is Life?," 500.

16. "Burke's Own Account," 43.

17. Burke, "On the Spontaneous Action of Radio-Active Bodies," 79.

18. "Action of Radium on Beef-Gelatine," 315. In fact, Burke had initially delayed publishing his results, as "so momentous a result as it seemed required careful confirmation, and much delay was also caused in taking the opinions of various men of science." Burke, "On the Spontaneous Action of Radium," 294.

19. Soddy had just become president at the previous meeting, on November 2. All subsequent quotes from this meeting are from Burke, "The Spontaneous Action of Radium and Other Bodies."

20. "Burke's Own Interpretation," 535.

21. Burke, "The Spontaneous Action of Radium and Other Bodies," 35.

22. Burke, *Origin of Life*, plates facing 98, 100, 102, 108, 110, 112.

23. "Origin of Life: Momentous Discovery," 5.
24. Saleeby, "Science: The Origin of Life," 668.
25. Saleeby, "Origin of Life," 4.
26. "Origin of Life: Momentous Discovery," 5.
27. Burke, "On the Spontaneous Action of Radio-Active Bodies," 79.
28. "Radium and Vitality: 'Radiobes,'" 1738.
29. "Cambridge Radiobes," 11.
30. Burke, "Evolution of Life," 185.
31. Burke, "Origin of Life," 396–97; Burke, *Origin of Life*, 109–10.
32. The meaning of "half-life" at this time came via at least three distinct roots: the work of Soddy, Becquerel's study of phosphorescence, and the work of Pflüger, discussed below.
33. "Radium and Life," 10. Another newspaper asked, "Who is Burke? Is he a foreigner? In one of the continental papers, in a very capable editorial on the matter, his name is given as Johannes Butler Borksi, whilst the same article calls him Burke, later on." "Spontaneous Generation," 2.
34. Thomson et al., *History of the Cavendish Laboratory*, 159.
35. These "research students" were ordinarily awarded an M.A. degree after two years' residence and the presentation of a original thesis. According to Thomson, "after a few years' trial this was replaced by Ph.D. (Doctor of Philosophy), a new degree created by the University for their benefit." Thomson, *Recollections and Reflections*, 136–37.
36. Kim, *Leadership and Creativity*, 98.
37. Thomson et al., *History of the Cavendish Laboratory*, 213; Burke, "On the Phosphorescent Glow in Gases."
38. Burke's work is cited in Kim, *Leadership and Creativity*, 189.
39. Thomson et al., *History of the Cavendish Laboratory*, 213. By 1905, Burke had been engaged in such experiments for six years.
40. Saleeby, "Origin of Life," 4.
41. Kim, *Leadership and Creativity*, 132.
42. Kim, *Leadership and Creativity*, 121, 167; Thomson et al., *History of the Cavendish Laboratory*, 232.
43. A footnote in the text at this point reads, "In the same sense as the cell, although it may admit of being broken up into its constituent parts by exceptional means."
44. Burke, "Radio-Activity of Matter," 126, 129–31.
45. "Origin of Life: Mr. Burke Describes His Experiments," 5.
46. In fact, just the year before his sensational announcement (1904), he had patented a neon lamp, and he continued with his interest in luminescence for years to come.
47. Burke, "On the Spontaneous Action of Radio-Active Bodies," 79.
48. Quoted in Verworn, *General Physiology*, 306; emphasis added.
49. "Origin of Life: Momentous Discovery," 5.
50. "A Fascinating Theme," 3; "Origin of Life: Mr. Burke Describes His Experiments," 5.
51. "Cambridge Radiobes," 11.
52. "Origin of Life: Momentous Discovery," 5.

53. "Origin of Life: Eminent Men," 5.
54. "The Microbe's Ancestor," 6.
55. "Professor Burke's 'Radiobes,'" 6.
56. "A Fascinating Theme," 3.
57. Saleeby, "Science: The Origin of Life," 668.
58. Saleeby, "Radium the Revealer," 85.
59. Saleeby, "Radium and Life," 226–30.
60. "Origin of Life: Mr. Burke Describes His Experiments," 5.
61. Ibid.
62. Pycraft, "What Is Life?," 500.
63. Slocum, "Mr Burke and His Radiobes," 1011.
64. "Clue to the Beginning," 6813–14.
65. "Life's Secret," 5.
66. Burke, "The Spontaneous Action of Radium and Other Bodies," 37.
67. Burke, *Origin of Life*, vi and chap. 7.
68. "Cambridge Radiobes," 11.
69. Burke, *Origin of Life*, vi.
70. Ibid., 191.
71. "Origin of Life: Momentous Discovery," 5.
72. "A Fascinating Theme," 3.
73. "Life's Secret," 5; cf. "Origin of Life: Theology and the Radium Experiments," 5; and "Origin of Life: Well-Known Theologian on Mr. Burke's Experiments," 3.
74. Hutton, "Origin of Life," 5.
75. Pycraft, "What Is Life?," 500. Then again, he said, arguments against spontaneous generation "cannot, strictly speaking, it is true, be met by any contradiction." The contention was a metaphysical one. Burke, "Origin of Life," 389–90.
76. Burke, *Origin of Life*, vi.
77. The word "sensational" is an actor's category, and it was widely used to describe Burke's discovery. James Secord has analyzed the use of this terminology in describing the impact of Chambers's anonymously published *Vestiges of the Natural History of Creation* (1844). More work would need to be done to ascertain whether the word had equivalently "sensorial" connotations in this particular turn-of-the-century context. See Secord, *Victorian Sensation*.
78. Hutton, "Origin of Life," 5.
79. "Origin of Life: Eminent Men," 5.
80. Lodge, "What Is Life?," 668.
81. Brewster, "J. Butler Burke's Own Explanation," SM7.
82. Burke, "Evolution of Life," 185.
83. Burke, *Origin of Life*, 81. One of Burke's reviewers pointed out that Burke "uses 'life' in a very extended and unusual sense. He says, for example, that a flame is alive. . . . But then nobody supposes for a moment that a live flame is at all like a living plant. Mr. Burke attributes to his radiobes a sort of artificial life, midway between the life of a plant and the so-called life of a fire. He holds that we ought to recognize many different sorts of life. The life of flame is one sort, and the life of radiobes another, and the life of animals

and plants still a third. He even holds that radium itself is, to some degree, alive." Brewster, "J. Butler Burke's Own Explanation," SM7. Angela Creager has noted a similar dynamic in the later history of virology: "The rendering of viruses as experimental systems for understanding the 'secrets of life' was somewhat paradoxical; its advocates never claimed that viruses were alive." Creager, *Life of a Virus*, 185.

84. Dana, "Origin of Life," BR1.

85. Burke claimed to have produced something on the way to life, not life itself. Accordingly, he thought Bastian was claiming too much, too fast, and explicitly distanced himself from such work. "[Burke] denies the validity of the experiments of Bastian, (who also denies his) and does not think that there has yet been any actual demonstration of the production of living cells out of dead organic matter." Dana, "Origin of Life," BR1. Intriguingly, one author had noted as late as 1911 that Bastian might benefit from looking into the effects of radium on the spontaneous origin of life, as "there is at present much evidence of the importance of radium in earth-history." Hall, "The Great Enigma," 355. For more on Bastian's prominent role in the history of spontaneous generation, see Strick, *Sparks of Life*.

86. Viewing "nutrition" in this way is not as much of a stretch as it might seem, given a physicist operating under a Spencerian definition of life and with Burke's own understanding of an "organism" as "a structure, a nucleus and an external boundary or cell wall." "Burke's Own Interpretation," 534.

87. Ibid. Cf. "Origin of Life: Mr. Burke Describes His Experiments," 5.

88. "Origin of Life: Momentous Discovery," 5. Indeed, one London newspaper had reported just the year before Burke's experiments that "the biologists and physicists are getting into disputes over the definition of the boundary line separating their respective provinces, for the 'neutral strip' between them, once a *terra incognita*, is being explored from both sides." "Growth of Non-living Matter," 589.

89. "Origin of Life: Mr. Burke Describes His Experiments," 5.

90. "Origin of Life: Momentous Discovery," 5.

91. Burke, *Origin of Life*, 166.

92. "Origin of Life: Mr. Burke Describes His Experiments," 5.

93. Ibid.

94. Burke, *Origin of Life*, vi; and Burke, "Origin of Life," 397. Cf.: "Once there had been a continuous scale, but, unfortunately, the greater portion of these things had been eliminated. The most one could hope to do in the laboratory was to attempt to bridge over the gaps—fill them in completely we never should." Burke, "The Spontaneous Action of Radium and Other Bodies," 37.

95. "Origin of Life: Momentous Discovery," 5.

96. "The Microbe's Ancestor," 6; emphasis added.

97. As Strick has noted, "the very slipperiness of the terms [in the spontaneous generation debates] and the changing definitions of many of them are crucial to how the debates turned. It is only by tracking the usage of the terms closely throughout that we can see when the ambiguity of a term, rather than actual empirical observations, is a source of disagreement as well as when deliberate changes in usage represent rhetorical strategies." *Sparks of Life*, 12.

98. "Wonders of Radium Explained," X4.

99. "Generation by Radium," 1; Hale, "Has Radium Revealed the Secret of Life?," 7; "The Microbe's Ancestor," 6.

100. "A Radium Product That Seems to Live," 1. This despite the fact that *New York Times* a day before had rapidly conflated two meanings of "artificial life"—Burke's and Loeb's—and in so doing had removed the epistemological subtlety of Burke's position; see "Generation by Radium."

101. "A Cell-Killer as a Cell-Maker," 6.

102. "Professor Burke's 'Radiobes,'" 6.

103. B. C. A. W., "Origin of Life," 187.

104. "Action of Radium on Beef-Gelatine," 315.

105. "Origin of Life" (*Times Literary Supplement*), 123.

106. "Action of Radium on Beef-Gelatine," 315.

107. Hutton, "Origin of Life," 5.

108. "Previous Experiments," 5.

109. Nevertheless, he concluded, "I shall watch with interest Mr. Burke's discovery in the hope that it may shed some light on the terra incognita which separates the organic from the inorganic." Best, "Origin of Life," 3.

110. Saleeby, "Science: The Origin of Life," 668.

111. Saleeby, "Radium and Life," 226.

112. "Origin of Life: Mr. Burke Describes His Experiments," 5.

113. "A Cell-Killer as a Cell-Maker," 6.

114. "Ramsay was *the* expert on disintegration," Badash has noted, even though at times his research "was incompetent . . . [and his] public statements . . . misleading, for he was prone to exaggeration and self-glorification." Badash, *Radioactivity in America*, 29.

115. Shenstone, "Origin of Life," 409; see also Shenstone, *New Physics and Chemistry*, 359–60.

116. "Ramsay, Radium, and Burke," 215.

117. Ramsay, "Can Life Be Produced by Radium," 556.

118. Ramsay, "Radium and Its Products," 57. Transmutation via radioactivity and the production of radium was often alluded to as the secret of the alchemists or the philosophers' stone at this time. George Darwin commented in 1905, for example, that "we are surely justified in believing that we have the clue which the alchemists sought in vain": Darwin, "Address of the President," 232. See also Saleeby, "Radium the Revealer," 85.

119. "Wonders of Radium Explained," X4.

120. Rudge took photographs from a few minutes after the initial contact of the radium with the gelatin to several days and "in some cases weeks later." Rudge, "Action of Radium," 380.

121. Ibid.; Rudge, "On the Action of Radium," 258–59. See also Burke, "The Spontaneous Action of Radium and Other Bodies," 33ff.

122. Rudge, "Action of Radium," 381.

123. Ibid., 382.

124. Ibid., 384.

125. Rudge's earliest criticisms of Burke's work were recorded at the same meeting of the Röntgen Society at which Burke presented his work, in

December 1905. See Burke, "The Spontaneous Action of Radium and Other Bodies," 37ff.

126. Loeb, *Organism as a Whole*, 39.

127. Pauly, *Controlling Life*, 115–16.

128. Loeb, *Organism as a Whole*, 210.

129. B. C. A. W., "Origin of Life," 187–89.

130. Burke, "The Spontaneous Action of Radium and Other Bodies," 38.

131. B. C. A. W., "Origin of Life," 190.

132. Pycraft, "What Is Life?," 500; Thomson, "Radiobes and Biogen," 1–3.

133. "What is to be made of this, for example? 'If more progress is not made in this borderland it is, as we fear, of the awe and dread which in these departments of knowledge each professor entertains towards each.'" Shadwell, "Origin of Life," 123.

134. B. C. A. W., "Origin of Life," 190, quoting Burke, *Origin of Life*, 187.

135. Thomson, "Radiobes and Biogen," 1–3.

136. In an even more damning criticism, he concluded, "Sentences could be quoted which sound well but are unintelligible when analysed closely. The volume must be looked at as we look at an impressionist picture of which we enjoy the general effect but which we dare not examine in detail. We must especially avoid searching for a satisfactory proof of the existence of the radiobe, to which Mr. Burke owes his latter-day reputation." Schuster, "New Books," 5.

137. Mallock, *Immortal Soul*, 25, 167.

138. Ibid., 169, 180, 262, 265.

139. Ibid., 271, 396, 416.

140. Yarker, "W. H. Mallock's Other Novels," 190.

141. Burke, "Artificial Cells," 355.

142. Ibid., 357–58.

143. "Origin of Life: Momentous Discovery," 5.

144. Burke, "Artificial Cells," 357.

145. Ibid., 355–56.

146. Ibid., 358–59.

147. Aleksandr Oparin later described Leduc as having been "allured by the wraith of external resemblance." Oparin, *Origin of Life*, 56.

148. Burke, "Artificial Cells," 359.

149. Burke, "On the Spontaneous Action of Radium," 294. Dubois claimed to have first announced his discovery in an academic lecture he gave on November 3, 1904; Burke's announcement was his May 25, 1905, letter to *Nature*. See Slocum, "Mr. Burke and His Radiobes," 1011–12.

150. "Some critics have suggested that these forms I have observed may be identified with the curious bodies obtained by Quincke, Lehmann, Schenck, Leduc and others in recent times, and by Rainey and Crosse more than half a century ago; but I do not think, at least so far as I can at present judge, that there is sufficient reason for so classifying them together. They seem to me to have little in common except, perhaps, the scale of being to which as microscopic forms they happen to belong." Burke, "On the Spontaneous Action of Radium," 294.

151. Ibid.

152. "Origin of Life: Mr. Burke Describes His Experiments," 5.

153. "Flotsam and Jetsam," 87.

154. Burke, "Evolution of Life," 178.

155. Ibid., 179, 188, 179–80. Burke soon included psychology under the purview of biology as well.

156. As one reviewer summed it up, "He contends that he has produced something which gives a clue to the origin of life, or rather to its nature; for if his argument be correct, life has no origin." "Origin of Life" (*Times Literary Supplement*), 123.

157. Moore later provided a similar argument in a different context: "Although Pasteur has conclusively proven that life did not originate in certain ways, that does not exclude the view that it arose in other ways. The problem is one that demands thought and experimental work, and is not an exploded chimera." Moore, *Origin and Nature of Life*, 163. William Ritter's later critique of origin-of-life work—that it created things that could not be living because they did not come from living things—is contradicted by Burke's argument that there are many possible modes of origin. See Ritter, *Unity of the Organism*. This question of the essential historicity of life would return with Gager's work on the effects of radium on plants (described in chap. 3).

158. Burke, "Evolution of Life," 176.

159. Burke, "Correspondence—The Origin of Life," 560.

160. Only fleeting references to Burke's sensational work have survived in laboratory correspondence. A letter to Rutherford from George F. C. Searle—held to be "the Laboratory's most influential teacher for several generations"—refers to Burke's experiments as having "caused a little amusement" at the Cavendish. This letter is dated September 14, 1905, which means it would have just followed Burke's summer of fame: Cambridge University Library, MSS ADD 7653 S50, cited in Kim, *Leadership and Creativity*, 166.

161. Burke, "Artificial Cells," 357.

162. Burke, "Some Fruitless Efforts to Synthesise Life," 304.

163. Thomson et al., *History of the Cavendish Laboratory*, appendix.

164. Kim, *Leadership and Creativity*, 171–72.

165. Soddy, "Radioactivity" (1907).

166. One commentator held, for example, that "an artificial specimen must be capable of a cell formation which then goes on to reproduce itself by cell division or mitosis, and then continues to do so independently for successive generations. If it can do nothing of the kind, then this supposed living substance is a delusion." Thomson, *What Is Physical Life?*, 58.

167. Saleeby, "Radium and Life," 229.

168. "Life by Chemical Action," 8.

169. For an earlier iteration of this dynamic from the history of electricity, see Secord, "Extraordinary Experiment."

170. "Scientists Discuss the Origin of Life," 5.

171. See Brooke, "Wöhler's Urea and Its Vital Force?"; and Kohler, "History of Biochemistry."

172. The historian John Farley had hinted that there was a continuity between spontaneous generation and origin-of-life research, but stopped short

of giving a full account. Strick has added further details and noted that there is a "deceptive impression of discontinuity between 1880 and 1920, the time when Oparin commenced theorizing on the origin of life, that has dominated histories of the subject": Strick, *Sparks of Life*, 191. Along these lines, this chapter is a further effort to show how Burke's use of radium and his creation of radiobes were crucial in the reorientation of his own discourse from "spontaneous generation" to a new focus on the question of experimental access to the *historical*.

173. Burke, *Emergence of Life*, 4; Burke, *Mystery of Life*, 132.

174. Burke, "Evolution of Life," 189.

175. Ibid., 190.

176. Davis, "Review of Burke's *Emergence of Life*."

177. "Mr. J. B. Butler Burke," 6.

178. Burke, *Mystery of Life*, 132.

179. Clement, "Translator's Introduction," 11, 28.

180. Becquerel, "L'action abiotique," cited in "A Radio-Hypothesis of Life's Origin," 23.

181. Burke, *Emergence of Life*, 4.

182. Burke, *Mystery of Life*, 147.

183. "Mr. J. B. Butler Burke," 6.

CHAPTER 3

1. Moffet, "Wonders of Radium," 14.

2. Harding, "Dr. MacDougal's Botanical Feat," SM1.

3. MacDougal, "Trends in Plant Science."

4. MacDougal, "Heredity."

5. Ibid., 521.

6. Johannsen, "Genotype Conception of Heredity," 141. Johannsen remained interested in the work of MacDougal and Gager and visited the Brooklyn Botanic Garden in October or November of 1911.

7. Strick, *Sparks of Life*. See also his *Origin of Life Debate*.

8. MacDougal, "Origin of Species by Mutation" (*Torreya*, July 1902), 99–100; Korschinsky, "Heterogenesis und Evolution," 273. Intriguingly, both Burke and MacDougal were interested in producing artificial cells; MacDougal deployed artificial cell setups in his laboratory in order to better model osmotic and other processes in living cells.

9. Kellogg, *Darwinism To-Day*, 327.

10. A fascinating story remains to be told of the fuller history of "mutation"—from at least its transformation of earlier medieval discourses of the "marvels" and "wonders" associated with the preternatural to the nineteenth-century discourse of "sports" and "monsters." Radium, having been discovered at just this decades-long transition from talk of monsters to talk of mutants, experienced its own parallel discursive instability, as it was described alternatively as a "monstrosity of nature" or, as Soddy put it, that which underwent a process of "transmutation" itself. While a full history of the concept of biological mutation has yet to be written, for partial attempts see Mayr, *Growth of Biological Thought*; and Mayr, *Animal Species and Evolution*. As Mayr has

noted, "The term 'mutation' has had a tortuous history": *Animal Species and Evolution*, 168. Lock claimed a century ago that "perhaps the earliest use of the actual word 'mutation' in this sense is to be found in 'Pseudodoxia Epidemica' by Dr. Thomas Browne" of 1650, in the chapter "Of the Blackness of Negroes." Lock also claimed that "the actual observation of variations of this kind is of quite recent date, and their recognition is largely due to the exertions of Bateson": Lock, *Variation, Heredity, and Evolution*, 123. Others have found possible precursors to the mutation theory in the late nineteenth century in Meehan, "Hybrid Oaks," 55; and Kerner, *Die Abhängigkeit der Pflanzengestalt von Klima und Boden* (MacDougal et al., *Mutations*, 76). For more on the cultural history of mutation in the nineteenth and twentieth centuries, see Campos and Schwerin, *Making Mutations*.

11. See, for example, MacDougal, "Alterations in Heredity" (*Botanical Gazette*), 242. MacDougal viewed de Vries's mutation theory as being "a logical step from his earlier contribution to the subject of electrolytic dissociation of salts in solutions." He noted, in fact, that the "essential feature" of de Vries's theory was the existence of "separable, measurable characters." MacDougal, "Activities in Plant Physiology," 464.

12. Moore, "Deconstructing Darwinism."

13. "Or, to put it in the terms chosen lately by Mr. Arthur Harris in a friendly criticism of my views, 'Natural selection may explain the survival of the fittest, but it cannot explain the arrival of the fittest.'" De Vries, *Species and Varieties*, 825–26.

14. De Vries, *Species and Varieties*, 550.

15. De Vries, "Principles of the Theory of Mutation," 81.

16. Lock, *Variation, Heredity, and Evolution*, 313.

17. Gager, "Review of *The Mutation Theory*," 740.

18. Piper, "Botany," 895–96.

19. Theunissen, "Closing the Door," 241–42. For more on the reception of the mutation theory, see Allen, "Hugo de Vries and the Reception of the 'Mutation Theory,'" and Sharon Kingsland, "Battling Botanist."

20. Theunissen, "Closing the Door," 241.

21. Gager, *Heredity and Evolution in Plants*, 117.

22. Quoted in Gager, *Heredity and Evolution in Plants*, 115.

23. Gager, "Review of *The Mutation Theory*," 740; emphasis added.

24. Gager, *Heredity and Evolution in Plants*, 117–18.

25. Harding, "Dr. MacDougal's Botanical Feat," SM1.

26. Maienschein, *Transforming Traditions in American Biology*.

27. MacDougal, "Origin of Species by Mutation" (*Torreya*, May 1902), 65.

28. MacDougal, "Discontinuous Variation," 223.

29. MacDougal, "Trends in Plant Science," 490.

30. MacDougal, "Alterations in Heredity" (*Botanical Gazette*), 241–43.

31. Ibid., 242.

32. MacDougal et al., *Mutations*, 61.

33. MacDougal, "Mutation in Plants," 770.

34. MacDougal, "Discontinuous Variation," 224.

35. MacDougal et al., *Mutations*, 76.

36. "The use here of the name *Oenothera* for species belonging to the genus *Onagra* (Tournefort, Adanson, Spach) is according to the decision of Dr. J. N. Rose in a recent paper (1905), where he points out, and with good reason, that *Oenothera biennis* should be considered as the type of the genus." Anna Murray Vail, "Identity of the Evening Primroses," in MacDougal et al., *Mutations*, 66.

37. MacDougal, "Origin of Species by Mutation" (*Torreya*, June 1902), 83.

38. MacDougal, *Mutants and Hybrids*, 32.

39. MacDougal et al., *Mutations*, 3.

40. MacDougal, *Mutants and Hybrids*, 3.

41. MacDougal et al., *Mutations*, 61.

42. MacDougal also found *Raimannia odorata* to be a suitable test species.

43. MacDougal, "Alterations in Heredity" (*Botanical Gazette*), 242.

44. MacDougal, *Mutants and Hybrids*, 51.

45. Ibid., 33.

46. MacDougal, "Origin of Species by Mutation" (*Torreya*, June 1902), 83.

47. Ibid.

48. MacDougal et al., *Mutations*, 2.

49. MacDougal, "Heredity," 520.

50. Ortmann, "Facts and Interpretations in the Mutation Theory," 185.

51. Kingsland, "Battling Botanist," 486.

52. MacDougal, "Origin of Species by Mutation" (*Torreya*, July 1902), 98–99.

53. See, for example, Lock, *Variation, Heredity, and Evolution*, 146. For more on MacDougal, see Patricia Craig, "Daniel MacDougal."

54. de Vries, *Species and Varieties*, viii–ix.

55. Howorth, *Greatest Discovery*, 29–30.

56. Quoted in Pais, *Inward Bound*, 122.

57. Ibid.

58. Coen, "Scientists' Errors," 192.

59. Ibid.

60. See, for example: I. Bernard Cohen, "Scientific Revolution"; Ian Hacking, "Was There a Probabilistic Revolution 1800–1930"; and M. Norton Wise, "How Do Sums Count?"; in Krüger et al., *Ideas in History*.

61. Kingsland, "Battling Botanist," 492.

62. Pais, *Inward Bound*, 112.

63. Johannsen, "Genotype Conception of Heredity," 158.

64. Quoted in Thomson, *Heredity*, 90–91.

65. Webber, "Effect of Research," 598.

66. Thomson, *Heredity*, 93–94.

67. Bateson, *Materials for the Study of Variation*, 18. One reviewer later noted that "Mendelism, as expounded by Professor Bateson and his school, is the application of the atomic conception to organised life." "The Mendelian Theory," *Standard* (London), April 27, 1909.

68. As L. C. Dunn once noted, "There is little doubt that the most important methodological contributions to and from genetics have been logical and statistical ones by which theoretical models have been prepared for testing. . . .

This was a result of the discovery, itself due to statistical thinking, of the atomic and molecular organization of matter." Dunn, "Genetics in Historical Perspective," 76. Erwin Schrödinger would also later pick up on this theme: "The significant fact is the discontinuity. It reminds a physicist of quantum theory—no intermediate energies occurring between two neighbouring energy levels. He would be inclined to call de Vries's mutation theory, figuratively, the quantum theory of biology. We shall see later that this is much more than figurative. The mutations are actually due to quantum jumps in the gene molecule." Schrödinger, *What Is Life?*, 46. See also "The Quantum of Evolution" in chapter 5.

69. Darwin, "Address of the President," 228.

70. Of course, Darwin's creative leap was not uniformly embraced: as Edward B. Poulton noted, "I do not, of course, doubt that there is reality in the analogy between the evolution of States and of species, but it is not, I submit, close enough to justify the author's reasoning from one to the other." Poulton, "Theory of Natural Selection," 50.

71. Darwin, "Address of the President," 230.

72. Ibid., 228.

73. Ibid., 229.

74. Thomson, *What Is Physical Life?*, 179–80.

75. Conklin, "Problems of Evolution," 126, 127–28; emphasis added.

76. Conklin, "Mechanism of Evolution," 57.

77. Davenport, "Form of Evolutionary Theory," 451.

78. Conklin, "Mechanism of Evolution," 58. For more on the presence-absence theory, see Shull, "'Presence and Absence' Hypothesis"; and Swinburne, "The Presence-and-Absence Theory."

79. Muller, "Reversibility in Evolution," 263.

80. Davenport, "Form of Evolutionary Theory," 459.

81. Ibid., 463.

82. Ibid., 463–64.

83. MacDougal, "Trends in Plant Science," 489.

84. Richards, "Recent Studies," 289.

85. "If we except those actively working on radium, the belief in transmutation is for the most part confined to the American textbooks." "Radium and Helium."

86. de Vries, "The Aim of Experimental Evolution."

87. Gager, "Effects of Radium Rays on Plants," 1007.

88. Wilson, "Cell in Relation to Heredity and Evolution," 111.

89. Harding, "Dr. MacDougal's Botanical Feat," SM2.

90. Gates, "Mutation in *Oenothera*," 600, 662.

91. Gates, "Mutations and Evolution," 77.

92. MacDougal, *Mutants and Hybrids*, 80; cf. "The instability seems to be here as permanent a quality as the stability in other instances." de Vries, *Species and Varieties*, 543–45, 564. De Vries had found eight types of mutants, seven of which were constant while the eighth, *O. scintillans*, appeared only eight times and was the most unstable of them all, in a constant process of throwing off other mutants (Thomson, *Heredity*, 93). *O. scintillans* was first

named in de Vries, "Recherches Experimentales"; "Sur L'Origine des Espèces," and was said to "continuously and consistently g[i]ve rise to a variety of forms in its progeny, which included the parent and its mutants." MacDougal, "Heredity," 512.

93. MacDougal, "Heredity," 523.

94. Kingsland, "Battling Botanist," 487.

95. Richards, "Recent Studies," 291.

96. Morgan, *Experimental Embryology*, 32.

97. MacDougal, "Direct Influence of the Environment," 115, 128.

98. MacDougal, "Organic Response" (*American Naturalist*), 6.

99. See Fujimura and Clarke, eds., *The Right Tools for the Job*; Muriel Lederman and Richard M. Burian, "Introduction"; and Hull et al., eds., *PSA 1994*.

100. MacDougal, "Discontinuous Variation," 225.

101. MacDougal, *Mutants and Hybrids*, 31.

102. Gager, "Effects of Radium Rays on Plants," 989.

103. MacDougal, "Alterations in Heredity" (*Botanical Gazette*), 244.

104. Even Soddy had written in 1913 on the "biochemical effects of radioactivity": Soddy, "Radioactivity" (1913), 323.

105. MacDougal et al., *Mutations*, 62–64, 2–3.

106. "The discovery of the mutants in the seedling stage when only two or three small leaves are present is difficult for the first time, although after becoming accustomed to the typical forms and learning the aspect of the things to be looked for it is comparatively easy to recognize the better-known mutant types. Even then the mutants previously seen are much more readily distinguished than those known only by descriptions." MacDougal, *Mutants and Hybrids*, 31. See also Kohler's description of a similar dynamic among the drosophilists regarding the tacit knowledge required to identify mutations in the Morgan lab in *Lords of the Fly*. It was precisely this sort of subjective bias that Muller sought to avoid in his experimental technique (see chap. 5).

107. George Harrison Shull later complained, "The practice prevailing among taxonomists of ascribing a hybrid origin to a newly discovered form, which, in outward anatomical characters, is between two known species, is extremely pernicious and is not justified by facts obtained in cultural work." Shull, "The Fluctuations of Oenothera Lamarckiana and Its Mutants," in MacDougal et al., *Mutations*, 58.

108. MacDougal, *Mutants and Hybrids*, 90.

109. MacDougal, "Alterations in Heredity" (*Botanical Gazette*), 241. As MacDougal also noted, "The derivative of *Oenothera biennis*, first obtained in 1905, has now been tested to the fifth generation, hybridized with the parental form, and cultivated under the most diverse conditions. No reasonable doubt as to its character remains." "Alterations in Heredity" (*CIW Year Book*), 63. See also MacDougal, "Organic Response" (*American Naturalist*), 18.

110. MacDougal, *Mutants and Hybrids*, 31, 243.

111. MacDougal, "Discontinuous Variation," 210.

112. For more on the shifting meanings of mutation, see Campos and Schwerin, *Making Mutations*.

113. MacDougal, "Discontinuous Variation," 220.

114. Piper, "Botany," 895. Elsewhere, MacDougal noted that "De Vries assumes that any group of individuals which are independent, self-perpetuating and sufficiently distinct by taxonomic characters to meet the requirements of systematic botany constitutes a species irrespective of origin, and in the consideration of his results the importance of his conclusions is not lessened materially whether the forms with which he has dealt are considered as species or varieties so long as they are shown to consist of distinct and independent individuals capable of transmitting certain characters which are assumed to be constant within the limits of ordinary fluctuating variation." MacDougal, "Origin of Species by Mutation" (*Torreya*, May 1902), 66.

115. "It is not so much the extreme types of leaves which give to a plant its characteristic appearance and appeal to the systematist, as the type to which the majority of the leaves belong." MacDougal, *Mutants and Hybrids*, 44.

116. Ibid., 85.

117. MacDougal, "Organic Response" (*American Naturalist*), 36.

118. Ibid., 18.

119. MacDougal, "Heredity," 518–19.

120. MacDougal et al., *Mutations*, 78.

121. MacDougal, "Heredity," 516.

122. MacDougal et al., *Mutations*, 84.

123. Coen, "Scientists' Errors," 180.

124. Ibid.

125. Craig, "Daniel McDougal," 39, footnote 6; MacDougal et al., *Mutations*, 90.

126. MacDougal, "Alterations in Heredity" (*Botanical Gazette*), 256.

127. MacDougal et al., *Mutations*, 87.

128. After a stint as professor of botany at the University of Missouri (Columbia) from 1908 to 1910, Gager returned to New York City, this time to the Brooklyn Botanic Garden, where he remained as director until his death in 1943. "Gager, Dr. Charles Stuart."

129. Gager, *Effects of the Rays of Radium on Plants*, 76.

130. Ibid., 76–77. See, for example, Dubois, *La création de l'être vivant et de lois naturelles*, and Dubois, "Radioactivité et la vie." Gager even cites the priority dispute between Dubois and Burke, noting that Dubois claimed priority because similar effects could be produced with nonradioactive bodies.

131. Gager, *Effects of the Rays of Radium on Plants*, 77.

132. Ibid., 81–83.

133. Gager, *Heredity and Evolution in Plants*, 43. Such an emphasis on the historicity of life would later find a keen defense elsewhere as well, as in Ritter's *Unity of the Organism*. See also Moore's and Ritter's arguments in the context of Burke's findings, mentioned in chapter 2.

134. Gager, "Review of *The Mutation Theory*," 740.

135. Gager, *Effects of the Rays of Radium on Plants*, v. One can only wonder if Mr. Lieber was the inspiration for the name of Mallock's character Mr. Hugo, in *An Immortal Soul*, as mentioned in chapter 2.

136. Gager, "Influence of Radium Rays." Cold Spring Harbor director

Charles Davenport was interested in Gager's radium-coated rods, or "radium pencils," as a means to "remove electrical charge from paraffin ribbons." Davenport to Gager, December 5, 1909, APS Davenport, series 1, box 40, folder 1, "Gager, C. Stuart." Moreover, while scholars like Angela Creager have recently emphasized how the medicinal uses of radioactive materials in the late 1930s and the 1940s gave physicists of the time "a strong incentive to build bridges with physicians and biomedical researchers"—uniting cyclotrons with hospitals, for example—similar ties were already extant in the early days of radium-based experimental biological research. Researchers at the time regularly thanked both hospitals and colleagues in physics departments for their supplies, and the links between physics and biology—radium and life—were as real and institutional as they were metaphorical. Creager, "Tracing the Politics," 370.

137. "If the radium is in *a sealed tube*, it cannot make anything outside it *radioactive*. The effects you observe can only be ascribed to the penetrating rays passing through the glass viz. β and γ rays." Rutherford to Gager, June 2, 1907, BBG, box 3, folder 12, "Misc R."

138. Gager, "Effects of Radium Rays on Plants," 987.

139. Gager, *Plant World*, 113.

140. Gager, *Relation between Science and Theology*, 16–21.

141. Gager, *Heredity and Evolution in Plants*, 82.

142. Gager, "Influence of Radium Rays," 223–24.

143. Gager, "Some Physiological Effects of Radium Rays," 773.

144. Gager, *Effects of the Rays of Radium on Plants*, 22.

145. Ibid., 56. Marie Curie had even reported by 1903 that Friedrich Giesel had observed radium-treated leaves to "turn yellow and wither away." Curie, *Radio-Active Substances*. See also "Radium and Vegetation," *Leavenworth Times*, December 6, 1903, cited in Lavine, *First Atomic Age*.

146. Caspari, "Die Bedeutung des Radiums," 37.

147. Gager, "Influence of Radium Rays," 224.

148. Gager, *Effects of the Rays of Radium on Plants*, 253.

149. Gager, "Some Physiological Effects of Radium Rays," 778. Gager noted elsewhere that "the irregularities produced by radium rays in karyokinesis do not seem to call for any special explanation in addition to that suggested in discussing the abnormalities of tissues and organs in chapter 16. Such irregularities are only a morphological expression of physiological disturbance, and it may be seriously questioned whether we are justified in expecting the morphological appearance and behavior of chromosomes to explain things, any more than do variations in leaf-margins, or other purely structural facts. The problem of the causes of variation and inheritance lies deeper than morphology." Gager, *Effects of the Rays of Radium on Plants*, 272.

150. Gager, *Effects of the Rays of Radium on Plants*, 245, 237.

151. Gager, "Influence of Radium Rays," 224, 228, 230.

152. Ibid., 231.

153. Ibid., 228.

154. Gager, *Effects of the Rays of Radium on Plants*, 228. Gager's language is echoed by Alexander, who writes that a "large single dose [of

radiation] which does not kill also leaves a permanent mark which is revealed as premature ageing." . . . "Some investigators believe that atomic radiations hasten the onset of typical senile alteration and can be considered to accelerate aging." Alexander, *Atomic Radiation and Life*, 91–92.

155. Nordau, *Degeneration*, 552–53.

156. Gager, *Effects of the Rays of Radium on Plants*, 274. Cf. "Influence of Radium Rays," 232, and "Some Physiological Effects of Radium Rays," 763–64, 778.

157. "Correcting Nature."

158. "Sleeping Plants Wakened by Radium."

159. Gradenwitz, "Forcing Plants by Means of Radium," 77.

160. "Lorsqu'il s'agit d'une radiation existant à dose élevée dans le milieu ambient où évoluent les êtres vivants, on peut se demander l'effet de l'absence de cette radiation. C'est la *radioexpérimentation negative*." Guilleminot, *Rayons X et Radiations Diverse*, 121.

161. Gager, "Effects of Radium Rays on Plants," 999, citing Blaauw and van Heyningen, "The Radium-Growth-Response of One Cell."

162. Further exploration of this shifting context is called for: does the later success in "negative" radioexperimentation (in order to study what Gager would later call a "deradiation response") have more to do with changing notions of "background radiation" than it does with the simple possibility of a new kind of experiment? While an earlier generation may well have viewed the residual radiation of all things as potentially mutagenic, later successes may have had more to do with operational definitions of a background equivalent of zero when no new mutants appeared as a result of natural radiation. This would then have defined a "sufficiently" shielded environment.

163. Richards, "Recent Studies," 289.

164. "Have you tried soaking peony seeds in various solutions of acids and alkalies? I think you would be more apt to secure acceleration of germination in a manner that would be useable in practice by such methods, rather than by radium." Gager to L. C. Glenn, November 2, 1918, BBG, box 4, folder 5, "G."

165. Gager, "Effects of Radium Rays on Plants," 1004. For more about other later induced-mutation research in horticulture and agriculture, see Helen Curry, "Accelerating Evolution."

166. Richards, "Recent Studies," 298.

167. Gager, *Effects of the Rays of Radium on Plants*, 260.

168. Gager, "Effects of Radium Rays on Plants," 988.

169. "To have been a 'stimulating' factor which the lapse of more than twenty years has not effaced is an assurance that the efforts of those early days were not in vain." Charles W. Hargitt to Gager, October 21, 1916, BBG, box 3, folder 7, "Misc H–I–J."

170. W. E. Castle to Davenport, May 22, 1908, APS Davenport, series I, box 10, folder 5, "Castle, William E."

171. Chamberlin, "Introduction," 4.

172. Davenport to De Vries, March 22, 1904, APS Davenport, series I, box 93, folder 1, "Vries, Hugo de."

173. Davenport to De Vries, April 27, 1904, ibid.

174. Gager, "Effects of Radium Rays on Plants," 999.

175. In Secord's view, "sensational" acted simultaneously as a term of notability and sensory application in the case of the anonymously published evolutionary epic *Vestiges of the Natural History of Creation*: Secord, *Victorian Sensation*. See "Mind the Gap," in chapter 2, for a discussion of "sensation" in the context of Burke's findings.

176. B. C. A. W., "Origin of Life," 190.

177. Doyle, *On Beyond Living*.

178. MacDougal to Gager, December 17, 1908, and MacDougal to Gager, December 21, 1908, BBG, box 3, folder 9, "Misc M."

179. MacDougal to Gager, November 7, 1908, BBG, box 3, folder 9, "Misc M."

180. MacDougal to Gager, November 27, 1908, ibid.

181. MacDougal to Gager, December 7, 1908, ibid.

182. MacDougal, "Direct Influence of the Environment," 128.

183. MacDougal, "Trends in Plant Science," 487–95.

184. MacDougal to Gager, December 7, 1908, BBG, box 3, folder 9, "Misc M."

185. "Department of Botanical Research," 70.

186. Note that the statistical genotype-phenotype distinction was not proposed by Johannsen until 1911; it did not attain its current genetic expression and meaning until years after its initial populational meaning. Johannsen, "Genotype Conception of Heredity."

187. Gager, *Effects of the Rays of Radium on Plants*, 247.

188. Ibid., 247, 254.

189. Ibid., 255–56.

190. Ibid., 235–36.

191. MacDougal, "Origin of Species by Mutation" (*Torreya*, May 1902), 67.

192. Gager, *Heredity and Evolution in Plants*, 43.

193. "But though the mutation theory is a direct outgrowth of the hypothesis of intracellular pangenesis, it fortunately does not stand or fall with the latter, for no scientific theory ever had a firmer foundation in fact—in experimental evidence—than that of mutation." Gager, "Review of *The Mutation Theory*, vol. 2," 493.

194. Gager, *Effects of the Rays of Radium on Plants*, 237.

195. Pond, "Review," 810.

196. That is, apart from efforts that accidentally succeeded in artificially replicating a de Vriesian "mutating period"—though whether Gager's account left room for a "mutating period" is not fully clear. It is also unclear whether Gager's reworking may have unwittingly unleashed the bugbear of requiring joint mutating periods, a phenomenon that de Vries's mechanism of intracellular pangenesis was already able to account for but which was not yet clearly related to the rest of Gager's understanding.

197. MacDougal, "Organic Response" (*Science*), 97.

198. "Curious Modifications in Plant Life Possibly Due to Radium," 6.

199. Spillman saw this as a confirmation of sorts of his own earlier

explanation of "the interesting work of McDougall [*sic*] in which mutants were produced by chemical stimulants." "Mutation" at this time could be accounted for, at least in part, by such chromosomal irregularities. The very distinctions between gene-level and chromosomal-level effects that contributed to the fall of the mutation theory were, at least until the waning years of the mutation theory's popularity, well within the rubric of "mutation," as we will see in chapter 4. Spillman, "Mendelian Phenomena without De Vriesian Theory," 216.

200. Gager, *Heredity and Evolution in Plants*, 73.

201. White, "Heredity, Variation, and the Environment," 967. This chapter in Gager's *General Botany* was written by Orland E. White, curator of plant breeding and economic botany at the Brooklyn Botanic Garden.

202. Cleland, "Genetics of *Oenothera*."

203. Gates, "Mutations and Evolution," 26.

204. Sturtevant, *A History of Genetics*, 70.

CHAPTER 4

1. Thomson, *Heredity*, 98–99.

2. "The Law of Vast Numbers," 308.

3. Allen, *Thomas Hunt Morgan*, 105–6. According to Alfred Sturtevant, "Morgan's interest in genetics seems to have stemmed, at least in large part, from a visit to de Vries's garden in Holland (probably in 1900). In 1903 he wrote 'No one can see his experimental garden, as I have had the opportunity of doing, without being greatly impressed.'" Sturtevant, "Thomas Hunt Morgan," 290.

4. George Ledyard Stebbins's later reference to *Crepis* as the "plant *Drosophila*" is thus a historical twist from the moment when *Oenothera* was the model case for *Drosophila*. For more on E. B. Babcock's work on *Crepis*, including his relationship to the Morgan school, see Smocovitis, "The 'Plant *Drosophila*,'" 303ff. Others have attributed the term "plant Drosophala [*sic*]" to Nikolai Vavilov; see F. A. Varrelman to E. W. Sinnott, Oct 3, 1932, Sinnott Papers, Yale University, box 7, folder 135.

5. Allen, *Thomas Hunt Morgan*, 147–48; Sturtevant, "Thomas Hunt Morgan," 292. A few years later, Morgan even subjected some 31,168 flies to etherization before concluding that it seemed "highly probable therefore that ether has no specific effect in producing mutations in Drosophila ampelophila . . . and one is inclined to look elsewhere for a solution of the problem." Morgan, "Failure of Ether," 708, 710.

6. Fernandus Payne to Ernst Mayr, February 11, 1972, APS Genetics Collection, box 3, "Payne, Fernandus, to Ernst Mayr" folder. Comparing this with Kohler's account, it seems unclear whether Payne used X-rays or radium: "In a single batch of X-rayed flies he found several flies with wing defects that seemed to be inherited; however, he did not continue the experiment, apparently because the physicists at Columbia would not let him use their radium source." Kohler, *Lords of the Fly*, 38.

7. "Problems of Radiobiology with Emphasis on Radiation Genetics," lecture, Oregon State College, Biology Colloquium, April 21, 1951, APS Stern, box 34.

8. Allen, *Thomas Hunt Morgan*, 147–48.

9. "*Beaded Wings*.—In May, 1910, a number of flies, pupae, larvae and eggs of *Drosophila* were subjected to radium rays. One fly was produced, the marginal vein of whose wings was beaded." Morgan, "The Origin of Nine Wing Mutations in Drosophila," 497. See also Dexter, "Analysis of a Case of Continuous Variation," 716. According to Blakeslee, Morgan's "work was not followed up apparently because the numbers of mutants were small and the effects not specific." Blakeslee, "Twenty-Five Years of Genetics," 36.

10. Morgan to Davenport, June 11, 1910, APS Davenport, series I, box 72, folder 2, "Morgan, T. H."

11. Others have claimed that the white-eyed mutation came first, in January 1910. Green, "The 'Genesis of the White-Eyed Mutant.'" The *truncate* mutant also came from radium treatment. Carlson, *Mendel's Legacy*, 173.

12. MacDougal, "Organic Response" (*Science*), 97.

13. "Prof de Vries has kindly consented to take down the radium with him for your use. There is about 12 mgms in the . . .—pure $RaBr_2$. . . . In an experiment, place . . . containing Ra beads[?] on top of radium or radium on one side. The effective radiation will be mostly gamma, and a little β. It is not possible for use of α rays." Rutherford to Loeb, July 18, 1906, LOC, box 13, "Rutherford, Ernest."

14. Loeb and Bancroft, "Some Experiments," 782.

15. Spillman, "Notes on Heredity," 512.

16. Loeb and Bancroft, "Some Experiments," 781.

17. Loeb, in what Nathan Reingold has characterized as a "touchy but friendly moment over the question of priority" in the use of radium to produce mutations, wrote back that he "had not the slightest idea that you ever had worked with radium and still less, treated your flies with it." Moreover, he continued, "I was under the impression that your mutations had sprung up accidently just as De Vries's had and I got my first intimation that you had treated your flies in any way, through McDougal's article." Reingold, "Jacques Loeb, the Scientist," citing Loeb to Morgan, March 16, 1911.

18. As Morgan asked Loeb, "Doesn't that make you want to go?!" March 16, 1911, cited in Reingold, "Jacques Loeb."

19. Morgan, "Failure of Ether," 708.

20. "As a beginning student [in Morgan's lab], Sturtevant also tried to produce wing mutants using radium." Kohler, *Lords of the Fly*, 38; "Radium Experiment with 'Big Smooth Black' Fruit Flies," CIT Sturtevant, box 16, folder 1.

21. Kohler, *Lords of the Fly*; and Kohler, "*Drosophila* and Evolutionary Genetics."

22. Morgan instead attributed the failure to the effects of inbreeding; only in 1914 would he begin to make the association between irradiation and sterility. Muller, by contrast, would later design his experimental setup in such a way that he was *looking* for sterility in order to detect induced mutations.

23. Morgan, "Failure of Ether," 708–9. Like many biologists, Morgan was a novice when it came to radioactive phenomena ("emanations" came from radium, not X-rays). As Gager had already noted by this time, "The use of

the plural 'emanations' to designate all the rays and influences coming from radium has been somewhat common in biological papers. It has no warrant, is only confusing, and should be abandoned." Gager, "Some Physiological Effects of Radium Rays," 763.

24. Bridges and Morgan, *The Third-Chromosome Group*, 37. According to Curt Stern, "With this carefully worded conclusion the development of radiation genetics was arrested, to lead to birth only 17 years later." "Problems of Radiobiology," APS Stern, box 34.

25. Morgan to Blakeslee, May 27, 1935, CIT Morgan, box 1, folder 4; emphasis added.

26. Morgan, "Failure of Ether," 709–10.

27. Morgan to Blakeslee, May 27, 1935, CIT Morgan, box 1, folder 4.

28. Fernandus Payne to Ernst Mayr, February 11, 1972, APS Genetics Collection.

29. Morgan, "Failure of Ether," 711.

30. Carlson, *Mendel's Legacy*, 173.

31. Morgan, "Genesis of the White-Eyed Mutant," 92.

32. The introduction of chance into considerations of genetic phenomena is a well-known advance of the first third of the twentieth century, most often associated with a spate of remarkable work by population geneticists in that period (including "genetic drift"). Such considerations also influenced experimental studies of mutation.

33. Morgan, "Failure of Ether," 710. Kohler uses "breeder reactor" as a technical term from the production of nuclear energy; confusingly, this usage is not directly related to the "breeding" of organisms. Kohler, *Lords of the Fly*, 47.

34. Davenport's comment here, in the annual report of the Carnegie Institution's Department of Genetics, and apparently unrelated in tone and in its use of analogy to the surrounding text, serves as yet another surprising instance of the powerful resonances between radium and life. "Department of Genetics" (1922), 94.

35. "Again, studies made at this Station on the evolution of the chromosomal complex, especially in the flies, have led to the general conception that evolution has proceeded not primarily by modifications of the series of visible organisms whose evolution is the goal of our researches, but rather evolution has proceeded by changes in the 'germ-plasm,' the chromosomes, and that these changes have occurred in some cases apparently owing to its intrinsic properties—as radium changes into lead—and sometimes under the influence of intracellular change, such as are induced by hybridization, and sometimes, perhaps, by extreme conditions external to the germ-cell." "Department of Experimental Evolution and Eugenics Record Office" (1920), 107.

36. Gates, "Mutations and Evolution," 74.

37. For more on the complicated legacy of the de Vriesian mutation theory, see Dunn, "Genetics in Historical Perspective," and the various series of investigations conducted by Renner, Cleland, and Bradley. See especially Cleland, "Genetics of *Oenothera*"; see also Campos, "'Complex Recombinations.'"

38. Gager, "Present Status of the Problem." His review came just before a war-related "hiatus" in radium-based publications from 1915 to 1920.

39. Davenport to de Vries, March 2, 1916, APS Davenport, series I, box 93, folder 2, "Vries, Hugo de."

40. MacDougal et al., *Mutations*, 62. Gager and Blakeslee remained in frequent contact over the years, sharing experimental instrumentation (radium "needles") and visiting each other as time and circumstance allowed (even staying at each other's homes). There are also many commonalities between Blakeslee's work and that of E. B. Babcock, who spent much of the 1930s similarly interested in studying chromosomes to gain insights into plant evolution. As Smocovitis notes, Babcock inaugurated an "inventive and ambitious phylogenetic and evolutionary study of a plant genus that fully embraced available genetical knowledge—the first such study seriously attempted in plants." Smocovitis, "The 'Plant *Drosophila*,'" 314.

41. Blakeslee to de Vries, April 7, 1933, APS Blakeslee, "Vries, Hugo de," box 21; "Lebenslauf of A.F.B.," p. 3, APS Blakeslee, "Biographical Materials," box 25, folder 2.

42. On the occasion of de Vries's eighty-fifth birthday, Blakeslee wrote to him, "It is a pleasure to have known such a founder of modern genetics who has been an inspiration to my own work." Blakeslee to de Vries, May 24, 1933, APS Blakeslee, "Vries, Hugo de," box 21. And in a letter to de Vries's wife, Blakeslee recalled that "at the quarter centennial of the founding of the Brooklyn Botanic Garden I pointed out an instance of his wonderful prevision in suggesting in 1904, in an address at the dedication of our Department here, that attempts be made to induce mutations by the use of X-rays and radium. My own researches owe much to him. In a measure, I feel that I have been carrying on the torch which he has laid down." Blakeslee to Mrs. de Vries, May 23, 1935, APS Davenport, series I, box 93, folder 2, "Vries, Hugo de."

43. "Seventy-Five Years of Progress in Genetics," p. 15, APS Blakeslee, "Lectures, Papers, Etc.," box 23, folder 35.

44. "Lebenslauf of A.F.B.," p. 6, APS Blakeslee, "Biographical Materials," box 25, folder 2.

45. Ibid., 4; see also Blakeslee to de Vries, April 7, 1933, APS Blakeslee, "Vries, Hugo de," box 21.

46. "Lebenslauf of A.F.B.," p. 5, APS Blakeslee, "Biographical Materials," box 25, folder 2.

47. Sinnott, "Albert Francis Blakeslee" (*NAS Biographical Memoirs*).

48. "Lebenslauf of A.F.B.," p. 5, APS Blakeslee, "Biographical Materials," box 25, folder 2.

49. "I am not sure that we could find sufficient greenhouse space here for the experiments. It would not take, of course, more than eight or ten linear feet of bench space to do quite a bit of work." Gager to Blakeslee, February 7, 1923, APS Blakeslee, "Gager, C. Stuart," box 10, folder 4. For more on Blakeslee's greenhouse, see Kimmelman, "Mr. Blakeslee Builds His Dream House."

50. Blakeslee to Morgan, May 22, 1935, CIT Morgan, box 1, folder 4.

51. "Evolution to Order," radio broadcast "under the auspices of Science Service, over the Columbia Broadcasting System," Thursday March 24, 1938, APS Blakeslee, box 23, "Lectures, Papers, Etc.: Adventures in Science" folder.

52. "Lebenslauf of A.F.B.," p. 6, APS Blakeslee, "Biographical Materials," box 25, folder 2. Though not in all respects: as Blakeslee later wrote to a colleague, "some of these species [of *Datura*] give very poor germination—sometimes not over a tenth of one percent." Blakeslee to O. L. Inman, December 11, 1934, CSH Blakeslee, "Blakeslee, Albert—1934" folder. Blakeslee later became interested in animal polyploidy, but found this a considerably more difficult task, as animals were "functionally dioecious." Blakeslee to Emmeline Moore, November 15, 1937, CSH Blakeslee, "Blakeslee, Albert—1937" folder.

53. "Lebenslauf of A.F.B.," p. 5, APS Blakeslee, "Biographical Materials," box 25, folder 2. According to Sinnott, Blakeslee had encountered "one or two Jimson weeds which were different from the typical ones and had begun to study them" while at Storrs. Sinnott, "Albert Francis Blakeslee" (*APS Year Book*).

54. Blakeslee, "Globe Mutant."

55. Blakeslee and Avery, "Mutations in the Jimson Weed," 115.

56. Ibid., 115–20.

57. Ibid., 119. As Blakeslee later recounted in 1921, "It may be mentioned that the tetraploid datura was called 'New Species' before its tetraploid nature was suspected. It satisfied the requirements of an independent species. The pollen was relatively good, and the mutant formed a distinct race, self-fertile and fertile *inter se*, while practically sterile with the parent stock." Blakeslee, "Types of Mutations," 263.

58. "Lebenslauf of A.F.B.," p. 6, APS Blakeslee, "Biographical Materials," box 25, folder 2.

59. Sinnott, "Albert Francis Blakeslee" (*NAS Biographical Memoirs*), 9, 8.

60. Blakeslee and Bergner, "Methods of Synthesizing Pure-Breeding Types," 571.

61. Blakeslee and Avery, "Mutations in the Jimson Weed," 111.

62. "Department of Genetics" (1922), 95. Sinnott recalled that "it was characteristic of him, too, that he provided his collaborators with good support and left them free to work out the ideas, helping them in whatever way he could. In the true sense of the word the Datura program was a cooperative one. Its head was no dictator, but simply *primus inter pares*." Sinnott, "Albert Francis Blakeslee," 1. Demerec noted that although Blakeslee "was very generous about sharing the credit for research accomplishments with his collaborators and assistant, and most of his papers were published under joint authorship, Blakeslee adhered strictly to the policy that materials accumulated by him must be used only under his control and in his laboratory. This attitude imposed severe limitations on the general study of Datura, and prevented the development of other groups of investigators working with that plant. Datura research is consequently becoming a lost art. But Blakeslee's material still has rich potentialities for the solution of many problems in genetics and speciation, and could be used to good advantage." Demerec, "Albert Francis Blakeslee," 4.

63. Blakeslee, "Variations in *Datura*," 18.

64. Blakeslee et al., "Chromosomal Duplication," 388–90.

65. "Department of Genetics" (1921), 108.

66. Demerec, "Albert Francis Blakeslee," 1.

67. Blakeslee, "Mutations in Mucors," 278, 284.

68. Intriguingly, Blakeslee held that the "failure" of a particular mutation in the adzuki bean "to appear more than once in so large a number of individuals indicates that it is a variation genotypic in nature, since it could scarcely be attributed to the reappearance of a character through normal segregation nor be considered a mere modification induced by environmental factors." The sheer rarity of the mutation was an argument for its genotypic, rather than its chromosomal, basis. Blakeslee, "A Unifoliolate Mutation in the Adzuki Bean," 155.

69. Blakeslee, "Variations in *Datura*," 31; emphasis added.

70. Ibid., 27.

71. Blakeslee, "Types of Mutations," 255. Blakeslee later realized, of course, that duplication was not the only means of producing mutations. Following Calvin Bridges's work on nondisjunction, he acknowledged that there was room for a "rather novel study of trisomic, tetrasomic and pentasomic inheritance." Blakeslee, "Variations in *Datura*," 27.

72. Relating the existence of these chromosomal types to geographic distribution patterns also did much to help illuminate the evolutionary history of *Datura*. Sinnott, "Albert Francis Blakeslee"(*APS Year Book*), 394–98.

73. "Department of Genetics" (1921), 101.

74. "Department of Genetics" (1922), 93; emphasis added.

75. "Department of Experimental Evolution and Eugenics Record Office" (1920), 110.

76. "Department of Genetics" (1922), 93.

77. Davenport to de Vries, February 5, 1924, APS Davenport, series I, box 93, folder 2, "Vries, Hugo de."

78. "Department of Genetics" (1921), 103.

79. Ibid., 109.

80. "Department of Genetics" (1922), 93–94.

81. Blakeslee, "Variations in *Datura*," 27.

82. Wells, Huxley, and Wells, *Science of Life*, 594.

83. Thomson, *Outline of Natural History*, 716–17.

84. Morgan to Osborn, December 26, 1917, cited in Reingold, "Jacques Loeb, the Scientist," 125.

85. Getzendaner, "A Hypothesis of 'Valence,'" 428.

86. Stubbe, *History of Genetics*, 220–21.

87. MacDougal concluded that "this generalization, which is essentially of a physiological character, even when applied to inorganic substances, gave the basis for the researches upon descent which have been carried out with such notable results." MacDougal, "Hugo De Vries," 0–1.

88. Punnet, *Mendelism*, 62.

89. "New Light on Mutation," 355.

90. Gates, "Mutations and Evolution," 33.

91. Ibid., 218.

92. Loeb, *Organism as a Whole*, 241–44.

93. Hogben, *Nature of Living Matter*, 73, 77.

94. The article in question may have been what later became Blakeslee's "Globe Mutant."

95. Blakeslee to Shull, April 15, 1921, APS Blakeslee, "Shull, George H.," box 19, folder 3.

96. Shull to Blakeslee, April 17, 1921, ibid.

97. Neel to Stern, July 30, 1940, APS Stern, "Neel, James V.," box 19, folder 2. For more on the meanings of "mutation" in Neel's later work as a member of the Atomic Bomb Casualty Commission, see Lindee, "What Is a Mutation?"

98. Shull to Blakeslee, April 26, 1921, APS Blakeslee, "Shull, George H.," box 19, folder 3.

99. Blakeslee, "Types of Mutations," 254.

100. Ibid., 261; citing Waagen, "Die Formenreihe des *Ammonites subradiatus*," 185–86.

101. For a case of tetraploidy, for example, *not* to be considered a mutation was a significant alteration of de Vries's theory, as de Vries himself considered the origin of *Oenothera gigas* to be "the one absolutely typical case of species-formation in all my cultures." He prefaced his remark by saying, "Please tell Miss Lutz that I enjoyed her discovery of the double number of chromosomes in *Oenothera gigas* immensely." De Vries to Davenport, December 31, 1907, APS Davenport, series I, box 93, folder 1, "Vries, Hugo de." Blakeslee went on to insist, however, that "the occurrence of tetraploidy would therefore be no more a mutation than the doubling of chromosomes at the origin of the sporophyte from the gametophyte ferns." Blakeslee, "Types of Mutations," 262–63.

102. Blakeslee, "Types of Mutations," 262–63.

103. Ibid., 266.

104. Ibid., 262, 265–66.

105. Hurst, *Mechanism of Creative Evolution*. The term "transmutation" was revived two decades later for still another phenomenon: "In *Neurospora*, *Aspergillus*, and yeast, intragenic recombination occurs by a mechanism that can be interpreted as miscopying of small segments of genetic material. This process differs from conventional crossing over in that a single event does not result in reciprocal products and also by the fact that it does not necessarily lead to recombination of genetic markers close to and on opposite sides of the genes within which it occurs. It is proposed that this be called 'transmutation.'" Beadle, "The Role of the Nucleus in Heredity," 13, 25.

106. Dunn to Blakeslee, February 10, 1932, APS Blakeslee, "Dunn, Leslie C.," box 8, folder 1.

107. "Department of Genetics" (1921), 104.

108. Blakeslee even cited Muller's work on balanced lethals, which he said "strongly suggests that such of the Œnothera mutants as are not caused by chromosomal duplication are due to cross-overs from a balanced lethal condition." Blakeslee, "Types of Mutations," 257, 260, 262.

109. Gager to Hugo Lieber, January 12, 1909, BBG, box 3, folder 8, "Misc K–L"; cf. Gager to Lieber, November 13, 1908, ibid.

110. Gager and Blakeslee, "Induction of Gene and Chromosome Mutations," 424.

111. Ibid. As Blakeslee would later advise a researcher at the Smithsonian Institution's Division of Radiation and Organisms, "If you are interested in testing out the effects of different wave lengths of radium energy I think it would be very important, from a genetic standpoint, to get a good *Versuchsthier*." Blakeslee to Florence E. Meier, November 14, 1934, CSH Blakeslee, "1939" folder.

112. "Department of Genetics" (1921), 109.

113. Sinnott and Blakeslee, "Structural Changes," citing Blakeslee, "Types of Mutations," and Blakeslee, "Variations in *Datura*," 17.

114. Sinnott and Blakeslee, "Structural Changes."

115. Blakeslee to Halsey J. Bagg, March 15, 1921, BBG, box 9, folder 9, "C. S. Gager, Research, 1921–1928" folder. Gager's wife was Bertha Bagg Gager.

116. Blakeslee asked Gager this question some years later, in 1923, as they were testing the effects of radioactive soil on *Datura*, though the question was undoubtedly on both their minds when they began their first experiments in 1921. Gager's response was of a piece with his earlier findings—"I think there is no reason to expect any results from growing plants in soil containing radioactive residues other than a stimulation or acceleration of growth"—but also indicated that he would want to carry out experiments along those lines. Blakeslee to Gager, February 13, 1923, APS Blakeslee, "Gager, C. Stuart," box 10, folder 4; Gager to Blakeslee, February 15, 1923, ibid.

117. Among the two most common mutants of *Oenothera* were giant *Oenothera gigas* and dwarf *Oenothera nanella* forms.

118. "Dr. Blakeslee has now brought over from Cold Spring Harbor his plants of Jimson weed, and I shall be able to subject them to the radium treatment at any time when you can conveniently let me have the preparations." Gager to Bagg, April 14, 1921, BBG, box 9, folder 9, "C. S. Gager, Research, 1921–1928" folder.

119. Gager to Blakeslee, July 7, 1921, APS Blakeslee, "Gager, C. Stuart," box 10, folder 2.

120. Gager to Blakeslee, March 30, 1922, ibid., box 10, folder 3, and February 20, 1923, box 10, folder 4.

121. Gager to Blakeslee, June 10 and July 7, 1921, ibid., box 10, folder 2.

122. Blakeslee to Gager, September 18, 1922, ibid., box 10, folder 3.

123. Blakeslee to Gager, November 19, 1921, ibid., box 10, folder 2.

124. Gager to Blakeslee, December 6, 1922, ibid., box 10, folder 3.

125. Gager and Blakeslee, "Chromosome and Gene Mutations," 76.

126. Gager, *Effects of the Rays of Radium on Plants*, 256.

127. Gager and Blakeslee, "Chromosome and Gene Mutations," 76.

128. Blakeslee to Gager, December 3, 1923, APS Blakeslee, "Gager, C. Stuart," box 10, folder 4.

129. Gager and Blakeslee, "Chromosome and Gene Mutations," 75–76.

130. "Department of Genetics" (1929), 45.

131. Blakeslee, "Control of Evolution and Life Processes in Plants," 59.

132. Quoted in "Department of Genetics" (1922), 98.

133. Blakeslee also acknowledged, however, that some mutations were not expected to be mendelizing. Blakeslee to Gager, January 14, 1923, APS Blakeslee, "Gager, C. Stuart," box 10, folder 4.

134. Gager to Blakeslee, January 4, 1923, ibid.

135. Blakeslee to Gager, January 14, 1923, ibid. Twenty-five years later, Gager's work was all but unknown: "I am not personally familiar with the literature in this country dating back to Rusby and Gager on the effects of radioactivity on plants. I am acquainted with a considerable amount of work on the influence of x-rays and other penetrating radiation on the growth of plants." James Bonner, on behalf of Lewis J. Stadler, to Harold Arnold Wolff, February 23, 1948, UMC Anderson, folder 4.

136. Gager to Blakeslee, January 3, 1927, APS Blakeslee, "Gager, C. Stuart," box 10, folder 6.

137. Mavor, "Production of Non-Disjunction by X-Rays."

138. Blakeslee to Gager, January 14, 1923, APS Blakeslee, "Gager, C. Stuart," box 10, folder 4.

139. Ibid.

140. "Radium Experiment with 'Big Smooth Black' Fruit Flies," CIT Sturtevant, box 16, folder 1.

141. Blakeslee to Gager, January 14, 1923, APS Blakeslee, "Gager, C. Stuart," box 10, folder 4.

142. Blakeslee to Gager, February 25, 1927, ibid., box 10, folder 6.

143. "Department of Genetics" (1929), 45.

144. Gager and Blakeslee, "Chromosome and Gene Mutations."

145. Ibid., 75.

146. Gager and Blakeslee, "Induction of Gene and Chromosome Mutations," 424; Blakeslee, "Distinction between Primary and Secondary Mutants in Datura," 389.

147. Gager and Blakeslee, "Chromosome and Gene Mutations," 78.

148. Mavor, "Attack on the Gene," 358.

149. Gager and Blakeslee, "Chromosome and Gene Mutations," 79.

150. Blakeslee and Avery, "Methods of Inducing Doubling," 404, 408. Blakeslee began to use the language of "genetics engineering" after he turned to using colchicine as a mutagen in the mid-1930s. These efforts grew out of and were deeply steeped in his earlier work using radium as a mutagen.

151. There was some confusion four years later over why precisely this mutant was given that name. Gager to Blakeslee, March 4, 1927, APS Blakeslee, "Gager, C. Stuart," box 10, folder 6.

152. Blakeslee to Gager, February 5, 1923, ibid., box 10, folder 4.

153. Gager and Blakeslee, "Chromosome and Gene Mutations," 78. Blakeslee later noted that Buchholz had identified two other gene mutations: "an albino which is of little use to us, and *swollen* which acts curiously for a recessive," both of which "came from the earlier treatment by Gager." Blakeslee to Buchholz, May 9, 1929, APS Blakeslee, "Buchholz, John T.," box 4, folder 12.

154. Gager and Blakeslee, "Chromosome and Gene Mutations," 78. Recall that MacDougal had invented the concept of mutation frequency so as to avoid precisely this sort of dependency.

155. Ibid., 79.

156. Blakeslee to Gager, September 5, 1933, APS Blakeslee, "Gager, C. Stuart," box 10, folder 10. By 1927, Blakeslee had published some eighty articles.

157. Ibid.

158. "Department of Genetics" (1929), 45. By 1933, however, his tune had changed, and in a letter to Gager on September 5, he remarked that he had identified some of these same characteristics as a "series of types due to gene mutations" that "affect all parts of the plant." Blakeslee to Gager, September 5, 1933, APS Blakeslee, "Gager, C. Stuart," box 10, folder 10.

159. Demerec, "Albert Francis Blakeslee," 4.

160. Blakeslee to MacDougal, February 15, 1923, APS Blakeslee, box 14, "MacDougal, D. T."

161. Ibid.

162. Stadler, "Induced Mutations in Plants," 1264.

163. Lewis Stadler to Karl Sax, December 17, 1931, UMC Stadler, folder 5. Stadler also grappled with the nature of mutation in his work. Like Shull (who attempted to coin the term "anomozeuxis"), Stadler was one of many who struggled with the right words to describe the phenomena they were observing. In an early grant application, he indicated that he intended to complete "studies of the cytological effects of X-rays with special reference to the production of chromosomal mutations and aberrant types of chromosome behavior." And elsewhere, under the heading of "fallacious," he once listed "mutations a single class." In his typed lecture notes, he referred to "Mutation" as "In sense of 'gene mutation'" and "Chromosomal 'mutations' later," also noting, "Line drawn strictly between gene mutations and chromosomal aberrations." But on another occasion, when referring to "chromosome doubling or other chromosome irregularities," he wrote: "Strictly speaking, these are not mutations, but they are inherited variations of just as much practical value." Fascinatingly, Stadler would later even seek to relabel "the hypothetical transformation of a gene to an allelic form . . . as 'transmutations,'" presaging Hurst's coinage. Stadler, "Comparison of Ultraviolet and X-Ray Effects on Mutation."

164. Sinnott and Blakeslee, "Structural Changes"; Sinnott, "Albert Francis Blakeslee" (*NAS Biographical Memoirs*), 1; Demerec, "Albert Francis Blakeslee," 3–4; Smith, "Albert Francis Blakeslee," 307. Decades later, Jim Crow would still refer to Blakeslee's research as "a mountain of work on the jimsonweed, *Datura* . . . a most convincing demonstration that translocations were an important part of the evolutionary process." Crow, "Sixty Years Ago: The 1932 International Congress," 299.

165. Spillman had proposed four distinct types of "variation": the Mendelian recombination of characters; fluctuation due to the environment; the discontinuous hereditary "irregularities in the distribution of chromosomes . . . amenable to the action of natural selection" (or, as he also labeled it, in light of the new understanding of what was going on cytologically with *Oenothera*, "de Vriesian mutation"); and finally, what he called "fundamental change

in . . . the germ plasm," which he believed to be "by far the most important type of evolutionary change." W. J. Spillman, "Mendelian Phenomena without De Vriesian Theory," 216. Ten years later, the corn geneticist Edgar G. Anderson wrote to the drosophilist Alfred H. Sturtevant in 1920, "You have no statement in the introductory part of this chapter regarding the meaning of the term mutation. There are several conceptions of mutation and that held by the Oenothera people is not quite the same as yours." Anderson to Sturtevant, March 16, 1920, UMC Anderson, folder 45.

166. Muller, "Artificial Transmutation of the Gene."

167. Blakeslee to Gager, July 31, 1927, APS Blakeslee, "Gager, C. Stuart," box 10, folder 6.

168. It took Gager a little over a month to give his blessing for Buchholz to continue the work. "The worst that could have happened (if I had not approved) was that you would both have been arrested and fined, but I should probably have been too busy to have pushed the matter!" Gager to Blakeslee, September 2, 1927, APS Blakeslee, "Gager, C. Stuart," box 10, folder 6. Buchholz and Blakeslee turned to using both X-rays and radium, as well as heat, in their experiments. By 1936 they were publishing further results: "Lebenslauf of A.F.B.," p. 6, APS Blakeslee, "Biographical Materials," box 25, folder 2.

169. Blakeslee, "Twenty-Five Years of Genetics," 35.

CHAPTER 5

1. Allen, *Thomas Hunt Morgan*, 171. See also works by Carlson, especially "Unacknowledged Founding." For a more popular take on Muller, see Schwartz, *In Pursuit of the Gene*.

2. At the age of about 17, Muller had even written a short story entitled "In the Cause of Science," in which—after quoting a description of Swedenborg's vorticle theory of the universe and its claims that all time and space "radiate" from an "infinitesimal finite point" (see Clay, "Radium," and chap. 1)—the protagonist heads off in search of a rumored strontium mine. (Muller records the protagonist as generally having had "queer ideas of amusement.") Strontium is not naturally radioactive, although the young Muller seems to have had other ideas in mind. "In the cause of science," 1907, LL, series VI, box 1.

3. "Autobiographical notes [prepared for Vavilov]," 1936, LL, series II, box 1. Lock's glossary definition of "mutation" defines it as "the sudden origin of a new species at a single step": Lock, *Variation, Heredity, and Evolution*, 327.

4. Carlson, *Mendel's Legacy*, 204, footnote 212. Carlson also notes that during these years, due to financial exigencies, "Muller had to work as a Wall Street clerk, rush to work eating a sandwich on the subway to teach classes at night in English as a second language for immigrants, and tuck in the time to go to his classes and study at Cornell for his master's degree," in addition to supporting his mother.

5. Lock, *Variation, Heredity, and Evolution*, 225; Allen, *Thomas Hunt Morgan*, 308.

6. Lock, *Variation, Heredity, and Evolution*, 23.

7. Ibid., 19, 127.

8. Ibid., 156.

9. "Autobiographical notes [prepared for Vavilov]," 1936, p. 7, LL, series II, box 1.

10. Probably a course taught by Pegram, during the period 1907–1912. See Carlson, *Genes, Radiation, and Society*, 33.

11. "Basis of the theory of the gene: The experimental evidence concerning the properties of the gene," LL, series II, box 1. This piece, dated December 23, 1936, was one of Muller's efforts against Lysenkoism.

12. "Some recent work in heredity. Draft/notes for Rice Institute lecture," ca. 1916, LL, series II, box 1.

13. Handwritten draft manuscript with Muller's own strikethroughs: "Applications and prospects. Discussion of eugenic views and human evolution," 1916, p. 17, LL, series II, box 1. Cf. Muller's remark regarding Calvin Bridges: "United we stand—divided we might all fall but Bridges." "An Episode in Science," lecture given at the Biological Laboratory of the Brooklyn Institute, Cold Spring Harbor, July 25, 1921, APS Davenport, series I, box 72, "Muller, H. J." Muller seemed to like bridges in general; as he later told James Neel, regarding another genetical matter, "This ratio is arrived at by way of two 'bridges.'" Muller to Neel, December 7, 1956, APS Stern, "Neel," folder 6.

14. "The Essential Facts of Heredity," ca. 1917–1918, LL, series II, box 1.

15. "An Episode in Science," pp. 4–5. A similar account appears in Russian under the title "Results of a Decade of Genetic Research on Drosophila." In a 1947 lecture at Oak Ridge, Muller acknowledged that as a youth he had read Soddy "on the revolutionary possibilities of the control of atomic energy." CSH Muller, box 6, folder 7.

16. Elsewhere, to make the point, he asked his audience whether they would choose to save a beaker full of human eggs for a limited time or the whole universe without humans in it indefinitely. "Recent Findings in Heredity," ca. 1916, LL, series II, box 1.

17. Ibid., 23. Muller's mode of description drew on a long-standing discursive tradition in hereditary theory that spoke of an "organizing principle" of a nucleated cell that "*radiates*" influence in every direction. See, for example, Freke, *On the Origin of Species by Means of Organic Affinity*, 29.

18. "Recent Findings in Heredity," p. 35; emphasis added.

19. "Lecture, re: a general survey of the gene," ca. 1927, LL, series II, box 1; emphasis added. Blakeslee would make similar comments in 1949: "Chromosomes are the most precious material in the world and it is worth our while to learn all we can about them." "Seventy-Five Years of Progress in Genetics," p. 18, APS Blakeslee, "Lectures, Papers, Etc.," box 23, folder 35.

20. Mendel's own choice of word for the hereditary factors, it should be noted in passing, was *elemente*. See chapter 1 for more on the connection between living atoms and atoms of life.

21. "It may be calculated from these experiments that a large proportion of the genes in *Drosophila* must have a stability which—at a minimum value—is comparable with that of radium atoms. Radium atoms, it may be recalled, have a so-called 'mean life' of about two thousand years." Muller, "Mutation," 109.

22. Ibid.

23. Schultz, "Radiation and the Study of Mutation in Animals," 1239.

24. Muller, "Quantitative Methods in Genetic Research," 417.

25. Comfort, *Tangled Field*, 44.

26. Getzendaner, "A Hypothesis of 'Valence,'" 429.

27. Muller, "Variation," 44–45.

28. Wells, Huxley, and Wells, *Science of Life*, 475, 477. By 1929, Wells et al. were convinced that the "living atom" tradition, at least as concerned species, was faulty. Species were not "a natural unit at all like an atom, or a quantum." The tradition could still apply with great utility, however, to the Mendelian theory (387–88).

29. As Herbert Spencer Jennings described the situation just prior to Muller's 1927 announcement, "A few genes mutated, as a few of the atoms of radiant metal disintegrate; in both cases no outside agent appeared to be at work." Jennings, *Genetics*, 347. See chapter 1, as well as the works of the Russian mutation theorist S. I. Korschinsky (Korzhinskii), who had once noted in his *What Is Life* (1900) that all organisms contain "some deep, secret force . . . poured throughout the organic world, glimmering in every molecule of the plasm and blazing as a flame in human reason . . . it is life." Korzhinskii, "Chto takoe zhizn?," 56–57, cited in Daniel Todes, *Darwin without Malthus*, 72, 189, footnotes 46–47.

30. Muller, "The Gene as the Basis of Life," 919; emphasis added.

31. "Lecture, re: a general survey of the gene," ca. 1927, LL, series II, box 1.

32. Muller, "Need of Physics," 210.

33. Like radium before them, genes also became increasingly intimately associated with fundamental questions of thermodynamics. As Muller would later note, "It is only the Maxwell demon of selection inherent in gene duplication, that is, the differential multiplication of the mutations, which brings order out of mutations' chaos despite itself." Muller, "Gene," 30. Elsewhere, in an even more dramatic echo of this discourse, he claimed that "we fooled the demon god inside of life, and took it unawares, and hit the gene." "Lecture, re: a general survey of the gene," ca. 1927, LL, series II, box 1, 13–14.

34. Needham, *Order and Life*, 26.

35. Huxley, *Evolution*, 48, 27.

36. Muller to Huxley, March 27, 1918, LL, series I, "J. S. Huxley," "1917–1918" folder.

37. "The newer biology. Lecture at Rice Institute," 1916–1918, LL, series II, box 1.

38. "Data notebook: Temperature induced mutations," July–August 1919, LL, series VI, box 1; emphasis added.

39. "Data notebook: Chem[istry]," September 1921–October 1922, p. 35, LL, series VI, box 2.

40. Muller, "Present Status of the Mutation Theory," 6.

41. "The Methods of Genetics in Their Application to Problems of Life and Evolution," ca. 1942, p. 11, LL, series II, box 2.

42. Muller, "The Gene," 12.

43. Muller, "Nature of the Genetic Effects," 392; emphasis added.

44. Muller, "Measurement of Gene Mutation Rate," 281.

45. Muller, "Present Status of the Mutation Theory," 6.

46. "The empirical question of whether such mutations had actually been produced, had already been the subject of much controversy, with the literature of which I had been familiarizing myself." "Autobiographical notes [prepared for Vavilov]," 1936, pp. 7–8, LL, series II, box 1.

47. "The Influence of X-Rays upon Heredity," no date, but after 1931, p. 5, LL, series II, box 7.

48. Muller, "Effects of Roentgen Rays," 307.

49. "Old literature on radiation" or "Early x-ray literature (before 1926)," ca. 1933, LL, series II, box 1.

50. Ibid. Gager's work is here cited five times, at least once prominently in relatively large script, and another time with several lines of notes.

51. Muller, "Effects of Roentgen Rays," 312–13.

52. According to Jim Crow's recollection of what his graduate professor, J. T. Patterson, told him, "*Muller was completely convinced from a study of the literature* that X-rays could produce mutations. His fear in advance of the experiments was not that X-rays would fail to produce mutations, but that cell-killing effects might predominate and mask the mutagenic effects he was seeking." Crow, "Some Reflections on H. J. Muller"; emphasis added.

53. Muller, "Method of Evolution," 489. Not all of these efforts seemed to Muller to have led to success: already by 1920 he had written that "the oft-suggested possibility of artificially influencing the kind of mutation that occurs (cf. Stockard, Tower, MacDougal, Kammerer, Guyer) would seem to recede indefinitely, unless some unique method is found which does not merely consist in an acceleration or intensification of the ordinary process of mutation." Muller, "Further Changes in the White-Eye Series of Drosophila," 470.

54. As Carlson and others have frequently noted, Muller was keenly aware of issues of priority. Muller wrote, "S. Gager [*sic*] and A. F. Blakeslee in a paper published early in 1927 (before Muller's findings had been officially announced) reëxamined some of their earlier work on the Jimson weed, and found evidence in it that several gene mutations had appeared in progeny of plants that had been treated with radium. Later, Blakeslee and his co-workers produced numerous gene and chromosome changes in this organism. . . . Since 1928, additional species of plants, of widely different kinds (maize, wheat, cotton, primroses, snapdragons, etc.) have been used, with positive results, and also several other species of insects." Muller, "Effects of Roentgen Rays," 313.

55. "Elimination of the X-chromosome from the egg of D[rosophila] m[elanogaster] by x-rays," LL, series II, box 1.

56. "PROPOSED EXPERIMENTS WITH MIRABILIS jalapa," October 1922, LL, series VI, box 2, "Texas," folder 1.

57. Muller and Painter, "Cytological Expression of Changes," 193.

58. Muller referred to a prior "long succession of experiments on the effects of ionizing radiation on the hereditary material. It is true that certain suggestive results were reported." He cites J. Dauphin, Gager, Morgan, and Loeb and Bancroft, but explains away their contributions. He refers to the

work of Blakeslee and Gager in more detail, saying they "were able to show that many so-called 'chromosome mutants' had indeed been produced by the radiation, but most of these were types having an entire but normal extra chromosome, brought about by nondisjunction, an already known effect of radiation. . . . There was however one case of a structurally changed chromosome, and two cases of recessive visible mutations." Muller, "Nature of the Genetic Effects," 391.

59. Muller and Altenburg, "Study of the Character," 213.

60. Without "a thorough course of inbreeding beforehand, or else to have run at least half a dozen different pairs of parallel lines of the control and treated series, and to have obtained a definite difference in the same direction between the two lines of *each* pair," Muller argued, "it can be proved by the theory of 'probable error' that the differences observed may have been a mere matter of random sampling among genic differences originally present." Muller, "Variation," 46.

61. Muller to J. S. Huxley, December 13, 1919, LL, series I, "J. S. Huxley," "1919–1920" folder. Muller wrote elsewhere, "There is, to be sure, enough work to show that the real mutations are 'rare'—whatever that term may mean; but, so far as an approximate quantitative determination of the rate of factor change is concerned, it is not possible, from the published work, to determine even its general order of magnitude. Some special scheme of crossing is required for this purpose." Muller, "Rate of Change"; see also Carlson, *Genes, Radiation, and Society*, 112.

62. Muller, "Variation," 44.

63. Muller, "Measurement of Gene Mutation Rate," 283–84, see also 281.

64. As he would later write in 1927, much of the earlier research into induced mutation "has been done in such a way that the meaning of the data, as analyzed from a modern genetic standpoint, has been highly disputatious at best; moreover, what were apparently the clearest cases have given negative or contrary results on repetition." Muller, "Artificial Transmutation of the Gene," 84.

65. Moreover, he concluded, "it would be absurd and scholastic to try to classify mutations according to the nature of their effects. A mutation can do practically anything that life can do—or at least a little of it, for life is built out of mutations." Muller, "Method of Evolution," 488.

66. Muller's constriction of the meaning of mutation to the level of the gene was the logical extension of Edward M. East's own earlier reduction of the word "mutation" from large-scale de Vriesian mutations "to any inherited variation, however small." It also paralleled the genically oriented work of population geneticists such as R. A. Fisher in the 1920s. East, "Mendelian Interpretation," cited in Comfort, *Tangled Field*, 42.

67. Muller, "An Oenothera-Like Case in Drosophila," 621.

68. de Vries to Loeb, June 25, 1919, LOC, box 3, "De Vries."

69. "An Episode in Science," APS Davenport, series I, box 72, "Muller, H. J." folder.

70. Muller, "Effects of Roentgen Rays," 307. Elsewhere, Muller indicated that the reasons for his later success "in obtaining more conclusive results . . .

lay in the great developments which both genetic technique and genetic theory, based on studies of nonirradiated material, had by that time undergone. These made discriminations between mutagenesis, on the one hand, and both environmentally induced 'modifications' and genetic effects of inbreeding, on the other hand, more precise, and also made the analyses into different classes of heritable changes more informative." Muller, "Nature of the Genetic Effects," 391.

71. Muller, "Effects of Roentgen Rays," 307; emphasis added.

72. Compare with MacDougal's comment that while identifying mutants is "difficult for the first time," it becomes easier over time (MacDougal, *Mutants and Hybrids*, 31); Blakeslee's ability to "pick out . . . mutants unerringly" (Sinnott, "Albert Francis Blakeslee" [*NAS Biographical Memoirs*], 9, 8); and Kohler's description of the role of observation in his *Lords of the Fly*, mentioned in chap. 3. For another interpretation of the "mutant gaze," see Campos, "Mutant Sexuality: The Private Life of a Plant." See also Muller's reference to the "personal equation" in identifying mutations later in this chapter.

73. Muller to J. S. Huxley, December 13, 1919, LL, series I, "J. S. Huxley," "1919–1920" folder. See also Carlson, *Genes, Radiation, and Society*, 112. Curiously, Muller seemed to think that mutations were just as rare as MacDougal did, despite being able to identify many more. As MacDougal had noted in 1902, "Mutations are enormously rare in comparison with the fluctuating variations described above, and this very rarity has led to an underestimation of their value in the origin and development of species, according to de Vries's conclusions." MacDougal, "Origin of Species by Mutation" (*Torreya*, July 1902), 99. See also Muller, "Data notebook: Temperature induced mutations," July–August 1919, LL, series VI, box 1.

74. "Autobiographical notes [prepared for Vavilov]," 1936, p. 7, LL, series II, box 1.

75. Muller, "Measurement of Gene Mutation Rate," 301.

76. Muller to Huxley, June 28, 1920, LL, series I, "J. S. Huxley," "1919–1920" folder.

77. Muller to Huxley, May 4, 1922, LL, series I, "J. S. Huxley," "1921–1925" folder.

78. "Autobiographical notes [prepared for Vavilov]," 1936, pp. 7–8, LL, series II, box 1.

79. Muller, "Variation," 46.

80. Ibid., 46–47.

81. Muller, "Mutation," 106.

82. Muller, "Method of Evolution," 489–90.

83. Ibid., 490–91. No devil, but perhaps a demon. See the discussion of thermodynamic considerations in Chapter 1. See also Keller, "Molecules, Messages, and Memory."

84. Carlson, *Genes, Radiation, and Society*, 84, footnote 224.

85. Later researchers such as James Neel, however, were to view even X-rays as a "rather gross and crude" tool that probably acted "through a dozen different channels both directly and indirectly on the chromosomes." Neel to Stern, January 11, 1941, APS Stern, "Neel, James V.," folder 2.

86. Muller to Altenberg, ca. October 22, 1924, CSH Muller, box 1, folder 2.

87. Just what role Muller thought radiation could play in his experiments is complicated. Although Carlson has noted that Muller was "stimulated" by J. W. Mavor's discovery in the early 1920s that X-radiation affected the frequency of crossing over, it was not initially obvious to Muller that radiation would be successful as a mutagen, despite the deep association he found between life and radiation. Carlson argues that Muller was not entirely committed to a plan of radiation mutagenesis until later. Carlson, *Genes, Radiation, and Society*, 137–39.

88. Muller to Altenburg, November 12, 1924, LL, series I "Edgar Altenburg," 1919–1929 folder.

89. Carlson, "Legacy of Hermann Joseph Muller"; see also Carlson, "Unacknowledged Founding," footnote 39.

90. E. G. Anderson had already been actively using X-rays for some years in the early 1920s to follow up on Mavor's discoveries of nondisjunction. Anderson to R. A. Emerson, June 1, 1923, Cornell University, Department of Plant Breeding Records, box 1, "E. G. Anderson" folder. See also letters from January 21, 1924, and November 3, 1925.

91. As early as 1904, Whetham had remarked that the "superficial similarity between Becquerel rays and Röntgen rays" had "proved for the most part misleading." Nevertheless, he concluded, "the relations between the two branches of the subject are so intimate that it is impossible to study satisfactorily the phenomena of radio-activity without a knowledge of the results previously and simultaneously reached" in the investigation of X-rays. Whetham, *Recent Development of Physical Science*, 200.

92. Lord, *Radium*, 67.

93. Ibid., 69.

94. Quoted in Robards, *Practical Radium*, 112.

95. "One of the essential differences between the 'X' rays and Radium is, that while the quality of the rays from a Crooke's tube varies considerably from time to time, the output from the Radium is quite constant." "Vital Effects of Radium," cited in Hotblack, *A New Activity?*, 20.

96. In fact, he thought, it was the β-rays that "appear to be those which are of great importance in the medical application of radium as a curative agent." Lord, *Radium*, 74, 50.

97. Soddy, *Science and Life*, 98. Cf. "The trouble with the use of X-rays up to this time has been that they are not as penetrating as the so-called 'gamma rays' of radium." "Improved X-rays for Cancer Work," 7.

98. Breit, "Radium—Lost and Found," SM35. By 1932, radium went for up to $70,000 per gram.

99. Biological researchers using radium regularly thanked both hospitals and colleagues in physics departments for their supplies at this time. This suggests that the links between physics and biology—radium and life—were as real and institutional as they were metaphorical, decades before the later medicinal uses of radioactive tracers linked cyclotrons with hospitals; see Creager, "Tracing the Politics," 370; and Creager, *Life Atomic*.

100. Hessenbruch, "The Commodification of Radiation," 7, 9.

101. "The term γ-ray is used when the radiations are given off by radioactive substances, while they are called X-rays when produced in special high-voltage equipment." And again: "Radium gives off 2 MeV γ-rays (these are the same as X-rays), and it can therefore be used instead of these high-voltage machines as a source for high-energy radiation." Alexander, *Atomic Radiation and Life*, 1, 141.

102. Lavine, *First Atomic Age*, 244.

103. Muller, "The Effect of X-Rays and Radium on the Germ Plasm," CSH Muller, box 4, folder 13.

104. Blaauw and van Heyningen, "The Radium-Growth-Response of One Cell"; Gager, "Some Physiological Effects of Radium Rays," 763; Gager, *Effects of the Rays of Radium on Plants*, 269.

105. Packard, "Effect of Radium Radiations."

106. Levine, "Cytological Studies on Irradiated Tissues," 290; emphasis added.

107. Morgan, *Experimental Embryology*, 33.

108. In fact, by 1938, X-rays had so eclipsed radium that even Blakeslee, one of the most successful geneticists ever to experiment with radium, could marvel at the misplaced enthusiasm for radium demonstrated by an Argentinian Guggenheim fellow, commenting that "most of us who have used [radium] have discarded [it] on account of the difficulty of measuring it." Blakeslee to Henry A. Moe, January 14, 1938, CSH Blakeslee, "Blakeslee, Albert—January–June 1938" folder.

109. "Autobiographical notes [prepared for Vavilov]," 1936, pp. 7–8, LL, series II, box 1.

110. Muller, "Effects of Roentgen Rays," 306.

111. Muller, "Method of Evolution," 490–91. Newspaper reports describing Muller's experiments also readily equated the two for their readers: "The physical and chemical composition of the genes may be changed by X-rays or gamma-rays of radium." "Evolution Process Is Aided by X-Rays," 3. Later authors agreed: "From a biological point of view X- and γ-rays need be considered merely as a means of releasing high-energy electrons within the object which is being irradiated whatever its chemical constitution . . . only the energetic X- and γ-rays are able to strip off electrons from every type of atom to produce ions." Alexander, *Atomic Radiation and Life*, 141.

112. Kaempffert, "Darwin After 100 Years," SM10.

113. Muller, "Method of Evolution," 491.

114. Carlson, *Genes, Radiation, and Society*, 138; Muller to Huxley, June 20, 1924, LL, series I, "J. S. Huxley," "1921–1925" folder.

115. Muller to Hartman, May 6, 1927, CSH Muller, box 1, folder 21.

116. deJong-Lambert, *The Cold War Politics of Genetic Research*, 12.

117. Carlson, *Genes, Radiation, and Society*, 145.

118. Keller, "Physics and the Emergence of Molecular Biology," 398. Elsewhere, Keller has noted, "Indeed, after Rutherford's success in 1919 in inducing a transmutation of the elements, Muller pursued his own search for a means of inducing mutation with that precedent directly in mind, even enti-

tling his discovery of X-ray induced mutations 'Artificial Transmutation of the Gene.'" Keller, *Century of the Gene*, 152, footnote 10. As Carlson has noted, "Muller was aware of the parallel of his biological work to the transmutation of elements first induced artificially by Rutherford in 1919." Carlson, *Genes, Radiation, and Society*, 147. Loeb had proclaimed to Rutherford upon the announcement of this discovery, "How all biological work shrinks into insignificance when measured by the progress you have made." Loeb to Rutherford, June 10, 1919, LOC, box 13, "Rutherford, Ernest."

119. Muller, "Artificial Transmutation of the Gene." Cf. Muller, "The Production of Mutations by X-Rays."

120. Muller, "How Evolution Works," 13. Cf. Muller, "Method of Evolution," 491–92.

121. Muller, "How Evolution Works," 14.

122. Ibid., 491; Carlson, "Unacknowledged Founding." Elsewhere, Carlson notes that in his 1927 paper, Muller "claimed a 15,000% increase over spontaneous mutation frequency, the production of over 100 new mutations, the similarity of X-ray mutations to those obtained spontaneously, the susceptibility of sperm as well as eggs to X-ray mutagenesis, and the large number of fractional mutations or mosaics which implied a 'precocious doubling' of the sperm chromatid." Carlson, "H. J. Muller (1890–1967)."

123. Coulter, "X-Ray Mutations," 110.

124. Faber, "X-Rays Form New Life."

125. Wells, Huxley, and Wells, *Science of Life*, 1477.

126. Huxley, "Where Darwin's Theory Stands Today," SM5.

127. Muller, "Quantitative Methods in Genetic Research," 419.

128. "News and Views," 658.

129. Hanson and Heys, "Effects of Radium," 115–16.

130. Hanson and Heys, "Analysis of the Effects," 202.

131. "This Number Is Devoted to the Discussion of the Effects of X-rays," introduction.

132. Mohr, *Heredity and Disease*, 190; Hanson and Winkleman, "Visible Mutations," 277.

133. With exceptions, of course. Demerec would soon complain, "I think that Muller's work must be repeated. His paper given at the Congress made me just mad. Big conclusion are drawn, but the data given are so meager that very little could be taken as proven." Demerec to Anderson, December 20, 1927, UMC Anderson, folder 17. Sturtevant would also complain, "I was very doubtful about Muller's claims for X-ray production of mutation, until I saw Weinstein's results . . . [but] the data look O. K. to me if taken in connection with Muller's sketchy account." Sturtevant to Anderson, January 21, 1927, UMC Anderson, folder 45.

134. Carl G. Hartman to Muller, August 2, 1927, CSH Muller, box 1, folder 21. L. C. Dunn would also later describe this work as Muller's "big thunder": "The Reminiscences of L. C. Dunn (1961)," Columbia Center for Oral History, Columbia University, 346. According to Carlson, "overnight, Muller had become a world figure, and the excitement of man's first alteration of living matter strengthened the hope he had long nourished that this would

also provide the power for man's control of his own evolution." Carlson, "Unacknowledged Founding," 163.

135. Sinnott and Dunn, *Principles of Genetics*, 192.

136. Broadcast, "Open University, A second level course; S299/6—Genetics: The Phage School; Tape No.: BLN16FW802," April 21, 1976, CIT Delbrück, box 32, folder 1.

137. Hollaender, *Radiation Biology*, v; "Genetic Eff. of Rad. on Man," 1953, APS Stern, "Stern, C.," "Radiation Genetics," box 36, folder 1.

138. Crow and Abrahamson, "Mutation Becomes Experimental."

139. Carlson, *The Gene*, xi, 3, 252; emphasis added.

140. Sapp, *Genesis*, 141. Sapp's sequence echoes that of Ernest Everett Just: "Moreover, it was soon learned, both for animals and for plants, that mutations can be induced not only by Roentgen rays but also by radium and by ultra-violet rays." Just, "On the Origin of Mutations," 61.

141. In more recent work, Carlson has softened his account considerably, saying that "what was original was not the idea of trying it, but of proving it." In the process, however, Carlson misconstrues the nature of Blakeslee's findings, declaring that "Blakeslee, like Morgan, could not really tell if he had induced any mutations. Over the years, the contradictory results would not be resolved until Muller (and independently, a year later, L. J. Stadler) showed that X rays were indeed mutagenic." Carlson, *Mendel's Legacy*, 173, footnote 34.

142. Carlson, *Genes, Radiation, and Society*, 153; Muller to Altenburg, July 14, 1928, LL, series I, "Edgar Altenburg," "1919–1929" folder.

143. "The Secret of Life," 52.

144. Neel to Stern, July 30, 1940. Neel also wrote to Stern in early July 1940, saying that the study of chromosomal aberrations had taken center stage: "CSH is swell. Learning a lot. Practically everybody here in the fly group—which I know best—is interested in the chromosomal rearrangement problem." APS Stern, "Neel," folder 2.

145. Hunt, "Dr. Muller and the Million Human Time-Bombs."

146. "Lecture, re: a general survey of the gene," ca. 1927, LL, series II, box 1.

147. "Evolution to Order," radio broadcast "under the auspices of Science Service, over the Columbia Broadcasting System," Thursday March 24, 1938, APS Blakeslee, "Lectures, Papers, Etc.: Adventures in Science," box 23, folder 2.

148. Goodspeed, "The Effects of X-Rays and Radium on Species of the Genus *Nicotiana*," 254.

149. Muller, "Further Studies on the Nature and Causes of Gene Mutations," 213.

150. Dunn, "Genetics in Historical Perspective," 43–44.

151. "Genetic Eff. of Rad. on Man," 1953, APS Stern, "Stern, C.," "Radiation Genetics," box 36, folder 1.

152. As Jack Schultz later noted, "Ideally, for the quantitative study of mutations *en masse*, it would be desirable to detect all types of variants. This would include the lethal effects, the ordinary visible mutants, and the very

slight types. Of these, the last present the greatest difficulty for measurement with any accuracy; there are at present no data available which permit even an estimate of their frequency. . . . For the 'visible' mutations, it is equally clear that criteria may differ from experiment to experiment, and notoriously from observer to observer. The frequency of occurrence of visible mutation might be determined with accuracy in an organism in which all the possible types of variation were known, and they could be distinguished from developmental accidents due to nongenetic influences. Even in *Drosophila melanogaster*, however, genetically the best studied of all animals, it is not possible to do this yet." Schultz, "Radiation and the Study of Mutation in Animals," 1211.

153. Blakeslee, "Mutations in Mucors."

154. Sinnott, "Albert Francis Blakeslee," *NAS Biographical Memoirs*, 9, 8.

155. Gager, *Plant World*, 99; Woods, "Botany and Human Affairs," 573.

156. As late as 1933, Blakeslee wrote, "I think it will be very difficult to separate the genetical from the cytological aspects since we are coming to the point where we cannot make clear a distinction we once thought between the gene and the chromosomal mutations." Blakeslee to Benjamin Duggar, February 18, 1933, APS Blakeslee, "Duggar, Benjamin," box 8, folder 3.

157. Ford, *The Study of Heredity*, 96.

158. "An Episode in Science," APS Davenport, series I, box 72, "Muller, H. J." folder.

159. Avery, Satina, and Rietsema, *Blakeslee*, vii.

160. Muller, "Genetic Variability," 499. De Vries would not agree. He had written to Loeb a few months earlier, saying, "I have wholly retired from active business, but am still cultivating my beloved Oenothera's and find that they are always true to their principles, despite the rather curious and vague objections of Morgan and others. But I do not like to be taken into public discussions." De Vries to Loeb, March 17, 1919, LOC, box 3, "De Vries."

161. Blakeslee, "Twenty-Five Years of Genetics," 35.

162. Blakeslee, "Ideals of Science," 591.

163. "Seventy-Five Years of Progress in Genetics," p. 21, APS Blakeslee, "Lectures, Papers, Etc.," box 23, folder 35; emphasis added. A very similar sentence was published in Blakeslee, "Chromosomes, Chemical Stimulators, and Inhibitors of Normal and Abnormal Plant Growth," 43.

164. Blakeslee, "Control of Evolution and Life Processes in Plants," 59–60. Blakeslee would later note that Gager's "single experiment with radium emanation obtained an increase in non-disjunctional types and a couple of recessives out of a number so small as not to be surely significant statistically." Blakeslee, "Twenty-Five Years of Genetics," 36.

165. "Gager and Blakeslee now report mutations of various types following the treatment of *Datura* with radium rays." Coulter, "X-Ray Mutations," 111.

166. "Following my work on flies, Whiting has obtained positive results by the use of X-rays on wasps. Blakeslee, Buchholz and the others of this group have a mass of interesting results from X-rays and radium applied to the Jimson-weed, Datura, that extended the findings concerning lethal as well as visible mutations to dicotyledonous plants." Muller, "Method of Evolution," 495–96.

167. Stadler, "Mutations in Barley Induced by X-Rays and Radium." According to Carlson, Muller "often remarked to his classes that it was only luck that he, and not Stadler, was the first to come out in print with the clear evidence for radiation-induced mutations": Carlson, "H. J. Muller (1890–1967)." Muller worried in 1939 that Stadler might mistake his motives in other matters because of this matter of priority: Muller to Darlington, July 3, 1939, Darlington Papers, Oxford University, box 110, folder J.130. And a decade later, in 1949, he wrote, "We have done parallel work on the problem of mutations and their artificial production, he with plant and I with animal material": Muller to Loyalty Board, June 17, 1949, CSH Muller, box 1, folder 17. According to Jim Crow, Muller's "greatest contribution . . . was not the discovery of radiation mutagenesis. L. J. Stadler independently demonstrated this in barley at about the same time; radiation was in the air. Muller's great contribution was the invention of techniques that would make mutagenesis experiments feasible. Prior to Muller there was no systematic, objective, and quantitative way of measuring mutation rates." Crow, "Some Reflections."

168. Indeed, it was Muller's discussions with Stadler and others that would ultimately cause Muller to begin to question the viability of a firm distinction between point mutations and exceptionally minor chromosomal rearrangements, as well as the nature of the position effect.

169. Blakeslee to Morgan, May 22, 1935, CIT Morgan, box 1, folder 4.

170. Muller and Altenburg, "Study of the Character," 213.

171. "Elimination of the X-chromosome from the egg of D[rosophila] m[elanogaster] by x-rays" (1921, but these particular pages are dated around 1923), LL, series II, box 1.

172. On the occasion of his seventieth birthday, Muller was presented with a 600-page blank book as a "reverse *Festschrift*." Fascinatingly, the theme of "blanks"—from gunshots to empty spaces on the page—pervades Muller's work. In 1918 he had even sent a humorous "mad lib"–style letter full of blanks to Huxley, recounting a recent family picnic; see Carlson, *Genes, Radiation, and Society*, 106, 406.

173. Patterson, "X-Rays and Somatic Mutations," 261.

174. Muller, "Method of Evolution," 496.

175. "History of Genetics" (unpub.), XI-5, CIT Sturtevant, box 24, folder 1.

176. Dobzhansky to Demerec, April 2, 1934, APS Demerec, "Dobzhansky, Theodosius," box 7, folder 3.

177. "We cannot get irradiation done here and ordinarily whenever I want to get flies irradiated I have to take a trip to Greenfield, 17 miles away, with them, so this would spare me that trip." Muller to Stern, February 1, 1944, APS Stern, "Muller, Hermann J.," box 18, folder 5. Muller preferred radium especially for "our low intensity work because of the greater ease with which it can be used to give low intensities constantly over a long period." He emphasized again, however, "that there is in fact no quantitative or qualitative difference between x-ray + radium effects, for a given total dose in *r* units." Muller to Stern and W. P. Spencer, October 12, 1943, APS Stern, "Muller, Hermann J.," box 18, folder 4.

178. Muller, "Heribert Nilsson's Evidence," 166.

179. "Proposals for Investigations Concerning the Genetic Effects of Radium," 1933, LL, series VI, box 2.

180. Hanson and Heys, meanwhile, had undertaken experiments to establish "just what elements in radium and X-rays are responsible for the results obtained." They found that the γ-rays alone, of the radium emanations, were sufficient to produce lethal mutations in *Drosophila*—just as earlier investigators had long suspected. Earlier investigators attributed this ability to the γ-rays' "penetrance"; Hanson and Heys attributed it to the impact of β-particles produced when γ-rays passed through living matter. Hanson and Heys, "Effects of Radium," 115–16.

181. "Proposals for Investigations Concerning the Genetic Effects of Radium," 1933, LL, series VI, box 2.

182. As G. Failla noted in 1936, "Ionization is the only thing we know of to which, rightly or wrongly, we may attribute all other effects of radiation. Hence, any attempt at a correlation of the biological effects of radiation requires of necessity a quantitative knowledge of the ionization in the living materials studied. But this essential information is very difficult to obtain." Failla, "Ionization and Its Bearing on the Biological Effects of Radiation," 106.

183. "Biological Effects of X-Rays," 382.

184. Hanson and Heys, "Analysis of the Effects," 212.

185. "Lecture, re: a general survey of the gene," ca. 1927, LL, series II, box 1.

186. Blaauw and van Heyningen, "Radium-Growth-Response," 403–17.

187. Muller, "The Production of Mutations by X-Rays," 724; emphasis added.

188. Schultz, "Radiation and the Study of Mutation in Animals," 1250, citing Muller, "Radiation and Genetics."

189. "Dr. Hermann Muller is Dead," *New York Times*, April 6, 1967, 39.

CHAPTER 6

1. That is, there is more work to be done in making such language central to historical analysis, rather than simply deploying it for its radiant descriptive charm.

2. "Radium was to claim its victims too," Abraham Pais has noted, "but the story of its early physiological impact is incomparably less harrowing than that for X-rays. I know of no lethal injuries caused by exposure to radioactive material in the pre–World War I years." Pais, *Inward Bound*, 99.

3. Breit, "Radium—Lost and Found," SM35.

4. Marie Curie died of aplastic anemia, believed to have been caused by her long-term exposure to ionizing radiation. Lavine, *First Atomic Age*, chap. 4.

5. Emery, "Roll of Martyrs to Science Is Increasing," XX5; "How the Men Who Devote Themselves to Humanity Are Often Overlooked," E6; "New Radium Disease Found"; and "Death Stirs Action on Radium 'Cures,'" 12. According to Pais, "The bad years for radium came in the late nineteenth-twenties (radium was more plentiful by then) when numerous reports appeared of persons severely injured or killed by internally deposited radio-

active substances." Pais, *Inward Bound*, 100. For more on the longer history of radium in American life, see Lavine, *First Atomic Age*; Clark, *Radium Girls*; and Mullner, *Deadly Glow*.

6. Hotblack, *A New Activity?*, 36. As Lavine has suggested, "Radium, more and more directly personified in the press the more 'evil' it became, lashed out against its captors, burned those who embraced it, and killed its own 'mother.' Where x rays had *technicians*, it had *handlers*, and when it was lost, the 'radium hound' was called in to track the prey." Lavine, *First Atomic Age*, 179.

7. Ibid., 41.

8. For more on the commercialization of radium in American industry, see Rentetzi, "The U.S. Radium Industry."

9. Letter dated August 4, 1938, LL, series I, "Correspondence (General)," "August 1938" folder.

10. Gager to W. C. Curtis, February 27, 1928, BBG, box 9, folder 9, "C. S. Gager, Research, 1921–1928."

11. W. C. Curtis to Gager, March 10, 1928, ibid.

12. Communication No. 2, March 8, 1928, "To the group of investigators interested in securing a fund for support of research upon the 'Effects of Radiations upon Organisms,'" ibid.

13. APS Stern, "National Committee on Radiation Protection and Measurements," folder 1. For more on the history of radiological protection, see Hacker, *The Dragon's Tail* and *Elements of Controversy*; Walker, *Permissible Dose*; and Jolly, "Thresholds of Uncertainty." Lavine has suggested that radium and X-rays had different regulatory profiles: "Radiation had almost always involved an external agency to regulate it, and this became more and more true as the years wore on. Exactly the opposite was true with radium . . . the distribution of radium and other radioactive substances (or their imitators) tended to be decentralized. . . . Radium . . . was a western phenomenon, conjuring up alpine vistas and mountain springs, and was scarcely discussed for long in any context without a nod to its geological origins. It resided in a geographic region, while x rays were better understood as belonging merely to specific rooms, such as the doctor's office or the physics laboratory." Lavine, *First Atomic Age*, chap. 4.

14. As Muller had written to Stern, "Hanson's work . . . was done at my suggestion and in consultation with me." Muller to Stern, January 14, 1947, APS Stern, "Muller, Hermann J.," box 18, folder 8. Reference is made to Hanson's 1928 abstract in *Anatomical Record* 41, 99, and to his more detailed joint publication with Heys, "An Analysis of the Effects of the Different Rays of Radium in Producing Lethal Mutations in Drosophila."

15. The chairman of the executive committee of Cancer Control Organisation for Edinburgh and South-East Scotland wrote to Muller on March 7, 1938, "I have taken steps to arrange for 70 mgms. of radium being made available for your use." Muller acknowledged receipt on April 5. See also letters from F. A. E. Crew to Muller, in LL, series I, "Correspondence (General)," "March 1938," "April 1938," and "July 1938" folders.

16. "April 28, 1939. Muller's *Genetics and Society*," LL, series I, "Correspondence (General)," "April 16–31, 1939" folder.

17. Letter dated April 5, 1938, LL, series I, "Correspondence (General)," "April 1938" folder. Muller to Stern and W. P. Spencer, October 12, 1943, APS Stern, "Muller, Hermann J.," folder 4.

18. Frank Thone, Science Service Editor in Biology, wire report, November 1, 1946, LL, series VII, box 1, "1946 Nobel—misc" folder.

19. Demerec to Blakeslee, July 6, 1940, APS Demerec, "Blakeslee," box 3, folder 5.

20. As Blakeslee wrote to a colleague, "These grants run out, however, July 1 and, since the Radiation Committee, from which he [Cartledge] has received a single grant terminating June 30, has officially in its publications stated that its support from now on would be for study of the physical side of radiation rather than the biological side, there are no more funds from this source." Blakeslee to Ernst Bessey, May 28, 1936, CSH Blakeslee, "Blakeslee, Albert—1936" folder.

21. "Lebenslauf of A.F.B.," p. 6, APS Blakeslee, "Biographical Materials," box 25, folder 2. Blakeslee also noted that "Dr. Demerec and Dr. Kaufman are using the Xray machine nearly every day to induce mutations in Drosophila. They analyze the mutations by the changes in the salivary chromosomes." Blakeslee to W. D. Coolidge, General Electric Laboratories, Schenectady, July 5, 1939, CSH Blakeslee, "Blakeslee, Albert—1939" folder.

22. W. C. Curtis to Gager, February 18, 1928, BBG, box 9, folder 9, "C. S. Gager, Research, 1921–1928" folder.

23. F. A. E. Crew to Muller, July 8, 1938; R. Lamy to Muller, July 7, 1938; and K. MacKenzie to Muller, July 15, 1938, LL I, "Correspondence (General)," "July 1–15 1938" folder.

24. Bacq and Alexander, *Fundamentals of Radiobiology*, 2.

25. "Improved X-rays for Cancer Work," 7.

26. Israel Klein, "Science Tries to Equal Radium's Terrific Power by Electricity," *NEA Service* (January 5, 1928), quoted in Lavine, *First Atomic Age*, 238.

27. Brunngraber, *Radium*, 398, 408.

28. "Giant Cancer Tube Tried Out," *Los Angeles Times*, March 25, 1931, A1, quoted in Lavine, *First Atomic Age*, 238.

29. "Atom-Bomb By-Product Promises to Replace Radium as Cancer Aid," 1.

30. Calabrese and Baldwin, "Radiation Hormesis," 57; Curry, "Accelerating Evolution," chap. 6.

31. "New Elixir Found for Plant World," 17. Compare with Alexander in 1957: "Radium gives off 2 MeV γ-rays (these are the same as X-rays), and it can therefore be used instead of these high-voltage machines as a source for high-energy radiation. With the development of atomic energy, man-made isotopes have become available in large quantities and radium is no longer the only source of high-energy γ-rays." Alexander, *Atomic Radiation and Life*, 141. Lavine has suggested that radium and X-rays "developed very distinct identities over the course of their first half-century: by the 1940s, radium's profligate energy seemed even weirder and wilder (and more sinister) than they had at first glance in 1903, while the x-ray had been steadily 'tamed' until the threats it had first suggested in 1896 seemed remote indeed to most Americans." Lavine, *First Atomic Age*, 22.

32. As Angela Creager has noted, "Radiation itself was a critical tool, but it was more than a tool. In part this was an effect of novel technologies: radiation was produced in ever-increasing amounts by the new instruments of nuclear physics, from cyclotrons in the 1930s to reactors and nuclear weapons in the 1940s and 1950s." Creager, "Commentary on 'Tools'"; see also Creager and Santesmases, "Radiobiology in the Atomic Age"; Creager, "Tracing the Politics"; and Creager, *Life Atomic*.

33. Weart, *Nuclear Fear*, 172, quoting Ruth Ashton, "The Sunny Side of the Atom," CBS Radio, June 30, 1947.

34. Sinnott, "Albert Francis Blakeslee"(*NAS Biographical Memoirs*), 13; Calabrese and Baldwin, "Radiation Hormesis," 57; Cook, "Dr. Muller Receives Nobel Medicine Award," 325ff.

35. Calabrese and Baldwin, "Radiation Hormesis," 57.

36. Muller, "Radiation and Genetics," 240.

37. "Basis of the theory of the gene: The experimental evidence concerning the properties of the gene," LL, series II, box 1; Muller, "The Gene," 6–7. For more on material versus operational understandings of the gene, see Falk, "What Is a Gene?—Revisited."

38. Muller, "Radiation and Genetics," 240.

39. Cockerell, "Radiation and Genetics," 476.

40. Muller, "Radiation and Genetics," 241.

41. Muller tried to develop this theory of radium storage in a few different ways to see if he could get any mileage out of it, and he reported that experiments testing the possibility of innate radium storage in fruit flies were conducted by Mott-Smith, but had brought negative results. Muller was left to conclude that "natural mutations" had perhaps not so much to do with the absorption of radiation as he had first thought. We have, he said, "to accept as more probable the alternative that the great majority of natural mutations have as their primary cause some other process or processes than the absorption of radiation. Thus we seem to be obtaining a negative answer to the query which I raised in 1927 as to whether natural radiation fashions the building-blocks of evolution." Muller, "Radiation and Genetics," 242.

42. Burke, "Physics and Biology," 81.

43. Muller, "Need of Physics," 210.

44. "Inverted Synapsis of Genes as Evidence for the Periodic Character of their Mechanism of Attraction," by H. J. Muller and D. Raffel, November–December 1934, LL, series II, box 1.

45. Muller, "Need of Physics," 211.

46. Cf. "All such radiation originates at some gene locus, from which it travels outward at a presumably finite rate." Ibid., 4–5.

47. Ibid., 211.

48. Muller, "Method of Evolution," 498.

49. Even as late as 1936, in a letter to Stalin in which he proposed a new system of eugenics for the USSR, Muller noted, "Now by making step after step in this way, through several generations, a level is soon reached by great numbers which correspond with that of the genetically best equipped individuals of today, or which, by combining with the varied gifts of the latter, in sum

total even surpasses them. And this in turn supplies a kind of genetic tonic, as it were, a vitalizing element that diffuses out to mix with the whole population." "Letter to Stalin (1936)," LL, series II, box 1, republished as Glad, "Hermann J. Muller's 1936 Letter to Stalin." Muller frequently reiterated to varied audiences his interest in a sort of ultimate "intelligent control over biological evolution": "Work of the Department of Mutation and the Gene from September 1933 to December 1936," CSH Muller, box 5, folder 13.

50. "Department of Genetics" (1932), 33.

51. Brunngraber, *Radium*, 345.

52. Weart, *Nuclear Fear*, 49; see also Hollaender, "The Problem of Mitogenetic Rays," 919–59.

53. "The Effects of Roentgen Rays Upon the Hereditary Material (1933)," p. 18, LL, series II, box 1.

54. "Inverted Synapsis," p. 4, LL, series II, box 1.

55. Wainwright, "Historical and Recent Evidence for the Existence of Mitogenetic Radiation"; Stern, "Mitogenetic Radiation: A Study of Authority in Science."

56. Seifriz, "The Gurwitsch Rays," 307.

57. "Imprisoned in each of these cells is a force many thousand times more powerful than the pull of gravity, a force that seems to be the ultimate source of all life and energy." Stout, *Secret*, 4.

58. Pais, *Inward Bound*, 116. As Ralph Stayner Lillie put it in 1945, "In both physics and biology a 'nuclear' influence is generally conceived as one which originates in a small, centrally situated area and controls processes in the immediate surroundings." Lillie, *General Biology and Philosophy of Organisms* (Chicago: University of Chicago Press, 1945), quoted in Lehman, *Biology in Transition*, 97.

59. Hoffman, *The Life and Death of Cells*, 89, 99.

60. "Fission" in nuclear physics was explicitly coined with reference to biological phenomena. According to Elisabeth Rona, the American microbiologist William Arnold was at Bohr's laboratory in Copenhagen in January 1939 when he was asked to observe on the oscillograph the experimental splitting of atoms of uranium: "We could see the tall spikes produced by the energetic fragments of the splitting. [Otto] Frisch turned to me and asked, 'What do you call the splitting of bacteria?' I answered, 'Fission.' This term henceforth was used to describe the splitting of uranium into fragments." Rona, *How It Came About*, 44. In a Caltech press release in 1951, Delbrück would later reappropriate the term as if it had first come from physics: "'At the gene level we seem to have a binary fission mechanism,' or reproduction by splitting of parent gene." CIT Delbrück, box 3, folder 4.

61. Along these lines, even Kary Mullis's invention of the polymerase chain reaction (PCR) may be another echo of this legacy. As historian of molecular biology Michel Morange has noted, the name PCR was "chosen not at random but because of its reference to nuclear chain reactions. In both its spirit and the person of some of its founders, molecular biology is the descendant of the physics of the 1930s and 1940s." Morange, *A History of Molecular Biology*,

242. See also Morange's note on references to "deoxyribonucleic bombs," which "exploded" in Kary Mullis's head after his discovery: 319, footnote 32.

62. Muller, on gene evolution, Oak Ridge, April 8, 1947, CSH Muller, box 6, folder 7.

63. "For the problem of the evolutionary process is not just the problem of life. It is the problem of the cosmos itself." Kaempffert, "Darwin After 100 Years." Other examples abound.

64. Darlington, *Recent Advances in Cytology*, 246, 334.

65. "The central problem of biology—the essential 'secret of life'—lies in the genes themselves: their composition and structure, how they are autocatalytically synthesized, and how they may have first originated from lifeless matter." Plunkett, *Outlines of Modern Biology*, 667; see also Lehmann, *Biology in Transition*, 134.

66. Problems of Radiobiology with Emphasis on Radiation Genetics," lecture, Oregon State College, Biology Colloquium, April 21, 1951, APS Stern, box 34.

67. According to Max Delbrück, the theory "was a masterly summation of all the existent data on mutation, both radiation-induced and spontaneous." CIT Delbrück, box 22, folder 2. See Crowther, "Biological Action of X-Rays"; Lea, *Actions of Radiations on Living Cells*; Summers, "Physics and Genes"; and Summers, "Concept Migration."

68. Delbrück, "Radiation and the Hereditary Mechanism," 361–62.

69. Summers, "Concept Migration"; Creager, *Life of a Virus*, 192;

70. Stent, quoted in Summers, "Concept Migration."

71. For more on this work, see Sloan and Fogel, *Creating a Physical Biology*.

72. As Delbrück had noted, "It was all a matter of bridging physics and genetics at that time—there just weren't any people who could do that." Delbrück, "How It Was" (May–June 1980): 22.

73. Beyler, "Targeting the Organism," 267–68.

74. As early as 1931, Muller had "ruled out a target-theory measurement of gene size from x-ray induced mutations" in *Drosophila*: Carlson, "Unacknowledged Founding," 159. Indeed, Muller is said to have held target theory "in low regard" in light of the results of his own radiation work because its approach to measuring the size of a gene within a cell was based on the assumption of a one-to-one correspondence between incidents of ionizing radiation and mutation, when in fact no such correspondence existed. (Building on his earlier criticism of the approach from the results of his studies of one locus in 1931, Muller calculated in 1940 that "about 200 to 600 ionizations occurred per gene per mutation at higher doses.") Carlson, "H. J. Muller (1890–1967)," 20.

75. As Guido Pontecorvo noted in 1958, "Though this first application of physical ideas to a particular set of problems did not work out too well, the whole outlook in theoretical genetics has since been perfused with a physical flavour." And: "In the years immediately preceding WWII, something quite new happened: the introduction of ideas (not techniques) from the realm of

physics into the realm of genetics, particularly applied to the problem of the size, mutability, and self-replication of genes." Pontecorvo, *Trends in Genetic Analysis*, 2. Cf. "DNA Damage and Repair." For an early account, see Fleming, "Émigré Physicists and the Biological Revolution." See also Keller, "Physics and the Emergence of Molecular Biology."

76. Summers, "Concept Migration."

77. Bohr, "Light and Life."

78. As Delbrück would later say, "Such findings were vaguely reminiscent of the 'wholeness' of the atom, or the stability of the stationary states. The stability of the gene and the algebra of genetics suggested something akin to quantum mechanics." From a lecture entitled "Light and Life III," given at the inauguration of the New Carlsberg Laboratory, Copenhagen, September 27–28, 1976. Quoted in Shropshire, *Max Delbrück*, 94. But see McKaughan, "The Influence of Niels Bohr on Max Delbrück."

79. "Light and Life," *New York Times*, 70.

80. Delbrück to Pontecorvo, March 12, 1968, CIT Delbrück, box 17, folder 32. "Do you have a copy of Delbrück & Timofeef's write-up of our Copenhagen discussion on the mechanism of mutation? It was 5-pages of mimeographed sheets." Muller to Carlos Offermann, September 28, 1937, CSH Muller, David Muller series, box 5, folder 48.

81. Bohr, *Atomic Physics and Human Knowledge*, 2.

82. Ibid., 9.

83. Delbrück, "How It Was" (March–April 1980): 25.

84. Kay, "Quanta of Life." See also Kay, "Secret of Life"; Kay, "Conceptual Models and Analytical Tools"; Fischer and Lipson, *Thinking About Science*; and Roll-Hansen, "The Application of Complementarity to Biology."

85. Delbrück, "How It Was" (May–June 1980): 21ff.

86. Delbrück, "Problems of Modern Biology in Relation to Atomic Physics," chap. 8.

87. Stent, "Max Delbrück," 9.

88. Miller, "Max Delbrück, 1906–1981," 269.

89. Ernst-Peter Fischer, "Max Delbrück: A Physicist Who Looked at Biology" (talk delivered at the *Phycomyces* meeting at Cold Spring Harbor, August 1982), p. 6, CIT Delbrück, box 49, folder 4. As early as 1922, Muller himself had wondered whether bacteriophage might be, in its essence, a gene. Viruses such as phage, as molecules that could replicate and mutate, and existing in that half-alive realm between the organic and the inorganic, seemingly shared many features with radium. Much more could be said about the role of the virus as a half-living entity and the "importation of specific tools and concepts" to study it, "as well as on the analogies being made to other systems," as Creager has noted in the case of Wendell Stanley (*The Life of a Virus*, 317). In 1936, Blakeslee even suggested to Stanley that he try X-rays to induce mutations in his crystalline virus: Blakeslee to Stanley, January 21, 1936, CSH Blakeslee, "Blakeslee, Albert—1936" folder. See also Podolsky, "The Role of the Virus in Origin of Life Theorizing," and the discussion in chapter 1.

90. Schrödinger, *What Is Life?*, 34, 47, 63. Mark Adams has noted that when Schrödinger's work was translated into Russian, it "provoked great

interest." Soviet nuclear physicists perceived a "close analogy . . . between nuclear physics in the period 1900–1935 and the physico-chemical study of heredity as it was beginning to develop. . . . The analogy between molecular genetics and nuclear physics was repeated again and again by Soviet physicists: as the first-half-century was dominated by physics, the second would be dominated by molecular biology; as the study of the atomic nucleus had dominated the scientific scene, so too would the study of the nucleus of the cell." Adams, "Genetics and the Soviet Scientific Community," 223–24.

91. Delbrück to Bohr, April 14, 1953, CIT Delbrück, box 3, folder 29.

92. Inspired by Stanley's work on the crystallization of tobacco mosaic virus, Delbrück wrote "a short memorandum to himself entitled 'Riddle of Life'" in August 1937. Olby, *Path to the Double Helix*, 236. In his Nobel lecture of 1969, Delbrück proclaimed, "Molecular genetics, our latest wonder, has taught us to spell out the connectivity of the tree of life in such palpable detail that we may say in plain words, 'This riddle of life has been solved.'" Delbrück, "A Physicist's Renewed Look at Biology," 1312. See also Delbrück, "Riddle of Life (1937)," CIT Delbrück, box 36, folder 1. The terms "enigma," "riddle," and "secret"—so frequently used by an earlier generation to describe the mysteries of radium—were also frequently used to describe the phenomena of life prior to the emergence of "code" talk. For more on "code" language in the history of molecular biology, see Kay, *Who Wrote the Book of Life?*

93. Received accounts include McKaughan, "The Influence of Niels Bohr on Max Delbrück"; Kay, "Secret of Life"; and Roll-Hansen, "The Application of Complementarity to Biology."

94. White, "Heredity, Variation and the Environment." This chapter of Gager's *General Botany* was written by Orland E. White, the curator of plant breeding and economic botany at the Brooklyn Botanic Garden. White acknowledged that Muller's paper had been "freely used in writing this survey," and his wording was taken almost verbatim from Muller's paper "Mutation," published in 1923. Another commentator had also used Muller's analogy in 1929: "It has even been argued that all genes ought to be expected to change sooner or later, just as an atom of radium changes in the course of 2,000 years or so until it ends as a dross of lead. As the radium atom 'runs down,' so one might expect such an active chemical structure as the gene to 'run down,' although the 'life' of most genes is long—probably several thousand years at least." Popenoe, *The Child's Heredity*, 264.

95. "The Doctrine of the Gene," ca. 1936, p. 2, LL, series II, box 1.

96. Reprinted as Muller, "Genetic Nucleic Acid," 144 and as Muller, "Genetic Nucleic Acid: Key Material."

97. "The turnover or metabolism occurring in the other material of a cell represents, we might say, the fire that the genetic material keeps going outside itself, to get that other material to work for it, in the service of its own distinctive goal: its own survival and replication. Thus metabolism is not the essence of life but a kind of upper-level expression of life after the genetic material has succeeded in making for itself a workshop of protoplasm." Muller, "Genetic Nucleic Acid: Key Material," 14.

98. Blakeslee and Avery, "Methods of Inducing Doubling," 404, 408. For a

synthetic account of the history of colchicine in mutation breeding, see Curry, "Making Marigolds"; see also Curry, "Accelerating Evolution," chap. 4; and Goodman, "Plants, Cells, and Bodies." Muller was also one of the foremost proponents of a vision of genetic engineering: "We may after all make some headway on this sector of the biological battlefront," he noted in 1927. "We cannot make life—far from it; we probably can, however, *remake* it." "Lecture, re: a general survey of the gene," ca. 1927, p. 21, LL, series II, box 1.

99. Quoted in Comfort, *Tangled Field*, 93.

100. McClintock devised novel cytogenetic and phenotypic-level descriptions of mutation to capture other hereditary phenomena that Muller had been uncertain how to explain. As Comfort has argued, "McClintock . . . saw her theory as an *alternative* to the chemical-change-in-the-gene model. Muller's model posited autonomous, particulate genes, in which the mutations were 'true'—that is, chemical. McClintock's postulated an integrative genome in which genes acted in suites, controlled by regulatory elements. In her Carnegie report of that year, she wrote that her investigations 'cast doubt on interpretations that postulate a "true gene mutation," that is, a chemical change in a gene molecule,' and suggested that phenotypic change was rather the result of reversible inhibition and modulation of genes." Comfort, *Tangled Field*, 134. Muller, however, was not always able to recognize these theories as based on non-genocentric first principles of mutation.

101. Comfort, *Tangled Field*, 85. As Muller would later write, "What a relief it is for us mutation workers to know that the mutable genes are in a different class from . . . 'ordinary gene mutations,' after all." Muller to McClintock, October 27, 1948, quoted in Comfort, *Tangled Field*, 134.

102. "Evolution Process Is Aided by X-Rays," 3. Stadler argued that most mutations induced by X-ray treatment were "not representative of natural mutation in general. This special class may be wholly or largely made up of mechanical or extra-genetic change," resulting primarily in chromosomal derangements, not gene mutations. He also claimed that "the conclusion that mutations are the result of chemical changes within the gene is not inevitable. Certainly the extension of this conclusion to induced mutations in general is very questionable."

103. Delbrück, "A Physicist Looks at Biology," 188.

104. This idea might once have sounded much like Bateson's discredited presence-absence theory in genetics—that all mutational changes were losses from originary particulate hereditary wholes—but this theory had long been in disrepute; see chapter 3.

105. Muller, "Development of the Gene Theory," 77ff.

106. Luckey, *Hormesis with Ionizing Radiation*; Mattsonn and Calabrese, *Hormesis*; Calabrese and Baldwin, "Radiation Hormesis"; Upton et al., "The Health Effects of Low-Level Ionizing Radiation."

107. It is very difficult to make any precise claims about the birth of "radiophobia." The dangers of radiation had been known for decades, and yet at some point between the 1920s and the 1940s, popular and scientific conceptions began to change. Weart's literature review is the best explanation of this shift, but even he is justly wary of an overly simplistic answer, and attributes it

to a general period of transformation: Weart, *Nuclear Fear*, 53, 387. See also Boyer, *By the Bomb's Early Light*, and Creager, *Life Atomic*, 145.

108. Grobman, *Our Atomic Heritage*, 62, 73.

109. Muller, "Our Load of Mutations." See also Paul, "Our Load of Mutations Revisited."

110. For example, "The neutral rate of evolution would then be merely a function of the mutation rate, which was thought to be a random process analogous to radioactive decay." Dietrich, "Paradox and Persuasion," 105–6.

111. Keller, *Secrets of Life, Secrets of Death*, chap. 2; see also Masco, "Atomic Health."

112. Brunngraber, *Radium*, 387–88.

CONCLUSION

1. Adams, *Education*, 425–26.

2. McElheny, *Watson and DNA*, 70.

3. Watson later remarked, at a "Perspectives in Genetics" event at Harvard University on March 4, 2005, that Crick had marched into the Eagle pub in Cambridge on February 28, 1953, to announce that they had found the "secret of life." Crick never remembered saying this. Channeling Muller, Watson remarked, "I don't know whether Francis called it the secret of life. He had to say it. So I had taken a little liberty on putting those words in his mouth. As a highly intelligent person he couldn't be there without saying it, it was too good not to be true." Watson, personal communication. Crick did remember going home, however, "and telling my wife Odile that we seemed to have made a big discovery. Years later she told me that she hadn't believed a word of it. 'You were always coming home and saying things like that,' she said, 'so naturally I thought nothing of it.'" Crick, "Crick Looks Back on DNA," 667.

4. Creager, *Life of a Virus*, 12.

5. Besides which, as Lily Kay has argued, "the path to the double helix completely by-passed the program of radiation genetics." Kay, "Secret of Life," 497.

6. Are we, like Burke, "sometimes apt to be carried away by a flow of language which suggests rather than conveys" our meaning? Or is our tale of the secret of life, like Watson's, "too good not to be true"?

7. Rheinberger, "Experimental Systems," 77–78. For a clear and excellent summary of Rheinberger's most recent works, see Dörries, "Life, Language, and Science."

8. Riles, *The Network Inside Out*, 5. I am also reminded here of Michel Foucault's claim that there is a "whole backwash of history to which words lend their glow at the instant they are pronounced." Foucault, *Order of Things*, 315.

9. Rheinberger, "Experimental Systems," 69.

10. Novick, *That Noble Dream*, 8.

11. In some respects this book builds on the work of Everett Mendelsohn, who in his *Heat and Life* traced a nascent single connection out through its ramifications and transformations over time, examining the interplay of new and old and of thought and technique. As he says of the heat-life connection,

and I might say in some respects of the radium-life connection, "Furthermore, the scope of the problem broadens to such an extent . . . that we can no longer view the changes in a single theory, but become involved in the mutual influences of a number of interacting theories." Mendelsohn, *Heat and Life*, vii.

12. Quoted in Meyer, *Irresistible Dictation*, 305. Loy was referring to Gertrude Stein.

13. After all, radium was described as "constantly and without cessation throwing off from itself, at terrific velocity, particles of matter," just as Darwin described gemmules as having been "thrown off," and just as de Vries's particularly mutable species were said to "throw off" new varieties and species (see chaps. 1 and 3). So, too, the associations of radium and life "threw off" novel experimental systems.

14. For more on the later history of radiobiology, and how such radioactive traces led to radioactive "tracers," see Creager, *Life Atomic*.

Bibliography

ARCHIVAL SOURCES

The following manuscript collections were actively consulted in the course of this study. They are cited in the text by the depository abbreviation indicated on the left.

APS American Philosophical Society (Philadelphia, Pennsylvania)
- Albert Francis Blakeslee Papers
- Charles Benedict Davenport Papers
- Milislav Demerec Papers
- Genetics Collection
- Curt Stern Papers

BBG Brooklyn Botanic Garden (Brooklyn, New York)
- Charles Stuart Gager Papers

CIT California Institute of Technology (Pasadena, California)
- Max Delbrück Papers
- Thomas Hunt Morgan Papers
- Alfred H. Sturtevant Papers

CSH Cold Spring Harbor Laboratory (Cold Spring Harbor, New York)
- Albert Francis Blakeslee Papers
- Hermann J. Muller Papers

LOC Library of Congress (Washington, D.C.)
- Jacques Loeb Papers

LL Lilly Library, Indiana University (Bloomington, Indiana)
- Hermann J. Muller Papers

UMC University of Missouri, Columbia (Columbia, Missouri)
Edgar G. Anderson Papers
Lewis J. Stadler Papers

PRIMARY SOURCES

Ackroyd, William. "Radium and Its Position in Nature." *Nineteenth Century* 53 (May 1903): 856–64.

"Action of Radium on Beef-Gelatine." *Living Age* 246 (July 29, 1905): 314–16.

Adams, Henry. *The Education of Henry Adams: An Autobiography.* New York: Heritage, 1942.

Adams, Mark Boyer, "Genetics and the Soviet Scientific Community, 1948–1965." Ph.D. dissertation, Harvard University, 1973.

Alexander, Peter. *Atomic Radiation and Life.* Harmondsworth: Penguin Books, 1957.

Allen, Garland. "Hugo de Vries and the Reception of the 'Mutation Theory.'" *Journal of the History of Biology* 2 (1969): 55–87.

———. *Thomas Hunt Morgan: The Man and His Science.* Princeton: Princeton University Press, 1978.

Allyn, Robert S. "Costumes Treated with Radium Paint." *New York Times*, April 14, 1904, 6.

American Institute of Medicine. *Abstracts of Selected Articles on Radium and Radium Therapy.* New York: United States Radium Corporation, 1922.

"Artificial Biogenesis: A Step Forward." *Medical News* 87 (October 7, 1905): 702–6.

"Atom-Bomb By-Product Promises to Replace Radium as Cancer Aid." *New York Times*, April 22, 1948, 1.

Avery, Amos G., Sophie Satina, and Jacob Rietsema. *Blakeslee: The Genus Datura.* New York: Ronald Press Company, 1959.

Bacq, Z. M., and Peter Alexander, *Fundamentals of Radiobiology.* London: Butterworths Scientific Publications, 1955.

"Bacteria and Radio-Activity," *Knowledge and Scientific News* 1 (June 1904): 127.

Badash, Lawrence. "How the 'Newer Alchemy' Was Received." *Scientific American* 215 (August 1966): 89–95.

———. *Radioactivity in America: Growth and Decay of a Science.* Baltimore: Johns Hopkins University Press, 1979.

———. "Radium, Radioactivity, and the Popularity of Scientific Discovery." *Proceedings of the American Philosophical Society* 122 (1978): 145–54.

Bateson, William. *Materials for the Study of Variation Treated with Especial Regard to Discontinuity in the Origin of Species.* London: Macmillan, 1894.

B. C. A. W. [Sir Bertram Windle]. "The Origin of Life." *Dublin Review* 4 (July 1906): 187–92.

Beadle, George. "The Role of the Nucleus in Heredity." In *A Symposium on the Chemical Basis of Heredity*, edited by W. D. McElroy and B. Glass, 3–22. Baltimore: Johns Hopkins University Press, 1957.

Beale, Lionel S. *On Life and On Vital Action in Health and Disease.* London: Churchill, 1875.
Becquerel, P. "L'action abiotique de l'ultraviolet et l'hypotèse de l'origine cosmique de la vie." *Comptes rendus* 151 (1910): 86–88.
Best, Agar. "Origin of Life." *Daily Chronicle* (London), June 26, 1905, 3.
Beyler, Richard. "Targeting the Organism: The Scientific and Cultural Context of Pascual Jordan's Quantum Biology, 1932–1947." *Isis* 87 (June 1996): 248–73.
"Billiard Room Buzzings." *New York Times Magazine*, November 29, 1903, SM10.
"The Biological Effects of X-Rays." *Scientific Monthly* 27 (October 1928): 382.
Blakeslee, Albert F. "Chromosomes, Chemical Stimulators, and Inhibitors of Normal and Abnormal Plant Growth." *Proceedings of the National Cancer Conference* (1949): 42–49.
———. "Control of Evolution and Life Processes in Plants." In *Plant Growth Substances*, edited by F. Skoog, 58–66. Madison: University of Wisconsin Press, 1951.
———. "Distinction between Primary and Secondary Mutants in Datura." *Anatomical Record* 26 (December 1923): 389.
———. "The Globe Mutant in the Jimson Weed (*Datura stramonium*)." *Genetics* 6 (1921): 241–64.
———. "Ideals of Science." *Science* 92 (December 27, 1940): 589–92.
———. "Mutations in Mucors." *Journal of Heredity* 11 (1920): 278–84.
———. "New Jimson Weeds from Old Chromosomes." *Journal of Heredity* 25 (March 1934): 81–108.
———. "Twenty-Five Years of Genetics." *Brooklyn Botanic Garden Memoirs* 4 (1936): 29–40.
———. "Types of Mutations and Their Possible Significance in Evolution." *American Naturalist* 55 (1921): 254–67.
———. "A Unifoliolate Mutation in the Adzuki Bean." *Journal of Heredity* 10 (April 1919): 153–55.
———. "Variations in *Datura* Due to Changes in Chromosome Number." *American Naturalist* 56 (January–February 1922): 16–31.
Blakeslee, Albert F., and B. T. Avery, Jr. "Methods of Inducing Doubling of Chromosomes in Plants." *Journal of Heredity* 28 (1937): 393–411.
———. "Mutations in the Jimson Weed." *Journal of Heredity* 10 (March 3, 1919): 111–20.
Blakeslee, Albert F., J. Belling, and M. E. Farnham. "Chromosomal Duplication and Mendelian Phenomena in *Datura* Mutants." *Science* 52 (1920): 388–90.
Blakeslee, Albert F., and A. Dorothy Bergner. "Methods of Synthesizing Pure-Breeding Types with Predicted Characters in the Jimson Weed." *Science* 76 (1932): 571–72.
Blaauw, A. H., and W. van Heyningen. "The Radium-Growth-Response of One Cell." *Proceedings of the Royal Academy of Sciences at Amsterdam* 28 (1925): 403–17.

Blondlot, René. "Sur le renforcement qu'éprouve l'action exercée sur l'oeil." *Comptes rendus* 137 (November 23, 1903): 831–33.

Bohr, Niels. *Atomic Physics and Human Knowledge.* New York: Wiley, 1958.

———. "Light and Life," *Nature* 131 (1933): 421–23, 457–59.

Boltwood, B. B. "The Life of Radium." *Science* 42 (1915): 851–59.

Bono, James J. "Science, Discourse, and Literature: The Role/Rule of Metaphor in Science." In *Literature and Science: Theory and Practice*, edited by Stuart Peterfreund, 59–89. Boston: Northeastern University Press, 1990.

———. "Why Metaphor? Toward a Metaphorics of Scientific Practice." In *Science Studies: Probing the Dynamics of Scientific Knowledge*, edited by Sabine Maasen and Matthias Winterhager, 215–34. Bielefeld: Transcript, 2001.

"Bottling the Sunshine: Peculiar Results Obtained from the Use of Radium." *Brooklyn Daily Eagle* (June 18, 1901): 9.

Bottone, S. R. *Radium, and All About It.* London: Whittaker, 1904.

Boyer, Paul. *By the Bomb's Early Light: American Thought and Culture at the Dawn of the Atomic Age.* Chapel Hill: University of North Carolina Press, 1985.

Brandstetter, Thomas. "Imagining Inorganic Life." In *Imagining Outer Space: European Astroculture in the Twentieth Century*, edited by Alexander C. T. Geppert, 65–86. New York: Palgrave MacMillan, 2012.

Breit, Harvey. "Radium—Lost and Found." *New York Times Magazine*, February 13, 1944, SM35.

Brewster, E. T. "J. Butler Burke's Own Explanation of His Theories of the Origin of Life." *New York Times Magazine*, June 10, 1906, SM7.

Bridges, C. B., and T. H. Morgan. *The Third-Chromosome Group of Mutant Characters of Drosophila melanogaster.* Washington, DC: Carnegie Institution of Washington, 1923.

Brooke, John. "Wöhler's Urea and Its Vital Force?—A Verdict from the Chemists." *Ambix* 15 (1968): 84–114.

Brunngraber, Rudolf. *Radium: A Novel.* London: Harrap, 1936. (English translation.)

Burke, J. B. Butler. "Artificial Cells and Artificial Life: Further Experiments and Their Results—Misconception and Misrepresentation by the Public." *World's Work* (September 1907): 355–59.

———. "The Blondlot *n*-Rays." *Nature* 69 (February 8, 1904): 365.

———. "The Blondlot *n*-Rays." *Nature* 70 (June 30, 1904): 198.

———. "Correspondence—The Origin of Life." *National Review* 48 (September 1906–February 1907): 559–60.

———. *The Emergence of Life: Being a Treatise on Mathematical Philosophy and Symbolic Logic by which a New Theory of Space and Time Is Evolved.* London: Oxford University Press, 1931.

———. "The Evolution of Life or Natural Selection in Inorganic Matter." *Monist* (April 1908): 176–91.

———. *Mystery of Life.* London: E. Mathews & Marrot, 1931.

———. "On the Phosphorescent Glow in Gases." *Philosophical Magazine* 6 (1901): 242–56, 455–64.

———. "On the Spontaneous Action of Radio-Active Bodies on Gelatin Media." *Nature* (May 25, 1905): 79.
———. "On the Spontaneous Action of Radium on Gelatin Media." *Nature* (July 27, 1905): 294.
———. "The Origin of Life." *Fortnightly Review* (September 1, 1905): 389–402.
———. *The Origin of Life: Its Physical Basis and Definition.* London: Chapman and Hall, 1906.
———. "Physics and Biology." *Knowledge & Scientific News* (March 1907): 81.
———. "The Radio-Activity of Matter." *Monthly Review* 13 (November 1903): 115–31.
———. "Some Fruitless Efforts to Synthesise Life." *Knowledge* (August 1914): 304.
———. "The Spontaneous Action of Radium and Other Bodies on Gelatin Media." *Journal of the Röntgen Society* (December 1905): 33–40.
"Burke's Own Account of the Spontaneous Action of Radium on Gelatin Media." *Scientific American* 93 (July 15, 1905): 43.
"Burke's Own Interpretation of His 'Life-Originating' Experiments." *Current Literature* 39 (November 1905): 533–35.
Calabrese, E. J., and L. A. Baldwin. "Radiation Hormesis: Its Historical Foundations as a Biological Hypothesis." *Human and Experimental Toxicology* 19 (2000): 41–75.
"The Cambridge Radiobes." *New York Daily Tribune*, July 2, 1905, 11.
Campos, Luis. "'Complex Recombinations': Rethinking the Death of de Vries' Mutation Theory." Paper presented at International Society for the History, Philosophy & Social Studies of Biology, Exeter, UK, July 28, 2007. (Unpublished manuscript.)
———. "Mutant Sexuality: The Private Life of a Plant." In Campos and Schwerin, *Making Mutations*, 49–70.
———. "That Was the Synthetic Biology That Was." In *Synthetic Biology: The Technoscience and Its Societal Consequences*, edited by Markus Schmidt et al., 5–21. Dordrecht: Springer Academic Publishing, 2009.
Campos, Luis, and Alexander Schwerin, eds. *Making Mutations: Objects, Practices, Contexts.* Preprint 393. Berlin: Max Planck Institute for the History of Science, 2010.
Canales, Jimena, and Markus Krawjewski. "Little Helpers: About Demons, Angels and Other Servants." *Interdisciplinary Science Reviews* 37 (December 2012): 314–31.
Carlson, Elof Axel. *The Gene: A Critical History.* Ames: Iowa State University Press, 1989 (1966).
———. *Genes, Radiation, and Society: The Life and Work of H. J. Muller.* Ithaca, NY: Cornell University Press, 1981.
———. "H. J. Muller (1890–1967)." *Genetics* 70 (January 1972): 1–30.
———. "The Legacy of Hermann Joseph Muller: 1890–1967." *Canadian Journal of Genetics and Cytology* IX.3 (September 1967): 437–48.
———. *Mendel's Legacy: The Origin of Classical Genetics.* Cold Spring Harbor, NY: Cold Spring Harbor Laboratory Press, 2004.

———. "An Unacknowledged Founding of Molecular Biology: H. J. Muller's Contributions to Gene Theory, 1910–1936." *Journal of the History of Biology* 4 (1971): 149–70.

Caspari, W. "Die Bedeutung des Radiums und der Radiumstrahlen für die Medizin," *Zeitschrift für Diatetische und Physikalische Therapie* 8 (1904): 37.

Caufield, Catherine. *Multiple Exposures: Chronicles of the Radiation Age.* New York: Perennial Library, 1989.

"A Cell-Killer as a Cell-Maker." *New York Times*, June 23, 1905, 6.

Chamberlin, T. C. "Introduction." In *Fifty Years of Darwinism: Modern Aspects of Evolution*, edited by the American Association for the Advancement of Science, 1–7. New York: H. Holt, 1909.

Clark, Claudia. *Radium Girls: Women and Industrial Health Reform, 1910–1935.* Chapel Hill: University of North Carolina Press, 1997.

Clay, H. Clinton. "Radium." *New Church Review* 11 (April 1904): 226–52.

Cleland, Ralph. "The Genetics of *Oenothera* in Relation to Chromosome Behavior, with Reference to Certain Hybrids." *Zeitschrift für Induktive Abstammungs- und Vererbungslehre*, Supplement, I (1928): 554–67.

Clement, Mark. "Translator's Introduction." In Georges Lakhovsky, *The Secret of Life: Cosmic Rays and Radiations of Living Beings*, translated by Mark Clement, 5–30. London: William Heineman, 1939 (1925).

"A Clue to the Beginning of Life on the Earth." *World's Work* 11 (November 1905): 6813–14.

Cockerell, T. D. A. "Radiation and Genetics." *American Naturalist* 64 (1930): 474–76.

Coen, Deborah R. "Scientists' Errors, Nature's Fluctuations, and the Law of Radioactive Decay, 1899–1926." *Historical Studies in the Physical and Biological Sciences* 32 (2002): 179–205.

Cohen, I. Bernard. "Scientific Revolution, Revolutions in Science, and a Probabilistic Revolution 1800–1930." In Krüger et al., *Ideas in History*, 23–44.

Collins, A. Frederick. "Common Sense Applied to Radium: Some of the Extravagant Stories Told about the Metal and the Real Basis for Them." *New York Times Magazine*, May 8, 1904, SM4.

"Columbia St. Louis Exhibits." *New York Daily Tribune*, April 24, 1904, 6.

Comfort, Nathaniel. *The Tangled Field: Barbara McClintock's Search for the Patterns of Genetic Control.* Cambridge, MA: Harvard University Press, 2001.

Conklin, Edwin Grant. "The Mechanism of Evolution in the Light of Heredity and Development, II." *Scientific Monthly* 10 (January 1920): 52–62.

———. "Problems of Evolution and Present Methods of Attacking Them." *American Naturalist* 46 (March 1912): 121–28.

Cook, Robert C. "Dr. Muller Receives Nobel Medicine Award." *Journal of Heredity* 37 (1946): 325–26.

Corelli, Marie. *The Life Everlasting: A Reality of Romance*. New York: Hodder & Stoughton, George H. Doran, 1911.

Cornforth, Maurice. *Dialectical Materialism and Science*. London: Lawrence & Wishart, 1949.

"Correcting Nature: Novel Experiments with Trees and Flowers." *Standard* (London), April 30, 1907.

"Costly Particles of Radium, Exhibited to Physicians at a Meeting of the Academy of Medicine." *New York Times*, February 20, 1903, 6.

Coues, Elliott. *The Daemon of Darwin*. Boston: Estes and Lauriat, 1885.

Coulter, M. C. "X-Ray Mutations." *Botanical Gazette* 91 (March 1931): 110–12.

Craig, Patricia. "Daniel MacDougal: Engineer of Life." In *Centennial History of the Carnegie Institution of Washington*, vol. 4, *The Department of Plant Biology*, by Patricia Craig. Cambridge: Cambridge University Press, 2005.

Creager, Angela. "Commentary on 'Tools': Radiation, Health, and Heredity." In Campos and Schwerin, *Making Mutations*, 225–27.

———. *Life Atomic: Radioisotopes as Tools in Science and Medicine*. Chicago: University of Chicago Press, 2013.

———. *The Life of a Virus: Tobacco Mosaic Virus as an Experimental Model*. Chicago: University of Chicago Press, 2002.

———. "Tracing the Politics of Changing Postwar Research Practices: The Export of 'American' Radioisotopes to European Biologists." *Studies in the History and Philosophy of Biological and Biomedical Sciences* 33 (2002): 367–88.

Creager, Angela, and María-Jesús Santesmases. "Radiobiology in the Atomic Age: Changing Research Practices and Policies in Comparative Perspective." *Journal of the History of Biology* 39 (2006): 637–47.

Crick, Francis. "Crick Looks Back on DNA." *Science* 206 (November 9, 1979): 667.

Crookes, William. "The Emanations of Radium." *Nature* 67 (April 2, 1903): 522–24.

Crow, J. F. "Sixty Years Ago: The 1932 International Congress." *Genetics* 131 (1992): 761–68.

———. "Some Reflections on H. J. Muller." *Environmental Mutagenesis* 9 (1987): 349–53.

Crow, James, and Seymour Abrahamson. "Mutation Becomes Experimental." *Genetics* 147 (1997): 1491–96.

"Crowds Gaze on Radium: Flock to the Museum of Natural History to See a Little Capsule of the New Wonder of Science." *New York Times*, September 8, 1903, 3.

Crowther, J. A. "The Biological Action of X-Rays—A Theoretical Overview." *British Journal of Radiology* 11 (1938): 132–45.

"The Cure." *Wall Street Journal*, May 3, 1905, 1.

Curie, Eve. *Madame Curie: A Biography*. Translated by Vincent Sheean. New York: Pocketbooks, 1946.

Curie, Marie. *Pierre Curie*. Translated by Charlotte and Vernon Kellogg. New York: The MacMillan Company, 1932.

———. "Radio-Active Substances." Thesis presented to the Faculté des Sciences de Paris. London 1904. Reprinted from *Chemical News* 88 (1903): 85ff.

"Curious Modifications in Plant Life Possibly Due to Radium." *New York Times*, January 1, 1907, 6.

Curry, Helen. "Accelerating Evolution, Engineering Life: American Agriculture and Technologies of Genetic Modification, 1925–1960." Ph.D. dissertation, Yale University, 2012.

———. "Making Marigolds: Colchicine, Mutation Breeding, and Ornamental Horticulture." In Campos and Schwerin, *Making Mutations*, 259–84.

Dana, Charles Loomis. "The Origin of Life: Dr. Bastian's Return to His Old Views in a New Book—The Theory of John Butler Burke." *New York Times Saturday Review of Books*, July 7, 1906, BR1.

Darlington, Cyril. *Recent Advances in Cytology*. London: J. & A. Churchill, 1937.

Darwin, George. "Address of the President of the British Association for the Advancement of Science." *Science* 22 (August 25, 1905): 225–34.

Davenport, Charles B. "The Form of Evolutionary Theory That Modern Genetical Research Seems to Favor." *American Naturalist* 50 (August 1916): 449–65.

Davis, H. T. "Review of Burke's *Emergence of Life*." *Isis* 19 (1933): 214–17.

"Death Stirs Action on Radium 'Cures.'" *New York Times*, April 2, 1932, 12.

Degnen, M. L. "The Radio-Active Solar Pad." From *A Treatise on Radium and Its Therapeutic Uses*. Pamphlet. Los Angeles: Radium Appliance Company, 1917. Available at Early American Energy Medicine, version 2.0, accessed June 1, 2014, http://www.meridianinstitute.com/eaem/emrasp1.htm.

deJong-Lambert, William. *The Cold War Politics of Genetic Research: An Introduction to the Lysenko Affair*. Dordrecht: Springer, 2012.

De Kay, Charles. "Color Visions of the Kiowas." *New York Times Magazine*, September 1, 1901, SM13.

de la Peña, Carolyn. *The Body Electric: How Strange Machines Built the Modern American*. New York: New York University Press, 2003.

Delbrück, Max. "How It Was." *Engineering and Science* 43 (March–April 1980): 21–26.

———. "How It Was." *Engineering and Science* 43 (May–June 1980): 21–27.

———. "A Physicist Looks at Biology." *Transactions of the Connecticut Academy of Arts and Sciences* 38 (December 1949): 173–190.

———. "A Physicist's Renewed Look at Biology: Twenty Years Later." *Science* 168 (1970): 1312–15.

———. "Problems of Modern Biology in Relation to Atomic Physics." Vanderbilt University Lectures, 1944. Available at "Max Delbrück at Vanderbilt, 1940–1947," http://www.vanderbilt.edu/delbruck/documents/Problems_of_Modern_Biology-Full.pdf.

———. "Radiation and the Hereditary Mechanism," *American Naturalist* 74 (July–August 1940): 350–62.

Demerec, M. "Albert Francis Blakeslee." *Genetics* 44 (1959): 1–4.

"Department of Botanical Research." *CIW Year Book* 11 (1912): 70.

"Department of Experimental Evolution and Eugenics Record Office." *CIW Year Book* 19 (1920): 107–58.

"Department of Genetics." *CIW Year Book* 20 (1921): 101–56.

"Department of Genetics." *CIW Year Book* 21 (1922): 93–125.
"Department of Genetics." *CIW Year Book* 28 (1929): 33–66.
"Department of Genetics." *CIW Year Book* 31 (1932): 33ff.
Derrida, Jacques. *Margins of Philosophy.* Chicago: University of Chicago Press, 1982.
de Vries, Hugo. "The Aim of Experimental Evolution." *CIW Year Book* 3 (1904): 39–49.
———. *Intracellular Pangenesis.* Chicago: Open Court, 1910. Originally published as *Intracellulare Pangenesis* (Jena: Fischer, 1889).
———. *The Mutation Theory: Experiments and Observations on the Origin of Species in the Vegetable Kingdom.* Translated by J. D. Farmer and A. D. Darbishire. Chicago: Open Court, 1909, 1910. Originally published as *Die Mutationstheorie: Versuche und Beobachtungen über die Entstehung von Arten im Pflanzenreich* (Leipzig, Veit, 1901–1903).
———. "The Principles of the Theory of Mutation." *Science* 40 (July 17, 1914): 77–84.
———. "Recherches Experimentales"; "Sur L'Origine des Espèces." *Revue Générale de Botanique* 13 (1901).
———. *Species and Varieties: Their Origin by Mutation.* Chicago: Open Court, 1905.
Dexter, John S. "The Analysis of a Case of Continuous Variation in Drosophila by a Study of Its Linkage Relations." *American Naturalist* 48 (December 1914): 712–58.
"Diamond Rays Pierce Paper: Wonderful Properties of Radium Demonstrated—First Seen Here to Be on View at Museum of Natural History." *New York Times*, September 6, 1903, 10.
Dietrich, Michael. "Paradox and Persuasion: Negotiating the Place of Molecular Evolution within Evolutionary Biology." *Journal of the History of Biology* 31 (1998): 85–111.
"The Disintegration of the Atom." *Lancet* 164 (July 23, 1904): 232–33.
"DNA Damage and Repair." *Nature* 421 (January 23, 2003): 436–40.
"Does the Dead Body Possess Properties Akin to Radio-Activity?" *Lancet* 159 (April 26, 1902): 1201.
Dolbear, A. "Life From a Physical Standpoint." Biological lectures delivered at Marine Biological Laboratory, Woods Hole, MA, Summer Session, 1894. Boston: Ginn 1895.
"Do Men Radiate Light?" *New York Times*, June 14, 1903, 6.
Dörries, Matthias, ed. *Experimenting in Tongues: Studies in Science and Language.* Stanford: Stanford University Press, 2002.
———. "Life, Language, and Science: Hans-Jörg Rheinberger's Historical Epistemology." *Historical Studies in the Natural Sciences* 42 (February 2012): 71–82.
Doyle, Richard. *On Beyond Living: Rhetorical Transformations of the Life Sciences.* Stanford: Stanford University Press, 1997.
"Dr. Hermann Muller is Dead." *New York Times*, April 6, 1967, 39.
Dubois, R. "La création de l'être vivant et de lois naturelles." *La Revue des Idées* 2 (1905): 198.

Dubois, R. "Radioactivité et la vie." *La Revue des Idées* 1 (1904): 338.
Duggar, Benjamin, ed. *Biological Effects of Radiation*. New York: McGraw-Hill, 1936.
Dunn, L. C. "Genetics in Historical Perspective." In *Genetic Organization I*, edited by Ernst W. Caspari and Arnold W. Ravin, 1–90. New York: Academic Press, 1969.
East, Edward M. "A Mendelian Interpretation of Variation that Is Apparently Continuous." *American Naturalist* 44 (1910): 82.
Emery, Steuart M. "Roll of Martyrs to Science Is Increasing." *New York Times*, January 11, 1925, XX5.
Eve, A. S. "The Infection of Laboratories by Radium." *Nature* 71 (March 16, 1905): 460–61.
———. *Rutherford*. New York: MacMillan, 1939.
"Evolution Process Is Aided by X-Rays." *New York Times*, August 27, 1932, 3.
Faber, William M. "X-Rays Form New Life: Prof. Muller Talks on the Wonders of X Rays at Cleveland Eng. Society; X Rays Change Life; Prof. Muller's Experiments on Fruit Flies Speed Up Nature." *Integrator* (February 27, 1932), Clarkson College, Potsdam, New York.
"Facts about Radium: Its Occurrence and Isolation." *Lancet* 162 (October 3, 1905): 966.
Failla, G. "Ionization and Its Bearing on the Biological Effects of Radiation." In Duggar, *Biological Effects of Radiation*, 87–122.
Fajans, K., and W. Makower. "The Growth of Radium C from Radium B." *Philosophical Magazine* 23 (1912): 292–301.
Falk, Raphael. "What Is a Gene?—Revisited." *Studies in History and Philosophy of Biological and Biomedical Sciences* 41 (2010): 396–406.
Farley, John. *The Spontaneous Generation Controversy from Descartes to Oparin*. Baltimore: Johns Hopkins University Press, 1977.
"A Fascinating Theme: Mr. Herbert Burrows on Radium and Life." *Daily Chronicle* (London), June 26, 1905, 3.
Field, C. Everett. "Radium and Research: A Protest." *Medical Record* 100, October 29, 1921: 764.
Fischer, Ernst Peter, and Carol Lipson. *Thinking About Science: Max Delbrück and the Origins of Molecular Biology*. New York: W. W. Norton, 1988.
Fleming, Donald. "Émigré Physicists and the Biological Revolution." In *The Intellectual Migration: Europe and America, 1930–1960*, edited by Donald Fleming and Bernard Bailyn, 152–89. Cambridge, MA: Harvard University Press, 1969.
"Flotsam and Jetsam: Science and the Origin of Life." *Month* 107 (January 1906): 87.
Ford, E. B. *The Study of Heredity*. London, T. Butterworth, 1938.
Foucault, Michel. *The Order of Things: An Archaeology of the Human Sciences*. New York: Pantheon Books, 1971.
Freedman, Michael I. "Frederick Soddy and the Practical Significance of Radioactive Matter." *British Journal for the History of Science* 12 (November 1979): 257–60.

Freke, Henry. *On the Origin of Species by Means of Organic Affinity.* London: Longman, 1861.
"From the Lives of Players." *New York Times*, October 11, 1903, 21.
Fry, Iris. *The Emergence of Life on Earth: A Historical and Scientific Overview.* New Brunswick: Rutgers University Press, 2000.
Fujimura, Joan, and Adele E. Clarke, eds. *The Right Tools for the Job: At Work in Twentieth Century Life Sciences.* Princeton: Princeton University Press, 1992.
Gager, C. S. "The Effects of Radium Rays on Plants: A Brief Resume of the More Important Papers from 1901 to 1932." In Duggar, *Biological Effects of Radiation*, 987–1013.
———. *Effects of the Rays of Radium on Plants.* Memoirs of the New York Botanical Garden, vol. 4. New York: New York Botanical Garden, 1908.
———. *Heredity and Evolution in Plants.* Philadelphia, P. Blakiston's Son, 1920.
———. "The Influence of Radium Rays on a Few Life Processes of Plants." *Popular Science Monthly* 74 (1909): 222–32.
———. *The Plant World: Plant Life of Our Earth.* New York: The University Society, 1931.
———. "Present Status of the Problem of the Effect of Radium Rays on Plant Life." *Memoirs of the New York Botanical Garden* 6 (1916): 153–60.
———. *The Relation between Science and Theology: How to Think About It.* Chicago: Open Court, 1925.
———. "Review of *The Mutation Theory*, by Hugo de Vries." *Science* 31 (May 13, 1910): 740–43.
———. "Review of *The Mutation Theory*, vol. 2, by Hugo de Vries." *Science* 34 (October 13, 1911): 491–93.
———. "Some Physiological Effects of Radium Rays." *American Naturalist* 42 (December 1908): 761–78.
Gager, C. Stuart, and A. F. Blakeslee. "Chromosome and Gene Mutations in *Datura* Following Exposure to Radium Rays." *Proceedings of the National Academy of Sciences* 13 (February 1927): 75–79.
———. "Induction of Gene and Chromosome Mutations in *Datura* by Exposure to Radium Rays." *Anatomical Record* 24 (1923): 424.
"Gager, Dr. Charles Stuart." In *American Men of Science*, edited by J. McKeen Cattell and Jacques Cattell. New York: Bowker, 1938.
Gates, R. Ruggles. "Mutation in *Oenothera.*" *American Naturalist* 45 (1911): 577–606.
———. "Mutations and Evolution." *New Phytologist* 19 (1920): 26–34, 64–88.
Geison, Gerald. "The Protoplasmic Theory of Life." *Isis* 60 (1969): 273–92.
"Generation by Radium: Cambridge Professor Reported to Have Produced Artificial Life." *New York Times*, June 20, 1905, 1.
Getzendaner, F. M. "A Hypothesis of 'Valence' in Heredity and Evolution." *American Naturalist* 58 (September–October 1924): 428.
Glad, John. "Hermann J. Muller's 1936 Letter to Stalin." *Mankind Quarterly* 43 (Spring 2003): 305–19.

Goodman, Jordan. "Plants, Cells, and Bodies: The Molecular Biography of Colchicine, 1930–1975." In *Molecularizing Biology and Medicine: New Practices and Alliances, 1910s–1970s*, edited by Soraya de Chadarevian and Harmke Kamminga, 17–46. Amsterdam: Harwood, 1998.

Goodspeed, T. H. "The Effects of X-Rays and Radium on Species of the Genus *Nicotiana*." *Journal of Heredity* 20 (June 1929): 243–59.

Gradenwitz, Alfred. "Forcing Plants by Means of Radium: Treatment Must Be Applied During 'Period of Rest." *Scientific American Supplement* 74 (August 3, 1912): 77.

Gray, George W. *The Advancing Front of Science*. New York: Whittlesey House, 1937.

Green, M. M. "The 'Genesis of the White-Eyed Mutant' in *Drosophila melanogaster*: A Reappraisal." *Genetics* 142 (February 1996): 329–331.

Greinacher, H. "Ein neues Radium perpetuum mobile." *Verhandlungen der Physikalischen Gesellschaft zu Berlin* 13 (1911): 398–404, and *Deutsche-Mechaniker-Zeitung* (Berlin) 1911, 101–4.

Grobman, Arnold B. *Our Atomic Heritage*. Gainesville: University of Florida Press, 1951.

"The Growth of Non-Living Matter." *Independent* (New York) 59 (September 7, 1905): 589–90.

Guilleminot, Hyacinthe. *Rayons X et Radiations Diverse: Actions sur l'Organisme*. Paris: O. Doin, 1910.

Hacker, B. C. *The Dragon's Tail: Radiation Safety in the Manhattan Project, 1942–1946*. Berkeley: University of California Press, 1987.

———. *Elements of Controversy: The Atomic Energy Commission and Radiation Safety in Nuclear Weapons Testing 1947–1974*. Berkeley: University of California Press, 1994.

Hacking, Ian. "Was There a Probabilistic Revolution 1800–1930." In Krüger et al., *Ideas in History*, 45–55.

Haeckel, Ernst. *Kristallseelen: Studien über das anorganische Leben*. Leipzig: A. Kröner, 1925.

Hahn, O. "Muttersubstanz des Radiums." *Berichte der Deutschen Chemischen Gesellschaft* 40 (1907): 4415–20.

Hale, William Bayard. "Has Radium Revealed the Secret of Life?" *New York Times*, July 16, 1905, 7.

Hall, A. E. "The Great Enigma." *Academy* 2029 (March 25, 1911): 354–55.

Hammer, William. *Radium, and Other Radio-Active Substances*. New York: D. Van Nostrand, 1903.

Hampson, William. *Radium Explained: A Popular Account of the Relations of Radium to the Natural World, to Scientific Thought, and to Human Life*. New York: Dodd, Mead, 1905.

Hanson, Frank Blair, and Florence Heys. "An Analysis of the Effects of the Different Rays of Radium in Producing Lethal Mutations in Drosophila." *American Naturalist* 63 (May–June 1929): 201–13.

———. "The Effects of Radium in Producing Lethal Mutations in Drosophila Melanogaster." *Science* 68 (August 3, 1928): 115–16.

Hanson, Frank Blair, and Elvene Winkleman. "Visible Mutations Following

Radium Irradiation in Drosophila Melanogaster." *Journal of Heredity* 20 (June 1929): 277–86.
Haraway, Donna. *Crystals, Fabrics, and Fields: Metaphors of Organicism in Twentieth-Century Developmental Biology.* New Haven: Yale University Press, 1976.
Harding, John W. "Dr. MacDougal's Botanical Feat Threatens Evolution Theories." *New York Times*, December 24, 1905, SM1–2.
Harvie, David I. "The Radium Century." *Endeavour* 23 (1999): 100–105.
Hesse, Mary. "Models, Metaphors, and Truth." In *From a Metaphorical Point of View: A Multidisciplinary Approach to the Cognitive Content of Metaphor*, edited by Zdravko Radman, 351–72. Berlin: De Gruyter, 1995.
Hessenbruch, Arne. "The Commodification of Radiation: X-Ray and Radium Standards, 1896–1928." Ph.D. dissertation, University of Cambridge, 1994.
Hoffman, Joseph G. *The Life and Death of Cells.* Garden City, NY: Hanover House Books, 1957.
Hogben, Lancelot. *The Nature of Living Matter.* London: K. Paul, Trench, Trubner, 1930.
Hollaender, Alexander. "The Problem of Mitogenetic Rays." In Duggar, *Biological Effects of Radiation*, 919–59.
———. *Radiation Biology.* New York: McGraw-Hill, 1954.
Hotblack, Frank A. *A New Activity? A Treatise on Mrs. Dickinson's Discovery of a 'New Radio-Activity' (with some notes on radium).* London: Jarrolds, 1920.
Houllevigue, Louis. *La Matière: Sa Vie et Ses Transformations.* Paris: A. Colin, 1913.
Howorth, Muriel. *The Greatest Discovery Ever Made, from Memoirs of Professor Frederick Soddy.* London: New World Publications, 1953.
———. *Pioneer Research on the Atom: Rutherford and Soddy in a Glorious Chapter of Science; The Life Story of Frederick Soddy.* London: New World Publications, 1958.
"How the Men Who Devote Themselves to Humanity Are Often Overlooked." *New York Times*, February 15, 1925, E6.
Hull, D. L., M. Forbes, and R. M. Burian, eds. *PSA 1994*, vol. 1. East Lansing, MI: Philosophy of Science Association, 1994.
Humboldt, Alexander von. *Cosmos: A Sketch of the Physical Description of the Universe.* Vol. 1. Translated by E. C. Otté. Baltimore: Johns Hopkins University Press, 1997 (1845).
Hunt, Morton M. "Dr. Muller and the Million Human Time-Bombs." *Science Illustrated* 4 (May 1949): 46–48, 53–60.
Hurst, C. C. *The Mechanism of Creative Evolution.* Cambridge: Cambridge University Press, 1933.
Hutton, F. W., FRS. "The Origin of Life: Views of Sir William Ramsay and Other Scientists." *Daily Chronicle* (London), June 22, 1905, 5.
Huxley, Julian. *Evolution: The Modern Synthesis.* London: Harper & Brothers, 1942.
———. "Where Darwin's Theory Stands Today: Evolution Is a Fact, Says

Julian Huxley, and the Years Have Strengthened the Evidence Adduced by Darwin." *New York Times*, March 1, 1931, SM5.
"Improved X-Rays for Cancer Work: Harvard Physicists to Use Deeply Penetrating Type in New Laboratory; Hope to Displace Radium." *New York Times*, February 14, 1921, 7.
Jacob, François. *The Logic of Life: A History of Heredity*. New York: Pantheon Books, 1973.
James, William. *The Meaning of Truth*. New York: Longmans, Green, 1909.
———. *Pragmatism: A New Name for Some Old Ways of Thinking*. New York: Longmans, Green, 1907.
Jenkin, John G. "Atomic Energy Is 'Moonshine': What Did Rutherford Really Mean?" *Physics in Perspective* 13 (2011): 128–45.
Jennings, H. S. *Genetics*. New York: W. W. Norton, 1935.
Johannsen, Wilhelm. "The Genotype Conception of Heredity." *American Naturalist* 45 (March 1911): 129–59.
Jolly, J. Christopher. "Thresholds of Uncertainty: Radiation and Responsibility in the Fallout Controversy." Ph.D. dissertation, Oregon State University, 2003.
Just, Ernest Everett. "On the Origin of Mutations." *American Naturalist* 66 (1932): 61–74.
Kaempffert, Waldemar. "Darwin After 100 Years: Since the Great Naturalist Saw the Truth of Evolution Unfold, Many Discoveries Have Shed Light on Biology." *New York Times Magazine*, September 15, 1935, SM10.
Kamminga, Harmke. "Historical Perspective: The Problem of the Origin of Life in the Context of Developments in Biology." *Origin of Life and Evolution of the Biosphere* 18 (1988): 1–11.
———. "Studies in the History of Ideas on the Origin of Life From 1860." Ph.D. dissertation, University of London, 1980.
Kauffman, G. B., ed. *Frederick Soddy (1877–1956): Early Pioneer in Radio-Chemistry*. Dordrecht: D. Reidel, 1986.
Kay, Lily. "Conceptual Models and Analytical Tools: The Biology of Physicist Max Delbrück." *Journal of the History of Biology* 18 (1985): 207–46.
———. "Quanta of Life: Atomic Physics and the Reincarnation of Phage." *History and Philosophy of the Life Sciences* 14 (1992): 3–21.
———. "The Secret of Life: Niels Bohr's Influence on the Biology Program of Max Delbrück." *Rivista di Storia della Scienza* 2 (1985): 487–510.
———. "W. M. Stanley's Crystallization of the Tobacco Mosaic Virus." *Isis* 77 (1986): 450–72.
———. *Who Wrote the Book of Life? A History of the Genetic Code*. Stanford: Stanford University Press, 2000.
Keller, Alex. *Infancy of Atomic Physics: Hercules in His Cradle*. New York: Oxford University Press, 1983.
Keller, Evelyn Fox. *The Century of the Gene*. Cambridge, MA: Harvard University Press, 2000.
———. "Language in Action: Genes and the Metaphor of Reading." In Dörries, *Experimenting in Tongues*, 76–88.
———. "Molecules, Messages, and Memory: Life and the Second Law." In *Re-*

figuring Life: Metaphors of Twentieth-Century Biology, 43–78. New York: Columbia University Press, 1995.
———. "Physics and the Emergence of Molecular Biology: A History of Cognitive and Political Synergy." *Journal of the History of Biology* 23 (1990): 398–409.
———. *Secrets of Life, Secrets of Death: Essays on Language, Gender, and Science*. New York: Routledge, 1992.
Kellogg, Vernon L. *Darwinism To-Day*. New York: H. Holt, 1907.
Kerner, Anton. *Die Abhängigkeit der Pflanzengestalt von Klima und Boden*. Innsbruck, 1869.
Kevles, Bettyann. *Naked to the Bone: Medical Imaging in the Twentieth Century*. New Brunswick: Rutgers University Press, 1997.
Kim, Dong-Won. *Leadership and Creativity: A History of the Cavendish Laboratory, 1871–1919*. Boston: Kluwer, 2002.
Kimmelman, Barbara. "Mr. Blakeslee Builds His Dream House: Agricultural Institutions, Genetics, and Careers 1900–1915." *Journal of the History of Biology* 39 (2006): 241–80.
Kingsland, Sharon. "The Battling Botanist: Daniel Trembly MacDougal, Mutation Theory, and the Rise of Experimental Evolutionary Biology in America, 1900–1912." *Isis* 82 (1991): 479–509.
Klotz, Irving M. "The N-Ray Affair." *Scientific American* 242 (1980): 168–75.
Kohler, Robert E. "*Drosophila* and Evolutionary Genetics: The Moral Economy of Scientific Practice." *History of Science* 29 (1991): 335–75.
———. "The History of Biochemistry, A Survey." *Journal of the History of Biology* 8 (1975): 275–318.
———. *Lords of the Fly:* Drosophila *Genetics and the Experimental Life*. Chicago: University of Chicago Press, 1994.
———. "Systems of Production: *Drosophila*, *Neurospora*, and Biochemical Genetics." *Historical Studies in the Physical and Biological Sciences* 22 (1991): 87–130.
Korschinsky (Korzhinskii), S. I. "Heterogenesis und Evolution." *Naturwissenschaftliche Wochenschrift* 14 (1899): 273.
Korzhinskii, S. I. "Chto takoe zhizn?" In *Pervyi universitet v Sibiri*. Tomsk, 1889.
Krüger, Lorenz, Lorraine J. Daston, and Michael Heidelberger, eds. *Ideas in History*. Vol. 1 of *The Probabilistic Revolution*. Cambridge, MA: MIT Press, 1987.
Kuckuck, Martin. *Die Lösung des Problems der Urzeugung*. Leipzig, 1907.
Lakhovsky, Georges. *The Secret of Life: Cosmic Rays and Radiations of Living Beings*. Translated by Mark Clement. London: William Heineman, 1939 (1925).
Lakoff, George, and Mark Johnson, *Metaphors We Live By*. Chicago: University of Chicago Press, 1980.
Landa, Edmund R. *Buried Treasure to Buried Waste: Rise and Fall of the Radium Industry*. Golden, Colorado: Colorado School of Mines Press, 1987.
Lavine, Matthew. *The First Atomic Age: Scientists, Radiations, and the American Public, 1895–1945*. New York: Palgrave Macmillan, 2013.

"The Law of Vast Numbers." *Journal of Heredity* 2 (1911): 308.
Lea, D. E. *Actions of Radiations on Living Cells.* Cambridge: Cambridge University Press, 1946.
Le Bon, Gustave. *The Evolution of Matter.* London: Walter Scott, 1907.
Lederman, Muriel, and Richard M. Burian, eds. "Introduction." *Journal of the History of Biology* 26 (1993): 235–37.
Leff, Harvey S., and Andrew F. Rex. *Maxwell's Demon: Entropy, Information, Computing.* Princeton: Princeton University Press, 1990.
Lehman, Harry. *Biology in Transition: A Critical Inquiry.* Hicksville, NY: Exposition Press, 1978.
Levine, Michael. "Cytological Studies on Irradiated Tissues: I. The Influence of Radium Emanation on The Microsporogenesis of the Lily." *Proceedings of the 4th International Congress of Plant Sciences, Ithaca 1926* 1 (1929): 290.
"Life by Chemical Action." *New York Times*, June 22, 1905, 8.
"The 'Life' of Radium." *Literary Digest* 37 (July 4, 1908): 17.
"Life's Secret: Dean Fremantle on the Burke Experiments; Striking Letter; Discoveries a 'Gain to Religious Thought.'" *Daily Chronicle* (London), June 27, 1905, 5.
"Light and Life." *New York Times*, September 19, 1937, 70.
Lightman, Bernard. "The Evolution of the Evolutionary Epic." In *Victorian Popularizers of Science: Designing Nature for New Audiences*, 219–94. Chicago: University of Chicago Press, 2007.
Lillie, Ralph Stayner. *General Biology and Philosophy of Organisms.* Chicago: University of Chicago Press, 1945.
"Limit Their Size." *Wall Street Journal*, October 30, 1905, 1.
Lindee, M. Susan. "What Is a Mutation? Identifying Heritable Change in the Offspring of Survivors at Hiroshima and Nagasaki." *Journal of the History of Biology* 25 (1992): 231–55.
"'Liquid Sunshine' on Tap." *New York Times*, February 6, 1904, 6.
Lock, Robert Heath. *Variation, Heredity, and Evolution.* London: John Murray, 1907.
Lockyer, Norman. *Inorganic Evolution as Studied by Spectrum Analysis.* London: Macmillan, 1900.
Lodge, Oliver. "Radium and Its Lessons." *Nineteenth Century* 54 (July 1903): 78–85.
———. "What Is Life?" *North American Review* 180 (1905): 661–69.
Loeb, Jacques. *The Organism as a Whole: From a Physicochemical Viewpoint.* New York: G. P. Putnam's Sons, 1916.
Loeb, Jacques, and F. W. Bancroft, "Some Experiments on the Production of Mutants in Drosophila." *Science* 33 (May 19, 1911): 781–83.
Lorch, "The Charisma of Crystals in Biology." In *The Interaction between Science and Philosophy*, edited by Yehuda Elkana, 445–61. Atlantic Highlands: Humanities Press, 1974.
Lord, John Perceval. *Radium.* London: Harding Brothers, 1910.
"Lost Radium Tube: Dangers Allied to Research." *Standard* (London), December 29, 1911, 5.

Luckey, T. D. *Hormesis with Ionizing Radiation*. Boca Raton: CRC Press, 1980.
"'Macbeth' in 'Pure Radium.'" *Literary Digest* 68 (March 19, 1921): 30–31.
MacDougal, D. T. "Activities in Plant Physiology." *Scientific Monthly* 26 (May 1928): 464–67.
———. "Alterations in Heredity Induced by Ovarial Treatment." *CIW Year Book* 10 (1911): 63.
———. "Alterations in Heredity Induced by Ovarial Treatments." *Botanical Gazette* 51 (April 1911): 241–57.
———. "The Direct Influence of the Environment." In *Fifty Years of Darwinism: Modern Aspects of Evolution*, edited by the American Association for the Advancement of Science, 114–42. New York: H. Holt, 1909.
———. "Discontinuous Variation in Pedigree-Cultures." *Popular Science Monthly* 69 (September 1906): 207–25.
———. "Heredity, and the Origin of Species." *Smithsonian Report* 15 (1908): 505–23.
———. "Hugo De Vries." *The Open Court, a Quarterly Magazine* 19 (August 1905): 0–1.
———. *Mutants and Hybrids of the Oenotheras*. Washington, DC: Carnegie Institution, 1905.
———. "Mutation in Plants." *American Naturalist* 37 (November 1903): 737–70.
———. "Organic Response." *American Naturalist* 45 (January 1911): 5–40.
———. "Organic Response." *Science* 33 (January 1911): 94–101.
———. "The Origin of Species by Mutation." *Torreya* 2 (May 1902): 65–68.
———. "The Origin of Species by Mutation." *Torreya* 2 (June 1902): 81–84.
———. "The Origin of Species by Mutation." *Torreya* 2 (July 1902): 97–101.
———. "Trends in Plant Science." *Scientific Monthly* 52 (June 1941): 487–95.
MacDougal, D. T., A. M. Vail, and G. H. Shull. *Mutations, Variations, and Relationships of the Oenotheras*. Washington, DC: Carnegie Institution of Washington, 1907.
Macklis, Roger M. "Radithor and the Era of Mild Radium Therapy." *Journal of the American Medical Association* 164 (August 1990): 614–18.
Maienschein, Jane. *Transforming Traditions in American Biology 1880–1915*. Baltimore: Johns Hopkins University Press, 1991.
Mallock, W. H. *An Immortal Soul, A Novel*. New York: Harper, 1908.
"Man Competing with Radium." *Literary Digest* 44 (February 3, 1912): 209.
Masco, Joseph P. "Atomic Health, Or How Nuclear Fear Shaped American Notions of Death." In *Against Health: How Health Became the New Morality*, edited by Jonathan Metzl and Anna Kirkland, 133–56. New York: New York University Press, 2010.
Mattson, Mark P., and Edward J. Calabrese, eds. *Hormesis: A Revolution in Biology, Toxicology and Medicine*. London: Springer, 2010.
Mavor, J. W. 1925. "The Attack on the Gene." *Scientific Monthly* 21 (1926): 355–63.
———. "The Production of Non-Disjunction by X-Rays." *Science* 55 (1922): 295–97, and *Journal of Experimental Zoology* 39 (April 1924): 381–432.

Mayr, Ernst. *Animal Species and Evolution.* Cambridge, MA: Belknap Press of Harvard University Press, 1963.

———. *The Growth of Biological Thought.* Cambridge, MA: Belknap Press of Harvard University Press, 1982.

McElheny, Victor. *Watson and DNA: Making a Scientific Revolution.* Cambridge, MA: Perseus, 2003.

McKaughan, Daniel J. "The Influence of Niels Bohr on Max Delbrück: Revisiting the Hopes Inspired by 'Light and Life.'" *Isis* 96 (December 2005): 507–29.

Meehan, Thomas. "Hybrid Oaks." *Bulletin of the Torrey Botanical Club* 9 (1882): 55.

"The Mendelian Theory." *Standard* (London), April 27, 1909.

Mendelsohn, Everett. *Heat and Life: The Development of the Theory of Animal Heat.* Cambridge, MA: Harvard University Press, 1964.

Merrick, Linda. *The World Made New: Frederick Soddy, Science, Politics, and Environment.* Oxford: Oxford University Press, 1996.

Meyer, Steven. *Irresistible Dictation: Gertrude Stein and the Correlations of Writing and Science.* Stanford: Stanford University Press, 2001.

"The Microbe's Ancestor." *New York Times*, September 25, 1905, 6.

Miller, Julie Ann. "Max Delbrück, 1906–1981." *Science News*, April 25, 1981, 268–69.

Millikan, Robert A. "Radium, the Revolutionary Element." *Technical World* 1 (March 1904): 1–10.

Mirowski, Philip. *Natural Images in Economic Thought: Markets Read in Tooth and Claw.* Cambridge: Cambridge University Press, 1994.

Moffett, Cleveland. "The Wonders of Radium." *McClure's Magazine* 22 (November 1903): 3–15.

Mohr, Otto L. *Heredity and Disease.* New York: W. W. Norton, 1934.

Moore, Benjamin. *The Origin and Nature of Life.* New York: H. Holt, 1913.

Moore, James. "Deconstructing Darwinism: The Politics of Evolution in the 1860s." *Journal of the History of Biology* 24 (1991): 353–408.

Morange, Michel. *A History of Molecular Biology.* Cambridge, MA: Harvard University Press, 1998.

Morgan, T. H. *Experimental Embryology.* New York: Columbia University Press, 1927.

———. "The Failure of Ether to Produce Mutations in Drosophila." *American Naturalist* 48 (December 1914): 705–11.

———. "Genesis of the White-Eyed Mutant." *Journal of Heredity* 33 (1942): 91–92.

———. "The Origin of Nine Wing Mutations in Drosophila." *Science* 33 (March 31, 1911): 496–99.

Morus, Iwan. *Frankenstein's Children: Electricity, Exhibition, and Experiment in Early-Nineteenth-Century London.* Princeton: Princeton University Press, 1998.

"Mr. J. B. Butler Burke: Physicist and Scientific Author." *Times* (London), January 16, 1946, 6.

"Mr. Soddy's Views on Atomic Disintegration." *Standard* (London), December 21, 1907, 5.

Muller, H. J. "The Artificial Transmutation of the Gene." *Science* 66 (1927): 84–87.

———. "The Development of the Gene Theory." In *Genetics in the Twentieth Century: Essays on the Progress of Genetics During Its First Fifty Years*, edited by L. C. Dunn, 77–99. New York: Macmillan, 1951.

———. "The Effects of Roentgen Rays upon the Hereditary Material." In *The Science of Radiology*, edited by the American College of Radiology, 305–18. Springfield, Ill., Charles C. Thomas, 1933.

———. "Further Studies on the Nature and Causes of Gene Mutations." *Proceedings of the 6th International Congress of Genetics, I* (1932): 213–55.

———. "Further Changes in the White-Eye Series of Drosophila and Their Bearing on the Manner of Occurrence of Mutation." *Journal of Experimental Zoology* 31 (November 1920): 443–74.

———. "The Gene." Pilgrim Trust Lecture, read before Royal Society of London, November 1, 1945. *Proceedings of the Royal Society B* 134 (1947): 1–37.

———. "The Gene as the Basis of Life." *Proceedings of the 4th International Congress of Plant Sciences, Ithaca 1926* 1 (1929): 879–921.

———. "Genetic Nucleic Acid." *Graduate Journal* 7 (Spring 1962): 144.

———. "Genetic Nucleic Acid: Key Material in the Origin of Life," *Perspectives in Biology and Medicine* 5 (1961): 1–23.

———. "Genetic Variability, Twin Hybrids and Constant Hybrids, in a Case of Balanced Lethal Factors." *Genetics* 3 (1918): 422–99.

———. "Heribert Nilsson's Evidence against the Artificial Production of Mutations." *Hereditas* 16 (1932): 160–68.

———. "How Evolution Works." *Evolution* (February 1931): 12–16.

———. "The Measurement of Gene Mutation Rate in Drosophila, Its High Variability, and Its Dependence upon Temperature." *Genetics* 13 (1928): 279–357.

———. "The Method of Evolution." *Scientific Monthly* 29 (December 1929): 481–505.

———. "Mutation." *Eugenics, Genetics and the Family* 1 (1923): 106–12.

———. "The Nature of the Genetic Effects Produced by Radiation." In Hollaender, *Radiation Biology*, 351–473.

———. "The Need of Physics in the Attack on the Fundamental Problems of Genetics." *Scientific Monthly* 44 (1937): 210–14.

———. "An Oenothera-Like Case in Drosophila." *Proceedings of the National Academy of Sciences* 3 (October 1917): 619–26.

———. "Our Load of Mutations." *American Journal of Human Genetics* 2 (June 1950): 111–76.

———. "The Present Status of the Mutation Theory." *Current Science*, Special Number. (March 1938): 4–15.

———. "The Production of Mutations by X-Rays." *Proceedings of the National Academy of Sciences* 14 (September 15, 1928): 714–26.

———. "Quantitative Methods in Genetic Research." *American Naturalist* 61 (September–October 1927): 407–19.
———. "Radiation and Genetics." *American Naturalist* 64 (May–June 1930): 220–51.
———. "The Rate of Change of Hereditary Factors in Drosophila." *Proceedings of the Society for Experimental Biology and Medicine* 17 (1919): 10–14.
———. "Reversibility in Evolution Considered from the Standpoint of Genetics" *Biological Reviews* 14 (July 1939): 261–80.
———. "Variation Due to Change in the Individual Gene." *American Naturalist* 56 (1922): 32–50.
Muller, H. J., and Edgar Altenburg. "A Study of the Character and Mode of Origin of Eighteen Mutations in the Chromosome of Drosophila." *Anatomical Record* 20 (1921): 213.
Muller, H. J., and T. S. Painter. "The Cytological Expression of Changes in Gene Alignment Produced by X-Rays in Drosophila." *American Naturalist* 63 (May–June 1929): 193–200.
Mullner, Ross. *Deadly Glow: The Radium Dial Worker Tragedy.* Washington: American Public Health Association, 1999.
"Mystery of Radium: Sir William Ramsay Describes Its Nature." *New York Times*, December 23, 1903, 10.
Needham, Joseph. *Order and Life.* New Haven: Yale University Press, 1936.
"New Elixir Found For Plant World." *New York Times*, October 26, 1937, 17.
"New Light on Mutation." *Independent* (New York) 76 (November 20, 1913): 355.
"New Radium Disease Found: Has Killed 5; Women Painting Watch Dials in a Jersey Factory the Victims, Doctor Says . . . Cancer Called Incurable; Trust in Radium Is Unjustified, New York Physician Asserts." *New York Times*, May 30, 1925.
"New Rays Discovered: They Produce Better Photographs Than the X-Rays." *Los Angeles Times*, December 29, 1899, 14.
"News and Views." *Nature* 158 (November 9, 1946): 658.
Nordau, Max. *Degeneration.* New York: D. Appleton, 1895.
"Notes." *Electrician* 52 (January 8, 1904): 437–39.
Novick, Peter. *That Noble Dream: The "Objectivity Question" and the American Historical Profession.* Cambridge: Cambridge University Press, 1988.
Nye, Mary Jo. "N-rays: An Episode in the History and Psychology of Science." *Historical Studies in Physical Sciences* 11 (1980): 125–56.
"Odor and the New Radiation." *Literary Digest* 24 (May 17, 1902): 677.
Olby, Robert. *The Path to the Double Helix.* Seattle: University of Washington Press, 1974.
Oparin, Aleksandr. *The Origin of Life.* New York: Macmillan, 1938.
"The Origin of Life." *Times Literary Supplement*, April 6, 1906, 123.
"The Origin of Life: Eminent Men on the Remarkable Experiments of Mr. Burke." *Daily Chronicle* (London), June 21, 1905, 5.
"Origin of Life: Momentous Discovery by an English Scientist; 'Spontaneous Generation.'" *Daily Chronicle* (London), June 20, 1905, 5.

"The Origin of Life: Mr. Burke Describes His Experiments with Radium." *Daily Chronicle* (London), June 29, 1905, 5.
"Origin of Life: Theology and the Radium Experiments." *Daily Chronicle* (London), June 23, 1905, 5.
"Origin of Life: Well-Known Theologian on Mr. Burke's Experiments." *Daily Chronicle* (London), June 26, 1905, 3.
Ortmann, A. E. "Facts and Interpretations in the Mutation Theory." *Science* 25 (February 1, 1907): 185.
Osborn, Henry Fairfield. *The Origin and Evolution of Life: On the Theory of Action, Reaction and Interaction of Energy.* New York: Scribner's Sons, 1917.
Otis, Laura. *Müller's Lab.* Oxford: Oxford University Press, 2007.
Packard, Charles. "The Effect of Radium Radiations on the Development of Chætopterus." *Biological Bulletin* 35 (July 1918): 50–52.
Pais, Abraham. *Inward Bound: Of Matter and Forces in the Physical World.* New York: Oxford University Press, 1986.
Patterson, J. T. "X-Rays and Somatic Mutations." *Journal of Heredity* 20 (June 1929): 260–67.
Paul, Diane. "Our Load of Mutations Revisited." *Journal of the History of Biology* 20 (1987): 321–35.
Pauly, Philip. *Controlling Life: Jacques Loeb and the Engineering Ideal in Biology.* Berkeley: University of California Press, 1996.
Piper, C. V. "Botany in Its Relations to Agricultural Advancement." *Science* 31 (June 10, 1910): 889–900.
Plunkett, C. R. *Outlines of Modern Biology.* New York: H. Holt, 1930.
Podolsky, Scott. "The Role of the Virus in Origin of Life Theorizing." *Journal of the History of Biology* 29 (1996): 79–126.
Poincaré, Henri. "Éloge de Curie." *Comptes rendus* 143 (December 17, 1906): 989–98.
Pond, Raymond H. "Review of *Effects of the Rays of Radium on Plants*." *Science* 30 (December 3, 1909): 810–11.
Pontecorvo, G. *Trends in Genetic Analysis.* New York: Columbia University Press, 1958.
Popenoe, Paul. *The Child's Heredity.* Baltimore: Williams & Wilkins, 1929.
"The Popular Interest in Radium." *New York Times*, September 9, 1903, 6.
"A Possible Use for Radium." *Science* 18 (1903): 338.
Poulton, Edward B. "The Theory of Natural Selection From the Standpoint of Botany." In *Fifty Years of Darwinism: Modern Aspects of Evolution*, edited by the American Association for the Advancement of Science, 8–56. New York: H. Holt, 1909.
"Previous Experiments." *Daily Chronicle* (London), June 23, 1905, 5.
"Professor Burke's 'Radiobes.'" *New York Daily Tribune*, July 3, 1905, 6.
"A Prophet of Radium." *Standard* (London), April 27, 1909, 5.
Proumen, Henri. *Les Rayons X, Le Radium, Les Rayons N.* Verviers: A. Lacroix, 1905.
Punnet, R. C. *Mendelism.* New York: Wilshire, 1909.
Pycraft, W. P. "What Is Life?" *Academy* 70 (May 26, 1906): 500.

"The Radio-Active Substances." *New York Times*, August 15, 1903, BR7.
"Radio-Activity of the Animate as Well as the Inanimate." *Lancet* 163 (January 9, 1904): 104–5.
"A Radio-Hypothesis of Life's Origin." *Literary Digest* 86 (July 11, 1925): 23.
"Radium." *Electrician* 52 (December 11, 1903): 277.
"Radium." *New York Times*, February 22, 1903, 6.
"Radium." *Science* 18 (1903): 347.
"Radium and Helium: 'Lancet' Asserts Transmutation of Elements." *Standard* (London), July 27, 1907.
"Radium and Its Lessons." *Lancet* 161 (April 18, 1903): 1114–15.
"Radium and Life." *New York Daily Tribune*, June 25, 1905, 10.
"Radium and Vitality: 'Radiobes.'" *Lancet* 165 (June 24, 1905): 1738.
"Radium as a Preservative: Food Standard Committee Hears of Substitute for Chemicals." *New York Times*, May 15, 1904, 5.
"Radium Energised Wool: Russian Experiments." *Lancet* 164 (August 6, 1904): 389–90.
"The Radium Institute," *Lancet* 173 (March 13, 1909): 773–74.
"A Radium Product That Seems to Live: But Prof. Burke Does Not Say He Has Solved the Great Secret; Like Bacteria, Like Crystals; Meanwhile the Discoverer Calls the Minute Objects Radiobes, and Is Making Further Tests." *New York Times*, June 21, 1905, 1.
Rafferty, Charles W. *An Introduction to the Science of Radio-Activity.* London: Longmans, Green, 1909.
"Ramsay, Radium, and Burke." *Scientific American* 93 (September 16, 1905): 215.
Ramsay, William. "Can Life Be Produced by Radium?" *Independent* (New York) 59 (1905): 554–56.
———. "Radium and Its Products." *Harper's* 110 (December 1904): 52–57.
Raveau, C. "L'origine, la longévité, et la descendance du radium." *Revue Electrique* 6 (1906): 374–78.
"The Rays of Radium." *New York Times*, May 8, 1904, 6.
Reingold, Nathan. "Jacques Loeb, the Scientist: His Papers and His Era." *Quarterly Journal of Current Acquisitions* 19 (1962): 119–30.
"Religious Notices." *New York Times*, September 19, 1903, 9.
Rentetzi, Maria. "Packaging Radium, Selling Science: Boxes, Bottles, and Other Mundane Things in the World of Science." *Annals of Science* 68 (2011): 375–99.
———. *Trafficking Materials and Gendered Experimental Practices: Radium Research in Early 20th-Century Vienna.* New York: Columbia University Press, 2008.
———. "The U.S. Radium Industry: Industrial In-House Research and the Commercialization of Science." *Minerva* 46 (2008): 437–62.
"Revelations of Radium." *Review of Reviews* 28 (November 1903): 490.
Rheinberger, Hans-Jörg. "Experimental Systems: Historiality, Narration, and Deconstruction." *Science in Context* 7 (1994): 65–81.
———. "Experiment, Difference, Writing." *Studies in History and Philosophy of Science* 23 (1992): 305–31, 389–402.

———. *Toward a History of Epistemic Things: Synthesizing Proteins in the Test Tube*. Stanford: Stanford University Press, 1997.
Richards, A. "Recent Studies on the Biological Effects of Radioactivity." *Science* 42 (September 3, 1915): 287–300.
Riles, Annalise. *The Network Inside Out*. Ann Arbor: University of Michigan Press, 2001.
Ritter, William. *Unity of the Organism*. Boston: R. G. Badger, 1919.
Robards, Herbert. *Practical Radium: The Practical Uses of Radium in the Treatment of Obstinate Forms of Disease*. St. Louis: Nixon-Jones, 1909.
Roll-Hansen, Nils. "The Application of Complementarity to Biology: From Niels Bohr to Max Delbrück." *Historical Studies in the Physical and Biological Sciences* 30 (2000): 417–42.
Rona, Elisabeth. *How It Came About: Radioactivity, Nuclear Physics, Atomic Energy*. Oak Ridge: Oak Ridge Associated Universities, 1978.
———. "Laboratory Contamination in the Early Period of Radiation Research." *Health Physics* 37 (1979): 723–27.
Rudge, W. A. Douglas. "The Action of Radium and Certain Other Salts on Gelatin." *Proceedings of the Royal Society A* 78 (June 7, 1906): 380–84.
———. "On the Action of Radium and Other Salts on Gelatin." *Proceedings of the Cambridge Philosophical Society* 13 (1906): 258–59.
Russell, Israel C. "Research in State Universities." *Science* 19 (1904): 841–54.
Rutherford, Ernest. *Radioactive Transformations*. New York: Scribner's, 1906.
———. *Radio-Activity*. 2nd ed. Cambridge: Cambridge University Press, 1905.
———. "Radium—the Cause of the Earth's Heat." *Harper's* 110 (December 1904): 390–96.
Rutherford, Ernest, and Frederick Soddy. "Radioactive Change." *Philosophical Magazine* 5 (1903): 576–91.
Saleeby, C. W. "Origin of Life: Epoch-Making Discovery by an English Scientist." *Daily Chronicle* (London), June 20, 1905, 4.
———. "Radium and Life." *Harper's* 113 (July 1906): 226–30.
———. "Radium the Revealer." *Harper's* 109 (June 1904): 85–88.
———. "Science: The Origin of Life." *Academy* 68 (June 24, 1905): 667–68.
Salomons, Sir David Lionel Goldsmid-Stern. "The Wonders of Radium Explained in a Popular Manner." Delivered at the Tunbridge Wells Hospital, 1912.
Sapp, Jan. *Evolution by Association: A History of Symbiosis*. New York: Oxford University Press, 1994.
———. *Genesis: The Evolution of Biology*. New York: Oxford University Press, 2003.
"Says Radium Is Modern Miracle: British Scientist Finds That 'Mystery' Is Inadequate." *New York Times*, September 27, 1903, 30.
Schaffer, Simon. "The Nebular Hypothesis and the Science of Progress." In *History, Humanity and Evolution: Essays for John C. Greene*, edited by James Richard Moore, 131–64. New York: Cambridge University Press, 1989.
Schrödinger, Erwin. *What Is Life? The Physical Aspect of the Living Cell*. New York: Macmillan, 1945.

Schultz, Jack. "Radiation and the Study of Mutation in Animals." In Duggar, *Biological Effects of Radiation*, 1209–61.
Schuster, Arthur. "New Books: The Origin of Life." *Manchester Guardian*, April 27, 1906, 5.
Schwartz, James. *In Pursuit of the Gene: From Darwin to DNA*. Cambridge, MA: Harvard University Press, 2010.
"Science and Life." *Candid Quarterly* 1 (1914): 237–60.
"Scientists Discuss the Origin of Life: Sir Oliver Lodge Reiterates His Belief That It Does Not Arise in Matter." *New York Times*, September 17, 1913, 5.
"Scientist's Great Discovery." *Daily Chronicle* (London), June 20, 1905, 1.
Sclove, Richard. "From Alchemy to Atomic War: Frederick Soddy's 'Technology Assessment' of Atomic Energy, 1900–1915." *Science, Technology, & Human Values* 14 (Spring 1989): 163–94.
Secord, James. "Extraordinary Experiment: Electricity and the Creation of Life in Victorian England." In *Uses of Experiment: Studies in the Natural Sciences*, edited by David Gooding, Trevor Pinch, and Simon Schaffer, 337–83. New York: Cambridge University Press, 1989.
———. *Victorian Sensation: The Extraordinary Publication, Reception, and Secret Authorship of* Vestiges of the Natural History of Creation. Chicago: University of Chicago Press, 2000.
"The Secret of Life." *Time* (July 14, 1958): 52.
Seed, David. "H. G. Wells and the Liberating Atom." *Science Fiction Studies*, 30 (March 2003): 33–48.
Seifriz, William. "The Gurwitsch Rays." In *The Science of Radiology*, edited by the American College of Radiology, 412–27. Springfield, IL: Charles C. Thomas, 1933.
Shadwell, Arthur. "The Origin of Life." *Times Literary Supplement*, April 6, 1906, 123.
Sharp, Dallas Lore. "The Radium of Romance." *Atlantic Monthly* 122 (July 1918): 67–76.
Shaw, George Bernard. *The Doctor's Dilemma: With Preface on Doctors*. New York: Brentano's, 1909.
Shenstone, W. A. "The New Chemistry." *Cornhill* 19 (July–December 1905): 516–27.
———. *The New Physics and Chemistry: A Series of Popular Essays on Physical and Chemical Subjects*. London: Smith, Elder., 1906.
———. "The Origin of Life." *Cornhill* 21 (July–December 1906): 398–409.
Shropshire, Walter. *Max Delbrück and the New Perception of Biology, 1906–1981: A Centenary Celebration*. AuthorHouse, 2007.
Shull, George Harrison. "The 'Presence and Absence' Hypothesis." *American Naturalist* 43 (July 1909): 410–19.
Shull, George Harrison. "The Fluctuations of Oenothera Lamarckiana and Its Mutants." In MacDougal et al., *Mutations*, 18–64.
Singer, Charles Joseph. *A History of Biology to about the Year 1900*. New York: Abelard-Schuman, 1959.
Sinnott, Edmund W. "Albert Francis Blakeslee, November 9,

1874–November 16, 1954." *National Academy of Sciences Biographical Memoirs* (1955): 1–38.

———. "Albert Francis Blakeslee (1874–1954)." *American Philosophical Society Year Book* (1954): 394–98.

Sinnott, Edmund W., and Albert F. Blakeslee. "Structural Changes Associated with Factor Mutations and with Chromosome Mutations in Datura." *Proceedings of the National Academy of Sciences* 8 (February 15, 1922): 17–19.

Sinnott, Edmund W., and L. C. Dunn. *Principles of Genetics.* New York: McGraw-Hill, 1932 (1925).

"Sleeping Plants Wakened by Radium." *Literary Digest* 48 (April 11, 1914): 817.

Sloan, Philip. "Organic Molecules Revisited," *Buffon 88* (Paris: J. Vrin; Lyon: Institut interdisciplinaire d'études épistémologiques, 1992): 415–38.

Sloan, Philip, and Brandon Fogel, eds. *Creating a Physical Biology: The Three Man Paper and Early Molecular Biology.* Chicago: University of Chicago Press, 2011.

Slocum, John C. "Mr. Burke and His Radiobes." *The World To-Day* 9 (September 1905): 1011–12.

Smith, Harold. "Albert Francis Blakeslee." *Bulletin of the Torrey Botanical Club* 82 (July–August 1955): 305–8.

Smocovitis, V. B. "The 'Plant *Drosophila*': E. B. Babcock, the Genus *Crepis* and the Evolution of a Genetics Research Program at Berkeley, 1912–1947." *Historical Studies of the Natural Sciences* 39 (2009): 300–355.

Snelders, H. A. M. "Zijn vloeibare kristallen levende organismen?" *Gewina* 20 (1997): 129–42.

Soddy, Frederick. *The Chemistry of the Radio-Elements.* New York: Longmans, Green, 1911.

———. "The Evolution of Matter as Revealed by the Radio-Active Elements." *Manchester Memoirs of the Literary and Philosophical Society* 48 (March 16, 1904): 1–42.

———. "The Evolution of the Elements." *Chemical News* (August 24, 1906): 85–89.

———. *The Interpretation of Radium.* London: John Murray, 1909.

———. "The Life-History of Radium." *Nature* 70 (May 12, 1904): 30.

———. *Matter and Energy.* New York, H. Holt, 1912.

———. "The Parent of Radium." *Rivista di Scienza/Scientia Bologna* 5 (1909): 256–74, supp. 166–84; 6 (1909): 86–87; supp. 188–89.

———. "The Present Position of Radio-Activity." *Journal of the Röntgen Society* 2 (February 1906): 45–65.

———. "The Production of Radium from Uranium." *Philosophical Magazine* 9 (June 1905): 768–79.

———. "Radioactivity." *Annual Progress Report to the Chemical Society for 1904*, 1 (1905): 244–80.

———. "Radioactivity." *Annual Progress Report to the Chemical Society for 1906*, 3 (1907): 333–65.

———. "Radioactivity." *Annual Progress Report to the Chemical Society for 1912*, 8 (1913): 289–328.

———. "Radio-Activity." *Electrician* 52 (February 26, 1904): 724–25.

———. *Science and Life: Aberdeen Addresses*. London: J. Murray, 1920.

———. "Transmutation: The Vital Problem of the Future." *Scientia* 11 (1912): 186–202.

Spengler, Oswald. *The Decline of the West*. Abridged edition by Helmut Werner, English abridged edition by Arthur Helps, translated by Charles Francis Atkinson. New York: Oxford University Press, 1991 (1918).

Spillman, W. J. "Mendelian Phenomena without De Vriesian Theory." *American Naturalist* 44 (April 1910): 216.

———. "Notes on Heredity." *American Naturalist* 45 (August 1911): 507–12.

"Spontaneous Generation." *Christian Observer* 93 (August 2, 1905): 2.

Stadler, L. J. "The Comparison of Ultraviolet and X-Ray Effects on Mutation," *Cold Spring Harbor Symposium on Quantitative Biology* 9 (1941): 168–78.

———. "Induced Mutations in Plants." In Duggar, *Biological Effects of Radiation*, 1263–80.

———. "Mutations in Barley Induced by X-Rays and Radium." *Science* 68 (August 24, 1928): 186–87.

Stent, Gunther. "Max Delbrück." *Genetics* 101 (May 1982): 1–16.

Stern, Kurt H. "Mitogenetic Radiation: A Study of Authority in Science." *Journal of the Washington Academy of Sciences* 65 (1975): 83–90.

Stout, Wesley W. *Secret*. Detroit: Chrysler Corporation, 1947.

Strick, James. "From Aristotle to Darwin to Freeman Dyson: Changing Definitions of Life in Historical Context." In *Exploring the Origin, Extent, and Future of Life: Philosophical, Ethical, and Theological Perspectives*, edited by Constance M. Bertka, 47–60. Cambridge: Cambridge University Press, 2009.

———. *The Origin of Life Debate: Molecules, Cells, and Generation*. Bristol, UK: Thoemmes Press, 2004.

———. *Sparks of Life: Darwinism and Victorian Debates Over Spontaneous Generation*. Cambridge, MA: Harvard University Press, 2000.

Stubbe, Hans. *History of Genetics: From Prehistoric Times to the Rediscovery of Mendel's Laws*. Translated by T. R. W. Waters. Cambridge, MA: MIT Press, 1972.

Sturtevant, A. H. *A History of Genetics*. New York: Harper & Row, 1965.

———. "Thomas Hunt Morgan, September 25, 1866–December 4, 1945." *National Academy of Sciences Biographical Memoirs* 33 (1959): 282–325.

Summers, William C. "Concept Migration: The Case of Target Theories in Physics and Biology." Paper presented at the History of Science Society, Minneapolis, October 27, 1995.

———. "Physics and Genes: Einstein to Delbrück." In *Creating a Physical Biology: The Three Man Paper and Early Molecular Biology*, edited by Philip Sloan and Brandon Fogel, 39–60. Chicago: University of Chicago Press, 2011.

Swinburne, R. G. "The Presence-and-Absence Theory." *Annals of Science* 18 (1962): 131–45.

Terrall, Mary. "Salon, Academy and Boudoir: Generation and Desire in Maupertuis's Science of Life." *Isis* 87 (1996): 217–22.
Theunissen, Bert. "Closing the Door on Hugo de Vries' Mendelism." *Annals of Science* 51 (1994): 241–42.
"This Number Is Devoted to the Discussion of the Effects of X-Rays and Radium in Producing Changes in Genes and Chromosomes." *Journal of Heredity* 20 (June 1929): introduction.
Thomson, J. Arthur. *Heredity.* London: J. Murray, 1908.
———. *The Outline of Natural History.* London: G. Newnes, 1931.
———. "Radiobes and Biogen." *Nature* 74 (May 3, 1906): 1–3.
Thomson, J. J. *Recollections and Reflections.* New York: Arno Press, 1975 (1936).
Thomson, J. J. et al. *A History of the Cavendish Laboratory, 1871–1910.* London: Longman, Green, 1910.
Thomson, W. Hanna. *What Is Physical Life? Its Origin and Nature.* New York: Dodd, Mead, 1909.
Todes, Daniel. *Darwin without Malthus: The Struggle for Existence in Russian Evolutionary Thought.* New York: Oxford University Press, 1989.
"To Make Luminous Drinks From Radium: Drs. Kunz and Merton Discuss Uses of the New Mineral. Interiors of Patients May Be Lighted Up—Uses in Diagnosis and Cure of Disease." *New York Times*, January 14, 1904, 2.
Turner, B. B. "Ionium, the Parent of Radium." *American Chemical Journal* 39 (1908): 653–58.
Trenn, Thaddeus J. *The Self-Splitting Atom: The History of the Rutherford-Soddy Collaboration.* London: Taylor and Francis, 1977.
Upton, Arthur C. et al. "The Health Effects of Low-Level Ionizing Radiation." *Annual Review of Public Health* 13 (1992): 127–50.
Vail, Anna Murray. "Identity of the Evening Primroses." In MacDougal et al., *Mutations*, 65–74.
Venable, Francis P. *A Brief Account of Radio-Activity.* Boston: D. C. Heath, 1917.
Verworn, Max. *General Physiology: An Outline of the Science of Life.* London: Macmillan, 1899.
Waagen, Wilhelm Heinrich. "Die Formenreihe des *Ammonites subradiatus.*" *Benecke's Geognostiche Paläontologische Beiträge* 2 (1868): 185–86.
Wainwright, Milton. "Historical and Recent Evidence for the Existence of Mitogenetic Radiation." *Perspectives in Biology and Medicine* 41 (1998): 565–71.
Walker, J. Samuel. *Permissible Dose: A History of Radiation Protection in the Twentieth Century.* Berkeley: University of California Press, 2000.
Waters, Theodore. "Radium and Human Life." *Everybody's Magazine* 9 (1903): 328–33.
Weart, Spencer. *Nuclear Fear: A History of Images.* Cambridge, MA: Harvard University Press, 1988.
Webber, Herbert J. "The Effect of Research in Genetics on the Art of Breeding." *Science* 35 (April 19, 1912): 597–609.

Weeks, Mary Elvira. *Discovery of the Elements*. Easton, PA: Journal of Chemical Education, 1960.
Wells, H. G. *Tono-Bungay*. Lincoln: University of Nebraska Press, 1978 (1909).
———. *The World Set Free: A Story of Mankind*. London: Macmillan, 1914.
Wells, H. G., Julian S. Huxley, and G. P. Wells. *The Science of Life*. Garden City, NY: Doubleday, Doran, 1931.
"When Silence Is Golden." *New York Times*, May 8, 1904, 6.
Whetham, W. C. D. "The Life-History of Radium." *Nature* 70 (May 5, 1904): 5.
———. *The Recent Development of Physical Science*. London, John Murray, 1904.
White, Orland E. "Heredity, Variation, and the Environment." In *General Botany: With Special Reference to Its Economic Aspects*, by C. S. Gager. York, PA: Maple Press, 1926.
Wickham, Louis, and Paul DeGrais. *Radiumtherapy*. Translated by S. Ernest Dore. London: Cassell, 1910.
Wilder, Salem. *Life: Its Nature, Origin, Development, and the Psychical Related to the Physical*. Boston: Press of Rockwell and Churchill, 1886.
Wilson, Edmund B. "The Cell in Relation to Heredity and Evolution." In *Fifty Years of Darwinism: Modern Aspects of Evolution*, edited by the American Association for the Advancement of Science, 92–113. New York: H. Holt, 1909.
Wilson, Harold A. "Is Radium an Element?" *Nature* 70 (January 7, 1904): 241–42.
Wise, M. Norton. "How Do Sums Count? On the Cultural Origins of Statistical Causality." In Krüger et al., *Ideas in History*, 395–425.
Wolfe, Charles. "Endowed Molecules and Emergent Organization: The Maupertuis-Diderot Debate." *Early Science and Medicine* 15 (2010): 38–65.
"Women and Radium." *New York Times*, September 6, 1903, 6.
"Wonders of Radium Explained by a Professor of Johns Hopkins." *New York Times*, April 29, 1906, X4.
Wood, R. W. "The *n*-Rays." *Nature* 70 (September 29, 1904): 530–31.
Wood, R. W. "The Scintillations of Radium." *Science* 19 (1904): 195–96.
Woods, A. F. "Botany and Human Affairs." *Science* 81 (2111): 573.
Yarker, P. M. "W. H. Mallock's Other Novels." *Nineteenth-Century Fiction* 14 (December 1959): 190.

Index